Quenching and Distortion Control

Proceedings of the
First International Conference on Quenching and Control of Distortion
22-25 September 1992
Chicago, Illinois

Edited by
George E. Totten

Sponsored by
the Quenching and Cooling Committee
of the Heat Treating Technical Division
of ASM International®

Co-Sponsored by
The ASM Heat Treating Steering Panel
The IFHT (International Federation for Heat Treatment
and Surface Engineering) Committee on Quenching and Cooling
The IFHT Committee on Science and Technology Aspects of Quenching
and The ASM European Council

**The Materials
Information Society**

First printing, September 1992

This book is a collective effort involving hundreds of technical specialists. It brings together a wealth of information from worldwide sources to help scientists, engineers, and technicians solve current and long-range problems.

Comments, criticisms, and suggestions are invited, and should be forwarded to ASM International.

Library of Congress Cataloging Card Number: 92-73755
ISBN: 0-87170-455-2
SAN: 204-7586

Production Manager
Linda Kacprzak

Production Coordinator
Randall L. Boring

ASM International®
Materials Park, OH 44073-0002

Printed in the United States of America

Organizing Committee

Dr. George Totten, Chairman
Union Carbide Chemicals & Plastics Company, Inc.

Prof. George Krauss
Colorado School of Mines

Dr. Charles E. Bates, Co-Chairman
Southern Research Institute

Prof. Bozidar Liscic
University of Zagreb

Mr. Terrence D. Brown
Lindberg Corporation

Mr. Wim A.J. Moerdijk
Houghton Benelux B.V.

Mr. John A. Hasson
E.F. Houghton & Company

Mr. Sören Segerberg
Swedish Institute for Production Engineering Research

Dr. Maurice A.H. Howes
INFAC/IIT Gear and Bearing Center

PREFACE

Quenching is one of the least understood of the various heat treating technologies. This is a problem that continues to perplex heat treaters as they continue to institute programs to reduce production costs and improve product quality to assure global competitiveness and even continued existence. In a recent survey conducted by ASM International, various aspects of quenching including identification of reliable methods of periodic quench severity measurement and the development of technologies for cracking, distortion and residual stress prediction, were viewed by heat treaters as their most critical immediate needs.

The First International Conference on Quenching and Control of Distortion was held to obtain a global assessment of various aspects of current quenching technology. For the first time, an integrated view of quenching, distortion and residual stress was discussed in one forum to encourage an interdisciplinary approach to future technology development.

The conference speakers discussed current and future developments in quenchants, methods and limitations of property predictions, quench system and equipment technology, quenchants as heat transfer mediators, and sources of residual stress and distortion generation and their prediction and measurement.

Papers presented are compiled in this conference proceedings, which will serve as a formal report for presentation at ASM's Heat Treating Steering Committee. On the basis of a review of this report (proceedings), recommendations will be made regarding the best modes encouraging the future information dissemination, education and reduction to practice of this technology in heat treating workshops.

Putting together a conference of this magnitude is a result of the joint effort of the conference organizing committee. This committee included:

* Dr. Charles E. Bates - Southern Research Institute (conference co-chairman)
* Prof. George Krauss - Colorado School of Mines
* Sören Segerberg - Swedish Institute of Production Engineering Research
* Prof. Bozidar Liscic - University of Zagreb, Croatia
* Wim Moerdijk - Houghton Benelux B.V.
* Dr. Maurice Howes - IIT/ITRI
* John Hasson - E.F. Houghton Inc.
* Terry Brown - Lindberg Heat Treating Inc.

Finally, the conference organizing committee acknowledges tremendous effort, encouragement and patience by the ASM staff.

George E. Totten
Union Carbide Chemicals and Plastics Company Inc.
Tarrytown, New York
Conference Chairman

Table of Contents

Session 1: Quenchants and Quenching Technology I

Session 2: Quenchants and Quenching Technology II

Session 3: Processes/Equipment

Posters

Measurement and Evaluation of the Quenching Power of Quenching Media for Hardening

J. Bodin and S. Segerberg
Swedish Institute of Production Engineering Research
Göteborg, Sweden

ABSTRACT

First, distinction is made between the concepts cooling power and hardening power. Various methods of testing the cooling power are presented together with existing and proposed standards for testing of quenchants. A number of methods have been proposed for testing the hardening power, which take account of the material (steel) to be hardened. Some of these methods are mentioned.

The interrelation between the quenchant and steel being hardened has been described in a number of ways over the years. Recently, new ways of determining the hardening power, based on information from the cooling curve of the quenchant, as recorded in a standardised test, have been proposed. With knowledge of the hardening power of available quenchants, selection of the most suitable one for each application is simplified. A list of several commercial quenchants with their hardening power in relation to carbon and low-alloyed steels is presented.

QUENCHING IS THE MOST CRITICAL PART of the hardening process. The quenching process has to be designed so as to extract heat from the hot workpiece at such a rate as to produce the microstructure, hardness and residual stresses desired.

Basic research on the quenching process has been carried out since the beginning of this century. During the last decade, important progress has been made in developing deeper understanding of the quenching process and its influence on the properties of workpieces being hardened.

The rapid evolution of computer hardware both offers tools to facilitate testing and encourages development of software to predict the results of hardening. This development has stressed the importance of better understanding of the interrelation between the quenchant and the material being hardened as well as of standardised methods for testing quenching media.

Before entering into details about methods of testing and evaluation of quenchants, it is essential to make a distinction between the two concepts: *cooling power* and *hardening power* of quenchants. According to Metals Handbook (1), the cooling power is the "thermal response" of the quenchant or the rate of heat removal from a specimen, usually instrumented (see Chapter 2), while the hardening power is the "metallurgical response" of steel specimens or the ability of the quenchant to develop a specified hardness in a given material/section size combination (see Chapters 4 and 5).

Methods of testing the cooling power of quenching media

Grossmann et al (2) introduced *the H value*, the "heat transfer equivalent" (also called the "Grossmann hardenability constant" or the "quench severity factor") for the cooling power (also called the "severity of quench") of quenching media. The H value is defined as:

$$H = h/2k$$
where
h = the mean value of the heat transfer coefficient throughout the entire quench process
k = the thermal conductivity of the material

According to Grossmann, the H value for stationary water at room temperature is about 1.0 (cm^{-1}), while, consequently, slower-cooling quenchants have H values less than 1. Sometimes the H value is determined in comparison tests, where the relative cooling time between two temperatures, e.g. 700° and 550°C or the cooling rate at a relevant temperature, e.g. 700°C, is measured for stationary water and the quenchant of interest. Thus, the H value takes little or no consideration of the fact that both the heat transfer coefficient and the thermal conductivity vary with temperature. Therefore, a quenchant with a higher H value will not always produce a higher hardness when real steel workpieces are quenched. To be able to evaluate the result of hardening from the cooling power, it is obvious that the complete cooling process has to be considered, as in cooling curve (or thermal gradient) testing. This is dealt with in Chapter 4.

Cooling curve testing

This method is the most useful for testing the cooling power of quenching media and quenching systems, since the complete cooling process is recorded. The test is performed by quenching a test-piece (probe) with a thermocouple embedded at some point, usually at its geometric centre or at (or near) the surface, and monitoring the cooling process with a temperature-measuring device.

Figure 1 shows a typical cooling curve, where the three phases normally appearing when quenching in liquid media are clearly identified. In the figure, the cooling rate vs. temperature is also shown, as is common practice.

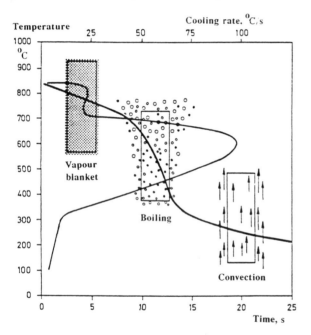

Fig. 1. Typical cooling curve for a quenchant with its three phases.

The probe *material* is usually austenitic stainless steel, nickel or nickel alloys or silver, all of which are free from transformation effects, or of carbon or alloyed steel. The probe *shape* is usually cylindrical. For many years, silver sphere probes were often used, but mainly for manufacturing reasons, cylindrical shapes are preferred today.

The cooling curves recorded are influenced, naturally, by the thermal properties of the probe material, the probe size and the position of the thermocouple in the probe. On the one hand, the sensitivity of measurement increases with increasing thermal conductivity of the probe material and decreasing probe size (when temperature is measured at the probe centre), or if temperature is measured at (or near) the probe surface. On the other hand, manufacture of the probe and design of the temperature monitoring system are more critical when probe sensitivity increases. Therefore, choice of probe material and size as well as type and position of the thermocouple have to be a compromise between sensitivity, manufacturing costs and probe life.

Cooling curve testing is now the subject of international standardisation and various national standards exist (see below). Equipment for testing, both stationary for laboratory use and portable for testing in the workshop, is available commercially.

International standards.

The International Federation for Heat Treatment and Surface Engineering (IFHT) and its "Scientific and Technical Aspects on Quenching" Sub-Committee, has engaged itself in preparing proposals for international standards as concerns quenchant testing. Different methods have been compared in round-robin tests performed in a number of countries. Based on this, proposals for standards are being submitted to the International Organization for Standardization (ISO).

The first proposal involving laboratory testing of *oils* (3) has been submitted to ISO and is now being reviewed by its TC 28 Technical Committee. It involves testing with a *ø12.5x60 mm cylindrical probe* made of *Inconel 600 (UNS N06600)*. Figure 2 shows the design of the test probe.

Testing is done from 850°C in a 2000 ml sample of stationary oil. Results of testing are expressed by (a) the temperature/time curve and the cooling rate/temperature curves, and (b) the following data read off from these curves:

- cooling time to 600°, 400° and 200°C
- maximum cooling rate
- temperature at which the maximum cooling rate occurs
- cooling rate at 300°C

The draft proposal is based on a specification elaborated by Wolfson Heat Treatment Centre (4). It has already reached widespread use, e.g. in Europe and the USA.

A few countries have standards for testing with silver probes (see below). Therefore, the IFHT Sub-Committee recommended that, as a *second-choice ISO standard* for testing of oils, testing with a *silver probe* according to the French national standard (see below) should be accepted. This will also be considered within the ISO committee.

As far as testing of *water-based polymer quenchants* is concerned, the IFHT Sub-Committee has organised round-robin tests which have been performed recently. The results are now being analysed. A draft proposal is expected shortly. In contrast to oils, testing of polymers has to be done with agitated quenchants in order to receive reproducible results. Therefore, the draft will also include a specification of the agitation system.

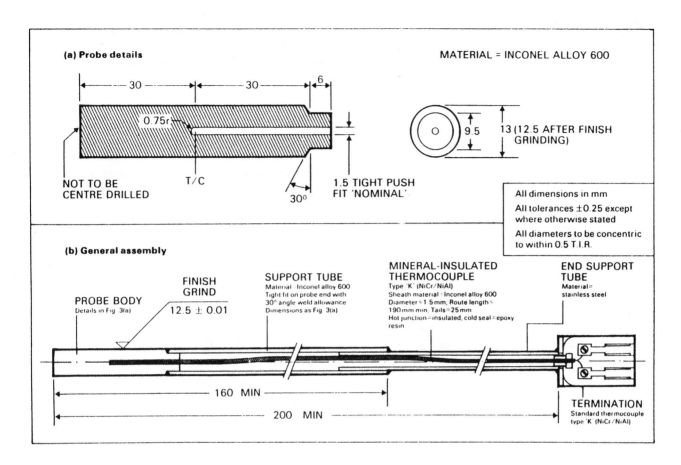

Fig. 2. Test probe according to the proposed international standard (ISO/DIS 9950 draft) (3).

Some national standards.

France. The French standard for testing of oils, NFT 60178 (5), is based on a ∅16x48 mm silver cylinder probe. Testing is done from 800°C in a 700 ml sample of stationary oil. In addition, there is a standard, NFT 60512, which classifies quenching fluids in product families and principal fields of application.

P.R. China. The Chinese standard GB 9449-88 (6) is identical with the French standard as concerns probe size and material, and probe and oil temperatures. The main difference in comparison with this standard is that a more sensitive probe thermocouple (0.5 mm instead of 1.0 mm dia.) is specified. The Chinese standard applies to both oil-based and water-based quenchants.

Japan. The Japanese standard JIS K 2242 (7) contains specification of heat treating oils and their classification as well as a method of testing their cooling performance. In this, a ∅10x30 mm silver cylinder equipped with a thermocouple at its surface is used, see Figure 3.

Testing is done from 810°C in a 250 ml sample of the oil. The result is expressed by the characteristic temperature (i.e. the transition temperature between film boiling and nucleate boiling) and the cooling time from 800° to 400°C derived from the cooling curve.

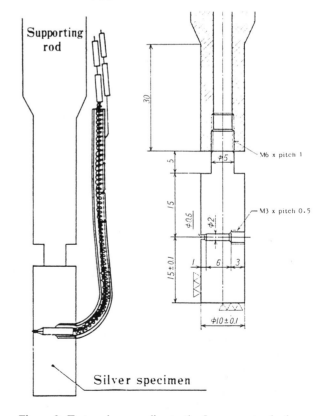

Figure 3. Test probe according to the Japanese standard JIS K 2242-1980 (7).

3

Recently, the standard has been subject to modification, mainly in order to lengthen probe life. Other probe designs have been considered, including those where the thermocouple is positioned in the centre of the probe. It has been decided, however, to stay with existing design, essentially in order not to impair sensitivity of measurement.

U.S.A. A draft specification has been elaborated recently by the ASM Quenching & Cooling Committee and is being discussed with the American Society for Testing and Materials (ASTM) for possible standardisation. The specification is similar to the ISO draft. Beside ASTM, some other competent U.S. organisations are discussing standardisation of test methods for quenchants.

Great Britain. Although not formally standardised, the Wolfson specification (4), which is similar to the ISO draft, has been accepted as a *de facto* standard within the British industry since its publication in 1982. In the first drafts from the mid-1970s, stainless steel was specified for the probe but later a change to Inconel 600 was made in order to increase the resistance to oxidation and corrosion.

Comments on the standards. The Wolfson method (and, thus, the ISO/DIS 9950 draft) can be characterised as an *engineering* method, where probe life, reproducability and, to some extent, similarity with quenching of steel have been considered the most important. The Japanese method in particular, and also the French, can be considered as *scientific*, where sensitivity has been considered the most important.

Thermal gradient testing

To be able to calculate the temperature within a workpiece being quenched, the heat transfer coefficient, or the heat flux, at the surface has to be determined. This can be calculated from measurements with thermocouples at or near the surface of the workpiece (i.e. cooling curves). However, as the heat flux from the workpiece is proportional to the *thermal gradient* at the surface, it can be determined directly by measuring the cooling curve at two points, at and near the surface. A method based on this concept has been developed by Liscic (8). A $\phi 50 \times 200$ mm stainless steel probe is used, having three thermocouples (at the surface, 1.5 mm below the surface and at the centre). The quenching intensity is represented by the heat flux density at the surface as calculated from the thermal gradient. The method is intended to be used also in workshop practice.

Other methods for testing the cooling power

The following methods have also been used to determine the cooling power of quenchants. These methods produce only small fractions of the information contained in cooling curves, as with the H value, and so their applicability is limited. Their advantage is the simplicity. However, one has be to very careful when interpreting results from measure-

ments with these methods. For further details see for example (1) and (9).

Houghton quench test (system Meinhardt). This method (10) is a simplified version of the cooling curve test. A test-piece of carbon steel, normally with thermocouple wire, which together with the steel probe and support tube forms the thermocouple, is immersed in the quenchant. The time to cool within a certain temperature interval is determined with a built-in stop watch. Thus, two points on the cooling curve can be determined in each test. The result is expressed as the time to cool between two fixed temperatures, normally from 700° to 400°C (or 300°C). (The probe thermocouple can, of course, also be connected to a temperature recording equipment so as to monitor the complete cooling curve, even if this is not the intended use). Figure 4 shows the equipment.

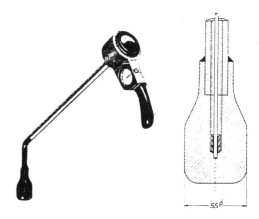

Fig. 4. Houghton quench test apparatus (system Meinhardt) (10).

The diameter of the probe is normally 55 mm, but can be chosen between 20 mm and 80 mm to correspond to different dimension of workpieces to be hardened. In addition to carbon steel, austenitic stainless steel can also be used as the probe material. In this case, a standard thermocouple must be used instead of the arrangement in Figure 4.

The Magnetic Test makes use of the fact that magnetic metals lose their magnetic properties above the Curie temperature. In this test a nickel sphere is heated to 835°C and quenched in a sample of the quenchant. The time required to cool to the Curie temperature, which is 355°C, is recorded. This time is a measure of the cooling power of the quenchant. By choosing other alloys with different Curie points, additional points on the cooling curve can be determined. The method, also known as the GM Quenchometer test, is standardised in the U.S.A. (ASTM 3520).

Hot Wire Test. In this test, a resistance wire is heated by means of an electric current in a sample of the quenchant. The current is increased steadily and the cooling power is indicated by the maximum steady current reading, as measured by an ammeter.

The main disadvantages are that comparison of quenchants is only made in the higher temperature range and that no information about the vapour phase characteristics and the transition temperature to boiling can be obtained (a stable vapour phase can not be achieved with a wire).

Interval Test. In this test, also known as the 5-s test, a heated bar of metal (usually stainless steel) is immersed for 5 s in a sample of the quenchant which is contained in an insulated container. The increase in temperature is noted. The process is repeated for a series of bars. Finally, a bar of identical size and material is fully quenched in a new sample of quenchant of the same volume. The cooling power is expressed as the rate between the average rise in temperature for the 5-s quench bars and the rise for the fully-quenched bar.

The method is simple and requires no special equipment. It can be used for determination of gross changes in cooling power. However, as with the interval test, measurement is only made in the high temperature range.

A similar method has been developed from the interval test, where consideration is taken of the cooling power at lower temperatures. Here, the rise in temperature is determined after various immersion times, or continuously.

Methods of testing the hardening power of quenching media

Immersion quench tests. These tests involve immersion of heated steel specimens in samples of the quenchant or in the quench tank itself. Some common practices and standards are presented below.

Cylindrical specimens. Testing with cylindrical specimens is common practice in most heat treatment shops. The dimensions and material of the specimens should be chosen with regard to the products to be hardened and the quenchant to be used. Variation in the hardening power within a quench tank can be monitored by placing specimens at various positions in a batch to be hardened.

Stepped, cylindrical specimens. Instead of using specimens with different diameters, stepped specimen with two (or more) diameters are sometimes used. The French automotive companies Renault and Peugeot, for example, used stepped specimen with two diameters made of 38C2 grade steel for a period of years (11). It is also obviously used in several other companies.

Wedge-shaped specimen. A method for testing with a wedge-shaped specimen has been developed by the French Association for Heat Treatment (ATTT). The method was submitted for standardisation both in France (NFT 60179 draft) (11) and internationally (12) in 1988.

The specimen is made of 38C2 grade steel (according to the French standard NF A 35-552). The shape of the specimen is shown in Figure 5. After quenching, the probe is cut into two halves, longitudinally, and the hardness is measured at prescribed points.

Modified Jominy end-quench test

The Jominy test (ISO 642 and ASTM A 255) is normally used for determination of the hardenability of steels. In this test, the quenching medium is water. According to (9), the same equipment can be used for evaluation of quenchants as well. In this case, specimen of known hardenability is used and the variables of the quenchant are changed. The quenchant has to be contained in a closed system in order to permit close control. The effect of the variable, whether composition, temperature, concentration or some other factor, is evaluated by the hardness pattern developed on the end-quench specimen.

The advantage would be that testing can be made in equipment that may already be on-site in the company. The main disadvantage would be that the quenching conditions are different from normal hardening in a quench tank (no stable vapour phase will be formed at the probe surface at high temperatures and convection will be greater). However, it could be a relevant method for testing of quenchants for induction hardening or for injection or spray quenching, where the vapour phase does not normally appear.

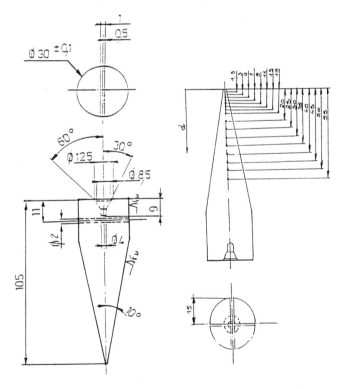

Fig. 5. (a) Wedge-shaped specimen proposed by ATTT, France. (b) Points where hardness is tested (11, 12).

5

Methods to interrelate quenching performance with the properties of steel workpieces being hardened

The interrelation between steel workpieces being hardened and the quenchant has been the subject of many investigations since the beginning of this century and more intensively since the 1930s. Names like Bühler, Grossmann, Bain, Wever, Rose, Peter, Tamura, Kobasko, Hougardy, Atkins, Beck, Thelning and their co-workers are only examples of people who have made important contributions to the understanding of the quench hardening process.

In this chapter, some examples from more recent engineering research are presented, where attempts were made to express the "hardening power" of the quenchant in a way relevant to the metallurgical response of steel workpieces being hardened. Before going into details about these, some comments are made concerning the possibilities and limitations of using CCT diagrams for evaluation of the quenching process.

Comments on the interpretation of cooling curves in relation to CCT diagrams

Continuous cooling transformation diagrams (CCT diagrams) are often used to evaluate the metallurgical response of different steels to cooling. Most CCT diagrams are constructed from tests with natural or linear cooling. In industrial quenching in liquids, the cooling rate varies with time and surface temperature of the workpieces. It has been shown by Shimizu and Tamura (13,14), Thelning (15), Loria (16) et al that a change in cooling rate during quenching, as in Figure 1, influences the transformation behaviour of the steel significantly. Shimizu and Tamura (13) also propose a graphical method to find the real transformation behaviour and the critical cooling rate in the CCT diagram, when the cooling rate is changed during continuous cooling.

Thelning (15) has constructed CCT diagrams representing cooling conditions prevailing at the surface and centre of steel bars quenched in water and oil. The diagrams show that the time to transformation is different for water and oil quenching and for the surface and the centre.

Allowance must also be made for the fact that the transformation behaviour depends on such factors as the austenising temperature before quenching, time at elevated temperature (for homogenising) and grain size of the steel.

Consequently, incautious superimposure of cooling curves on CCT diagrams may lead to considerable errors in prediction of the result of hardening.

Some methods proposed to correlate cooling curve information with the result of hardening

As mentioned in the introduction, computer models of the hardening process are now being developed at several places. A few of them use complete cooling curves or heat transfer data at the surface and work incrementally (in small time steps) to arrive at the final condition successively. Most of the models, however, seem to employ quite simple ways of describing the quenching part of the process and are therefore limited in their application. Below, some simpler methods of correlating the cooling curves with the result of hardening (mostly the hardness) are described.

Cooling rate at a certain temperature or temperature interval

A simple approach in trying to correlate hardening response with the cooling process is to consider the cooling rate at some selected temperature or cooling time in some temperature range. There is general agreement between researchers that a high cooling rate in the temperature region where the time to transformation to ferrite, perlite and bainite is shortest, is of decisive importance for the hardness and microstructure after hardening.

Table 1 gives examples of temperatures and intervals, where the cooling rate has been correlated to the hardening response, since the turn of the century. Obviously, the temperatures chosen has in some cases, at least, been directly related to the steel being considered.

Atkins and Andrews (17) and Murry (18) have constructed CCT diagrams with the abscissa graded in cooling rate at 800°, 750° or 700°C, depending on the steel composition, and cooling time between 700° and 300°C, respectively.

Figure 6 shows how hardness is correlated with the cooling time between 640° and 400°C for a low-alloy steel. Figure 7 shows similarly how hardness is correlated to the cooling rate at 550°C for carbon steel test-pieces. These figures show good correlation. However, such simple models can only be used with success within certain limits as concerns the cooling characteristics and test-piece dimensions.

Table 1. Temperature ranges and temperatures considered for studies of correlation between cooling time or rate and hardening response.

Researchers	Temp. range (°C)	Cooling rate at (°C)
Lechatelier (19)	700 -> 100	
Benedicks (20)	700 -> 100	
Mathews, Stagg (21)	650 -> 370	
Portevin, Garvin (22)	700 -> 200	
Grossmann, Asimow, Urban (2)	700 -> 300	
Murry (18)	700 -> 300	
Deliry, El Haik, Guimier (23)	640 -> 400	
Wever, Rose (24)	800 -> 500	
Kulmburg, Kornteuer, Kaiser (25)	800 -> 500	
Rogen, Sidan (26)	A_{C3} -> M_s	
Ives, Meszaros, Foreman (27)		700 and 200
Atkins, Andrews (17)		800, 750 or 700
Segerberg (28)		550

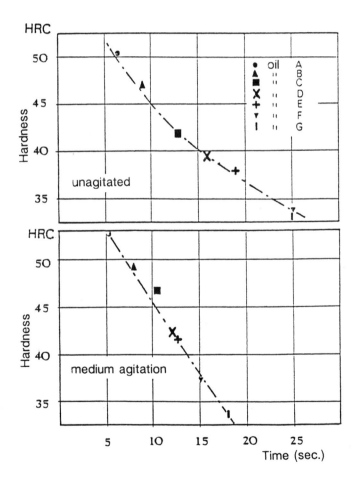

Fig. 6 - Correlation between hardness and cooling time from 640° to 400°C in ∅16x48 mm test-pieces of steel 38C2 (0.38 % C, 0.5 % Cr) (23).

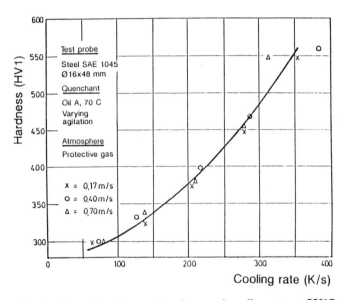

Fig. 7 - Correlation between hardness and cooling rate at 550°C for ∅16x48 mm test-pieces of SS 1672 (SAE 1045) steel (28).

The Tamura V value (29)

In this approach, the following steps are identified:

(1) Steels are classified roughly into four types according to the shape of their CCT diagrams (i.e. those for which non-martensitic structures are likely to be ferrite+perlite+bainite, perlite, perlite+bainite or bainite). For each type, a temperature range, X, can be defined, where rapid cooling is required to ensure martensitic transformation.

(2) For a given quenchant, the temperature for the start (T_c) and end (T_d) of boiling is determined by testing with the standardised silver probe (see Figure 3). These two temperatures are converted into those for steels from graphical data.

(3) A factor **V** is defined which is the ratio of the degree of overlap of the two ranges, X and T_c-T_d, see Figure 8.

(4) For a certain section size, the hardening response of a given steel in various quenchants can be predicted from experimentally-determined curves relating the **V** value with hardness.

Ts > Tc	Td > Tf		$V = \dfrac{Tc - Td}{X}$
Ts > Tc	Td < Tf		$V = \dfrac{Tc - Tf}{X}$
Ts < Tc	Td > Tf		$V = \dfrac{Ts - Td}{X}$
Ts < Tc	Td < Tf		$V = \dfrac{Ts - Tf}{X}$

Fig. 8. Calculation of the V value at the four different types of overlap between the boiling phase of the quenchant and the CCT diagram for the steel being hardened (29).

Figure 9 shows the correlation between surface hardness and the V value for steels of the four different categories, quenched in oils. Similar results were received for the center hardness and the hardening depth.

The IVF hardening power, HP (30, 31)

Based on measurements with the method proposed as an ISO standard for oils (probe according to Figure 2), a formula has been derived by regression analysis, where the hardening power for *oils* is expressed as a single value, **HP**:

$$HP = k_1 + k_2 T_{VP} + k_3 CR - k_4 T_{CP} \qquad (1)$$

where

T_{VP} = the transition temperature between the vapour phase and the boiling phase (in °C)

CR = the cooling rate over the temperature range 600° -> 500°C (in °C/s)

T_{CP} = the transition temperature between the boiling phase and the convection phase (in °C)

k_1, k_2, k_3, k_4 are constants

For *unalloyed steels* the formula is as follows:

$$HP = 91.5 + 1.34 T_{VP} + 10.88 CR - 3.85 T_{CP}$$

Hardness ratings of test-pieces of medium-carbon SS 1672 (SAE 1045) steel, hardened in immersion tests in several commercial oils on the Swedish market, have been compared with the HP values determined for these oils. The results are shown in Figure 10, which also shows HP values of some oils on the U.S. market. The definition of the three points on the cooling rate curve are also shown here.

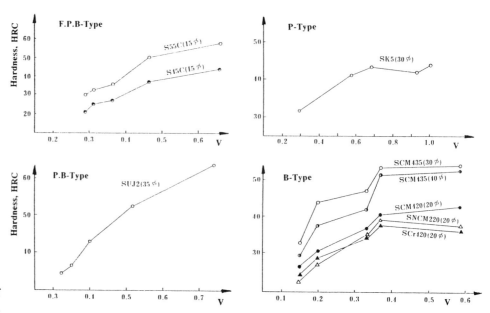

Figure 9. Relation between hardness of specimens and the V value of oils (29).

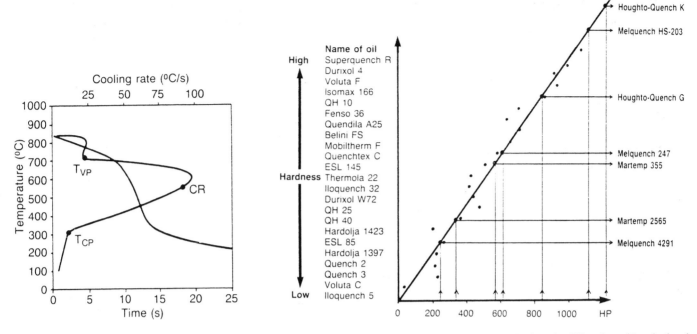

Fig. 10. Predicting the hardening power, HP, of oils (30): (a) definition of the three points used to calculate the HP value; (b) calculated values of HP matched to a straight line for quenching oils. Performance ranking applies only for hardening of unalloyed steel. HP values determined for some common U.S. oils are also included (right) for comparison (based on measurements made by the respective supplier).

For *alloyed steels*, the coefficients in Eq. (1) will be different, as will the ranking of oils.

For *polymer quenchants*, a similar approach has been made. However, since the vapour phase is often suppressed, or nonexistent, and the transition between the boiling and convection phases is not as pronounced for polymers, the hardening power has been formulated as follows:

$$HP = k_1\,CR_F + k_2\,CR_M - k_3$$

where

CR_F = the cooling rate at the ferrite/perlite nose (in °C/s)

CR_M = the cooling rate at the martensite start temperature (in °C/s)

k_1, k_2, and k_3 are constants

Figure 11 shows graphically the critical points on the temperature/cooling rate curve for polymers and the relationship between the hardness and the HP values.

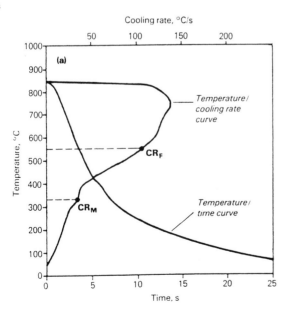
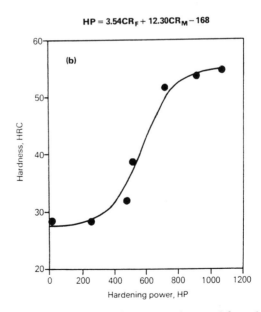

Fig. 11. Predicting the hardening power, HP, of polymer quenchants (31): (a) definition of the two points used for calculation of the HP value; (b) relationship between the measured hardness on ⌀16x48 mm test-pieces of medium-carbon SS 1672 (SAE 1045) steel and the HP value.

The Castrol/Renault hardening power, HP, and the Castrol Index, Ci (32)

From extensive tests on oils with probes to the French standard and the ISO draft, as well as with 16 mm dia. probes of Inconel and 38C2 steel with thermocouples at the centre, Deck et al (32) have developed formulas similar to the IVF formula for oils, but where the hardness achieved in workpieces of 38C2 steel is calculated directly. For the ISO draft probe, the formula is as follows:

$$HP \ (HRC) = 99.6 - 0.17 \ O'_2 + 0.19 \ V_{400}$$

where

O'_2 = the transition temperature between the boiling and convection phases (in °C)

V_{400} = the cooling rate at 400°C (in °C/s)

The formulas for the other Inconel probe and the steel probe are similar.

In the case of silver, the empirical laws described above could not be applied with sufficient accuracy. Based on thermokinetic reasoning, the following formula was developed, which in fact is independent of the probe material and size:

$$HP \ (HRC) = 3.1 \ Ci - 5.7$$

where $Ci = K' \ V_{max}/(O_{max} - O_l)$

and $K' = 11.5 \ \{(O_{max} - O_l)/V_{max}\}_{reference*}$

where V_{max} = the maximum cooling rate (in °C/s)

O_{max} = the temperature at which maximum cooling rate occurs (in °C)

O_l = the temperature of the quenchant (in °C)

* reference is made to values for an additive-free base oil, having an H value of 11.5 m⁻¹

As shown in Figure 12, the correlation between calculated HP values and hardness measured in steel test-pieces is quite good for both test probes.

HP (HRC)

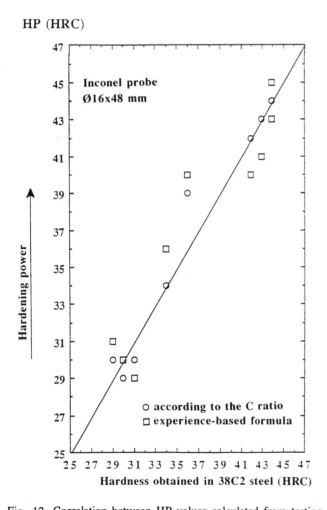

HP (HRC)

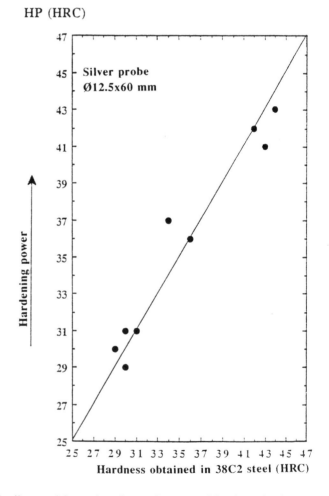

Fig. 12. Correlation between HP values calculated from testing with silver and Inconel probes and measured hardness in ⌀16x48 mm specimens in 38C2 steel (32).

10

Other similar methods

Totten et al (33) have made an excellent overview of some of the above and several other methods of correlating cooling curves with the results of hardening. These include work by Hilder (34) and Tensi (35) on critical transformation times, temperatures and rates, Liscic (36), Thelning (37) and Goryushin et al (38) on areas under the cooling time and rate curves, Evancho and Staley (39) and Bates (40) on quench factor analysis and Tamura and Tagaya (41) on master cooling curves.

In this context, work by Wünning and Liedtke (42) and Lübben et al (43) on the QTA method should be mentioned. In this method, the real cooling rate curve is approximated with an artificial one which is more convenient to work with in combination with computer prediction of the result of hardening.

Cooling curves vs. cooling performance in industrial hardening

Cooling curves are normally determined under controlled conditions in the laboratory, although equipment is available on the market today by which the cooling process can be monitored in real quench tanks. The cooling curve, as determined with a standard, or non-standard, test probe, reflects only the cooling characteristics at one point, or one circumference, of the probe. If the cooling curve is determined with a silver probe, it is also important to note that cooling curves determined with probes made of materials not having phase transformation, especially silver, are different from those of steel.

When quenching real workpieces in a tank, the heat transfer characteristics vary over the surface of the workpiece. The transition between the vapour phase and the boiling phase, for example, takes place at different times at different points on the surface. See, for example, Tensi et al (44) on wetting kinetics.

The workpieces in a batch also interact with each other, see e.g. Segerberg and Bodin (45), creating conditions around the workpieces which are difficult, if not impossible, to describe. It has also been found that oxidation of the workpiece surface will change the cooling performance. See, for example, Segerberg (46).

It is therefore of the utmost importance for researchers to work hand-in-hand with practical users of heat treatment in order to be able to transfer results from the laboratory to practice.

Summary and conclusions

The ongoing standardisation of methods for testing of quenchants for hardening will bring better tools to the heat treatment industry and to manufacturers of quenchants to monitor and understand the complex quenching process. The possibility of combining immense knowledge from research laboratories, acquired over a century, with practical experience from the workshop floor is better than ever, especially when taking into account the opportunities offered by the revolutionary development in computer technology and applications.

References

1. *Metals Handbook*, Ninth Ed., Vol. 4, Heat Treating, ASM International, U.S.A., 1981, 31
2. A. Grossmann, M. Asimow and S.F. Urban, in *Hardenability of Alloy Steels*. American Society for Metals, 1939, 124-190
3. *Industrial Quenching Oils - Determination of Cooling Characteristics - Laboratory Test Method,*. Draft international standard ISO/DIS 9950, International Organization for Standardization (submitted 1988)
4. *Laboratory Test for Assessing the Cooling Characteristics of Industrial Quenching Media*, Wolfson Heat Treatment Centre, Birmingham, England, 1982
5. *Drasticité des huiles de trempe. Essai au capteur d'argent*, NFT 60178, Association Francaise de Normalisation, Paris, France
6. *Test method for cooling properties of quenching media*, State Bureau of Standards of the People's Republic of China, Beijing, China
7. *Japanese Industrial Standard. Heat Treating Oils*, JIS K 2242-1980, Japanese Standards Association, Tokyo, Japan
8. B. Liscic, in *Proc. 6th Int. Congress on Heat Treatment of Materials. Chicago, U.S.A. 28-30 Sept. 1988*, ASM International, U.S.A., 1988, 157-166
9. *Quenching and Control of Distortion* (ed. H.E. Boyer and P.R. Cary), ASM International, U.S.A., 1988, 166
10. *Houghton Abschreckprüfgerät* (Houghton Quench Test Apparatus). Brochure from Houghton-Hildesheim, Germany
11. *Fluides de trempe. Determination de la sévérité de trempe d'une installation industrielle*, NFT 60179, Association Technique de Traitement Thermique, Paris, France, 1988
12. *Quenching Media. Determination of Quenching Severity of an Industrial Facility*, Draft international standard submitted by Association Technique de Traitement Thermique, France, to the International Federation for Heat Treatment and Surface Engineering (IFHT), 1988
13. N. Shimizu and I. Tamura, *Trans. ISIJ*, 1977, 17, 469-476
14. N. Shimizu and I. Tamura, *Trans. ISIJ*, 1978, 18, 445-450
15. K.E. Thelning, *Scandinavian J. of Metallurgy*, 1978, 7, 252-263
16. E.A. Loria, *Metals Technology*, Oct.1977, 490-492
17. M. Atkins and K.W. Andrews, in *BSC Report SP/PTM/6063/-/7/C/*
18. G. Murry, *Traitement Thermique*, 1976, 108, 47-54
19. H. Lechatelier, *Revue de Métallurgie*, 1904, 1
20. C. Benedicks, *J. Iron Steel Inst.*, 1908, 77
21. J.A. Mathews and H.J. Stagg, *Trans. ASME*, 1914, 36
22. A.M. Portevin and M. Garvin, *J. Iron Steel Inst.*, 1919, 99
23. J. Delixy, R. El Haik and A. Guimier, *Traitement Thermique*, 1980, 141, 29-33
24. F. Wever and A. Rose, *Stahl und Eisen* 1954, 74, 749
25. A. Kulmburg, F. Korntheuer and E. Kaiser, in *Proc. 5th Int. Congress on Heat Treatment of Materials, Budapest, Hungary, 20-24 Oct. 1986*, Scientific Society of Mechanical Engineers, Budapest, Hungary, 1986, 1730-1736
26. G. Rogen and H. Sidan, *Berg- und Hüttenmänn. Monatsh.*, 1972, 117, 250-258
27. M.T. Ives, A.G. Meszaros and R.W. Foreman, in *Heat Treatment of Metals*, 1988, 15, 11
28. S. Segerberg, *IVF-skrift 88804*, IVF - The Swedish Institute of Production Enginering Research, 1988
29. I. Tamura, N. Shimizu and T. Okada, *J. Heat Treating*, 1984, 3, 335-343
30. S. Segerberg, *Heat Treating*, Dec. 1988, 30-33
31. S. Segerberg, in *Heat Treatment of Metals*, 1990, 17, 67

32. M. Deck, P. Damay and F. Le Strat, in *Proc. ATTT 90 Internationaux de France du Traitement Thermique, Le Mans, 19-21 Sept. 1990*, Association Technique de Traitement Thermique, Paris, France, 49-70

33. G.E. Totten, M.E. Dakins and R.W. Heins, *J. Heat Treating*, 1988, 6, 87-95

34. N.A. Hilder, *Heat Treatment of Metals*, 1987, 14, 31

35. H.M. Tensi and E. Steffen, *Steel Research*, 1985, 56, 489

36. B. Liscic, *Härterei-Tech. Mitt.*, 1978, 33, 179

37. K.E. Thelning, in *Proc. 5th Int. Congress on Heat Treatment of Materials, Budapest, Hungary, 20-24 Oct. 1986*, Scientific Society of Mechanical Engineers, Budapest, Hungary, 1737-1759

38. V.V. Goryushin, V.F. Arifmetchikor, A.K. Tsretkov and S.N. Sinetskij, *Met. Sci. and Heat Treatment*, 1986, 709

39. J.W. Evancho and J.T. Staley, *Metallurgical Trans.*, 1974, 5, 43

40. C.E. Bates, *J. Heat Treating*, 1987, 5, 27

41. I. Tamura and M. Tagaya, *Trans. Japan Inst. of Metals*, 1964, 5, 67-75

42. J. Wünning and D. Liedtke, *Härterei-Tech. Mitt.*, 1983, 38, 149-155

43. Th. Lübben, H. Bomas, H.P. Hougardy and P. Mayer, *Härterei-Tech. Mitt.*, 1991, 46, 24-34, 155-171

44. H.M. Tensi, Th. Künzel and P. Stitzelberger, *Härterei-Tech. Mitt.*, 1987, 42, 125-132

45. S. Segerberg and J. Bodin, *in Proc. 3rd International Seminar: Quenching and Carburising, Melbourne, 2-5 Sept. 1991*, Institute of Metals & Materials Australasia Ltd. Parkville Vic 3052, Australia

46. S. Segerberg, *IVF-resultat 88605*, Sveriges Mekanförbund, Stockholm, 1988

Proceedings of the First International Conference on Quenching & Control of Distortion, Chicago, Illinois, USA, 22-25 September 1992

Quenchants: Yesterday, Today, and Tomorrow

J. Hasson
E.F. Houghton & Co.
Valley Forge, Pennsylvania

The art of quenching has been known for milleniums, and it has mystified humans for just as long. Prior to the 20th Century, quenching was a crude practice done in air, water, and animal or vegetable oils. In the early 1900's, the understanding of hardening or quenching crystallized.

This is a quote from a book published in 1914, "In order to successfully harden one must study the various parts, their design and hardness desired." With this understanding of hardening came the popular use of quenching oils based on mineral oil. Although taken for granted now, this was a significant event in assuring consistent metallurgical results in the quenching process. Animal and vegetable oils degraded quickly sometimes in a matter of days while mineral oil based quenchants lasted considerably longer - months and years, providing consistent results. We must keep in mind that the steels heat treated at that time in quenching oils contained considerable alloying elements. Water was used for quenching plain carbon steels.

Early researchers were aware of the different stages that occurred during the quenching process. Although the nomenclature of the zones or stages of quenching are different today, the basic understanding of what happens in each stage has been known since at least 1923. Attempts at recording cooling rates or curves are documented in Practical Metallurgy for Engineers, 1923 edition where an 11 pound test bar made from 1045 steel with a thermocouple embedded in the center is heated and quenched. The temperature change was measured in millivolts on a potentiometer and time measured by a stop watch.

Today, the equipment and techniques for measuring and analyzing cooling characteristics of quenchants is much more sophisticated. Using exotic alloys, powerful electronic hardware and specialized software, researchers can analyze the quenching power and effects of a quenchant or quenching system. The level of sophistication has reached the stage where programs have been developed to predict the physical properties of alloys quenched in a particular quenchant.

No matter how powerful the test equipment what we are measuring is the same stages of quenching observed by those early researchers known as the three stages of quenching. Because they are so fundamental to the quenching process, a review of this subject will help further discussions on the topic of quenching.

The first stage is the vapor blanket stage that begins at the instant of immersion of the hot part into the quenchant. The vapor blanket stage is a period of relatively slow cooling where the component is completely surrounded by the quench vapor. Heat is removed by radiation and conduction through the vapor film. As the component cools, the vapor blanket cannot be maintained and collapses into the boiling stage. During the boiling stage, the quenching fluid is in contact with the surface of the component and boils. Heat is removed rapidly from the workpiece by the heat of vaporization. As the component cools below the boiling point of the quenchant, the boiling stage ceases and the convection stage begins. The convection stage or the third stage of quenching is once again a stage of slow cooling. Heat is transported from the component by movement of the quenchant caused by heat conduction currents.

The convection stage for many quenchants occurs at temperatures which corresponds to the austenite to martensite transformation in many steel alloys. Therefore, the convection stage is a very important stage in the quenching mechanism for controlling and reducing distortion.

The three stages of quenching occur in all liquid quenching media. However, various quenching media will have characteristically different stages in the quenching mechanism. Besides the quenchant other parameters or factors influence the cooling characteristics of the various stages of quenching. Quenchant temperature and agitation of the quench bath have pronounced effects on quenching characteristics. Agitation has its greatest impact on the vapor blanket stage by physically dispersing the vapor surrounding the component. Agitation has a slight effect on the action of the boiling stage most likely due to the already violent action inherent to this stage. The effect of agitation on the convection stage is much less then its effect on the vapor blanket stage but more then its effect on the boiling stage. Though it may not be readily discernible by cooling rate analyses, agitation has an impact on the cooling rate of the convection stage. Practical experience has shown that altering the agitation during the convection stage has reduced distortion.

Temperature of the quenchant effects the three stages of quenching differently depending on the quenchant used.

Water based quenchants have increasingly longer vapor blanket stages as the bath temperature increases. This is due to the nature of water to form vapor (steam) as the temperature approaches the boiling point. The boiling stage and convection stage are not significantly effected by changes in bath temperature. Therefore, as the bath temperature of a water based quenchant increases its overall cooling rate decreases.

Oil based quenchants on the other hand have the opposite effect. Oil as it is heated has a proportional drop in viscosity. This drop in viscosity allows the quenchant to move more freely increasing the natural or mechanical turbulence which brakes the vapor blanket stage. The boiling stage of an oil quenchant, an already violent period, is not drastically altered by changes in bath temperature. The rate of cooling in the convection stage of an oil quenchant will become slower as the temperature of the bath increases. This is advantageous for a slower rate of cooling through austenite to martensite transformation generally reduces the tendency of distortion or cracking. In general, as the temperature of a quenching oil increases, the overall quenching rate increases.

The preceding discussion deals with the mechanics or what occurs in the quenching process, quenchant make-up or formulation to that discussion is secondary. However, the quenchant is very critical to the overall hardening process. In the late 1930's and early 1940's with World War II, two events occurred which forced a change in quenching oils. These events were; one, the shortage of alloying elements and two, the need for components with higher physical requirements, i.e., tensile impact, etc. In order to achieve this, the quenching oil had to be "faster."

Heat treaters for many years had a practice of adding salt to water quench baths to make the water quench "faster" or at least more uniform. This same principle applies to oil quenchants and quenchant researchers during this period began testing various organometallics and resins. These materials were added to base oils with positive results and the "modern" era of accelerated quenching oils began.

These materials or additives are often called "wetting agents" and a theory exists concerning their effect on the cooling characteristics of the quenching media. The resins or organo metallics are soluble in the oil and, at the moment of immersion, come in contact with the hot component. Instantaneously, the vapor blanket stage begins but remaining on the surface of the component are resin or organometallic crystals. These crystals carry with them, on a molecular scale, oil which superheats and explodes violently disrupting the vapor blanket stage hastening the boiling stage. The addition of wetting agents also makes the quenching action more uniform. Combine this with the shorter duration of the vapor blanket stage, the problems of insufficient hardness, spotty hardness and distortion of the component can be eliminated or reduced.

Since the beginning of the modern era of quenching, the need has continued for more specialized quenchants. Media needed to harden steel and cast irons and to solution treat aluminum alloys free of distortion while providing maximum mechanical properties. In the late 1940's, martempering oils were used in place of molten salt baths. Today the use of martempering oils is increasing. Industry is taking advantage of cost savings associated with martempering oils. Cost savings arising from near-net-shape forming and machining processes. This, in combination with the Global drive for improved quality is causing industry and quenchant researchers to investigate and develop improved products and processes.

Another development during the modern era of quenching is that of water based polymer quenchants. First introduced in the 1950's, these products filled the need in the growing surface hardening processes of induction and flame hardening. They provided the quenching characteristics which filled the void between those of water and the fastest quenching oils. This technology improved during the 1960's with introduction of glycol based quenchants used to replace quenching oils. In the early 1970's several factors positively influenced the use and development of polymer quenchants. Increased concern for plant and natural environment along with worker safety pressured many industries to adapt the use of polymer quenchants. In 1973 and 1974 with the shortage of petroleum oils and the subsequent rise in the cost of quenching oils along with their shortages gave industry the needed impetus to begin using polymer quenchants.

Today there are many different types of polymer quenchants used in the heat treatment of metals. Many of these are specialized products aimed to fill the need of specific industries. The polymer quenchant researcher has several chemical families to choose from for the testing and development of polymer quenchants. The understanding and knowledge of these materials has advanced to the point where the polymer molecules can be engineered to provide a particular characteristic; quenching or other to fill the requirements of a particular application.

Looking ahead, what does the future hold for quenchants? The answer to this lies with Government regulations, available resources, environment, industry need and many other areas. Research is underway to improve the characteristics of water soluble polymers, anhydrous polymers, vegetable oils and specially treated petroleum products as well as today's standard products. Whatever the course quenchants and hardening take, the aim is still the same as stated in Steel and Its Treatment (1914). In order to successfully harden one must study the various parts, their design and hardness desired. With the above quenching media a wide range of hardnesses can be secured and any condition satisfied if the bath is modified to suite the purpose.

References

1.   E. F. Houghton & Co., Steel and Its Treatment, Philadelphia, (1914)

2.   E. F. Houghton & Co., Practical Metallurgy for Engineers, Philadelphia, (1923 and 1952)

3.   Dicken, T. W., Heat Treatment of Methods, 1986.1, 6-8 (1986)

4.   Bates, C. E. and Totten, G. E., Heat Treatment of Metals, 1992.2, 45-48 (1992)

5.   Bodin, J. and Segerberg, S., Heat Treatment '84, London, England, May, 1984

Workshop Designed System for Quenching Intensity Evaluation and Calculation of Heat Transfer Data

B. Liscic, S. Svaic and T. Filetin
University of Zagreb
Zagreb, Croatia

ABSTRACT

Computer simulation of a quenching process that aims to predict precisely the resulting microstructures and the hardness distribution for components in workshop practice, needs first of all, as input, real and accurate heat transfer data to calculate cooling curves at every arbitrary point of the cross-section. A system has been developed for quenching intensity measurement, recording and evaluation and for calculation of heat transfer data for all quenchants, different quenching conditions and different quenching techniques. The main part of the hardware is the LIŠČIĆ-NANMAC cylindrical probe instrumented with multiple thermocouple assembly. Based on the TEMPERATURE GRADIENT METHOD the system enables to calculate and graphically present the heat flux on the surface of the probe during the whole quenching process. As an example, which conclusions can be drawn from it, an investigation of the influence of polymer-solution (PAG) concentration is presented. By using another software-module "PROBE", calculation of the real, temperature dependent heat transfer coefficient on the probe surface is possible. This heat transfer coefficient together with the surface temperature measured by the probe, enable calculating the cooling curves at every arbitrary point of the round bars cross-section. Examples for quenching in oil and in a polymer-solution (PAG) of high concentration are given. If, instead of the temperature dependent heat transfer coefficient, a constant (average) value of it is used for cooling curves calculation, the accuracy of such calculation is substantially decreased, as it is shown by error functions that are compared to measured temperature - time data.

IN PRACTICE, COMPUTER SIMULATION OF THE QUENCHING PROCESS of real components must take into account the following realities:

- Quenching conditions in practice (influence of other workpieces in the batch, actual flow rate and direction of the quenchant etc.) are substantially different from those prevailing in laboratory tests.

- Small diameter (usually 12.7 mm Dia) and small-mass specimens used in laboratory tests cool down to below 200°C in about 20-30 seconds when quenched in unagitated oil, while e.g. a cylindrical workpiece of 50 mm Dia for same quenching conditions, to be cooled in its center below 200°C, needs more than 600 seconds.

- When measuring and recording the temperature vs. time data within a cooling body, one should be aware of the damping effect i.e. the more the measuring point is shifted below the surface towards the core, the more real heat transfer phenomena, taking place on the surface, are damped. Clearly, Fig. 3a shows e.g. the difference between a temperature-time curve measured 1.5 mm below the surface of a 50 mm Dia bar, compared to the temperature-time curve measured on the same radius on the very surface, - during oil quenching.

Because of these realities the basic concept of the system described is the following:

a) To measure and record the temperature during the whole quenching process of a probe, which with its mass and shape resembles real components.

b) When investigating batch immersion quenching, the probe itself must be heated and quenched together with the batch of workpieces.

c) The temperature has to be measured as close as possible to the probe's surface.

When designing the system we have taken into account that nowadays many different quenching media under different quenching conditions (bath temperature, concentration, agitation rate etc.) are used, and different quenching techniques exist (direct immersion quenching, interrupted quenching, spray-quenching). There is a need, consequently, to compare the quenching intensity and thermal stresses developed among very many different quenching conditions that can be realized in order to optimize the quenching process in every specific case. One has to be aware that the temperature dependent heat transfer coefficient between a cooling body and the surrounding fluid, when quenching in an evaporable fluid, depends on many factors such as e.g.:

A. Factors which depend on the body itself:
  - Shape and size of the body
  - Position of the body (standing or lying)
  - Material of the body
  - Surface condition (roughness chemical state)

B. Factors which depend on the surrounding fluid:
  - Specific characteristics of the fluid (viscosity; wetting properties; spec. heat; boiling temperature; transition or Leidenfrost's temperature)
  - Temperature of the surrounding fluid
  - Pressure of the fluid
  - Flow rate and direction.

The combination of so many influential factors is, of course, different in every practical case. Therefore, no real heat transfer data can be calculated without having a temperature-time record taken near the surface of the probe in actual quenching conditions.

In selecting the mass, size and shape of the instrumented probe it was necessary to satisfy the following requirements:

- It should enable one-dimensional, symmetric heat flow in the cross-section where the thermocouples are placed.
- It should be applicable to all quenchants, to different quenching conditions and quenching techniques.
- It should be sensitive enough to indicate and record the change of every quenching parameter (concentration; bath temp.; agitation rate).
- It should maintain always the same surface condition.
- The dependence of the heat transfer coefficient on the shape and size of the probe should be as small as possible.
- It should have good durability and yield reproducible results.

Fig. 1 shows the layout of the system and its possibilities.

HARDWARE OF THE SYSTEM

The main part of the hardware is the probe shown in Fig. 2. It is a cylinder of 50 mm Dia and 200 mm length, made of AISI-304 having three thermocouples placed on the same radius in the mid-length cross-section. The outer thermocouple (T_N) measures the temperature on the very surface, the intermediate one (T_I) measures the temperature at the point

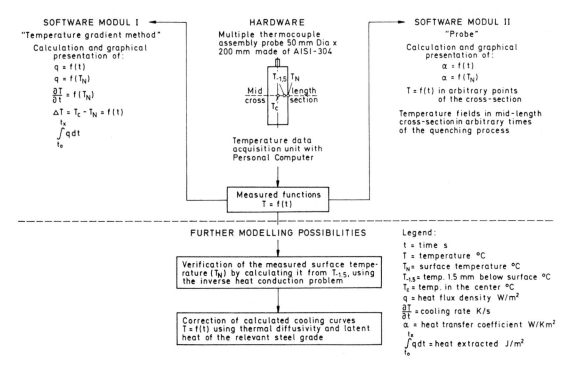

Fig 1. Workshop designed system for quenching intensity measurement, recording and evaluation and heat transfer data calculation

Within the figure:

SOFTWARE MODUL I
"Temperature gradient method"
Calculation and graphical presentation of:
$q = f(t)$
$q = f(T_N)$
$\frac{\partial T}{\partial t} = f(T_N)$
$\Delta T = T_c - T_N = f(t)$
$\int_{t_0}^{t_x} q\, dt$

HARDWARE
Multiple thermocouple assembly probe 50 mm Dia x 200 mm made of AISI-304

$T_{-1.5}$ T_N
Mid cross — length section
T_c

Temperature data acquisition unit with Personal Computer

SOFTWARE MODUL II
"Probe"
Calculation and graphical presentation of:
$\alpha = f(t)$
$\alpha = f(T_N)$
$T = f(t)$ in arbitrary points of the cross-section
Temperature fields in mid-length cross-section in arbitrary times of the quenching process

Measured functions
$T = f(t)$

FURTHER MODELLING POSSIBILITIES

Verification of the measured surface temperature (T_N) by calculating it from $T_{-1.5}$, using the inverse heat conduction problem

Correction of calculated cooling curves $T = f(t)$ using thermal diffusivity and latent heat of the relevant steel grade

Legend:
t = time s
T = temperature °C
T_N = surface temperature °C
$T_{-1.5}$ = temp. 1.5 mm below surface °C
T_c = temp. in the center °C
q = heat flux density W/m²
$\frac{\partial T}{\partial t}$ = cooling rate K/s
α = heat transfer coefficient W/Km²
$\int_{t_0}^{t_x} q\, dt$ = heat extracted J/m²

1.5 mm below the surface, and the inner thermocouple measures the temperature in the center of the cross-section. The probe has been developed in cooperation with, and is made by: NANMAC Corp. Framingham Centre, Mass, USA.

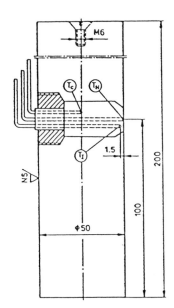

Fig. 2 The LIŠČIĆ-
NANMAC probe

Measurement of the surface temperature (of the body itself), that has until now been considered to be rather inaccurate, if at all possible, becomes possible and reproducible by using a new concept of a fast response thermocouple for transient thermal measurements (proprietary NANMAC Corp. - U.S. Pat. No 2, 829.185). A more detailed description of such thermo-couples is given in [1]. Specific features of the mentioned probe are:

- Extremely short response time of the surface thermocouple (10^{-5} s)
- The surface condition of the probe can been maintained constant by polishing the sensing tip of the thermocouple before each measurement.
- The body of the probe, made of an austenitic stainless steel, does not change in structure during the heating-quenching process, nor does it evolve or absorb heat because of phase changes.
- The size of the probe and its mass ensure a sufficient heat capacity and a symmetrical radial heat flow in the cross-sectional plane where the thermo-couples are located.
- The dependence of the average heat transfer coefficient during the boiling

stage is according to [2] , substantial only for diameters less than 50 mm.

For every test the probe is heated up to 850°C in a protective atmosphere furnace and transferred quickly (in less than 3 seconds) to the quenching bath and immersed. The temperature data acquisition unit consists of an interface having three analog-to-digital converters and three amplifiers connected to a personal computer.

THE TEMPERATURE GRADIENT METHOD

Fig. 3a shows temperature-time curves for all three thermocouples, measured and recorded by the probe, when quenching it in a mineral oil of 20°C, without agitation. Using the TEMPERATURE GRADIENT METHOD - see [1] [3] , which is based on the known physical rule that the heat flux within a body is equal to its thermal conductivity multiplied by the temperature gradient, the developed software - Module I, enables calculation and graphical presentation of the heat flux density between the point 1.5 mm below the surface and the surface itself vs. time (Fig. 3b) and vs. surface temperature (Fig. 3c) respectively.

Also, the temperature difference between the core and surface vs. time is calculated (Fig. 3d) thus characterizing the particular quenching process in respect to thermal stresses.

Graphical presentation of the heat flux density near the surface of the probe as shown in Fig. 3b enables the evaluation of the quenching intensity of a particular quenching process with respect to expected depth of hardening. For this evaluation two criteria should be used:
- the value of maximum heat flux density q_{max} W/m^2 , and
- the integral below the heat flux density curve for a certain period of time (representing the heat extracted) $\int_{to}^{tx} q\, dt$ J/m^2.

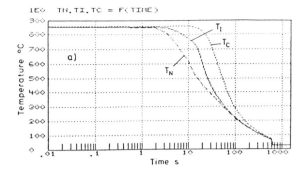

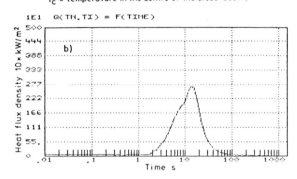

T_N = surface temperature
T_I = temperature 1.5 mm below the surface
T_C = temperature in the centre of the cross-section

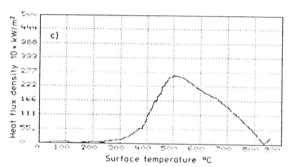

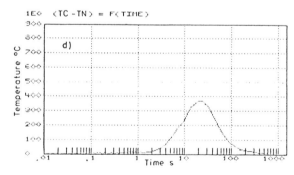

Fig. 3. Temperature v.s. time (a); heat flux density v.s. time (b); heat flux density v.s. surface temperature (c) and temperature difference core-surface v.s. time (d) - when quenching the probe in a mineral oil of 20°C, without agitation.

By comparing two quenching processes, the process which yields higher value of q_{max} and gives higher value of $\int_{t_o}^{t_x} qdt$ (t_x selected correspondingly) will certainly produce a greater depth of hardening . The heat flux density vs. surface temperature curve (Fig. 3c) enables evaluating the particular quenching process in respect to the risk of cracking due to transformation stresses that occur in the martensite formation range.

For this evaluation the following criterion should be used:
- the value of surface temperature at the moment when max. heat flux density occurs ($T_{q\ max}$).

The lower the value of $T_{q\ max}$, the higher is the risk that (especially for steels having high M_s temperature) tensile residual stresses occur on the surface, which increases the risk of cracking. It is a well known fact, that quenching in water (having lower $T_{q\ max}$ than oil) increases the risk of cracking, compared to oil quenching.

In order to demonstrate the sensitivity of the method to the change in some of quenching parameters an investigation of the influence of polymer-solution (PAG) concentration has been undertaken. The only variable was the polymer concentration, while all other quenching parameters have been kept constant (bath temperature 40°C, agitation rate 0,8 m/s). Fig. 4 shows the measured time-temperature curves, and calculated heat flux density vs. time curve when the probe was quenched in a polymer-solution of 5% concentration.

Fig. 5 shows the same for quenching in the same polymer solution of 15% concentration, and Fig. 6 for quenching in the same polymer solution of 25% concentration.

By comparing figures 4, 5 and 6, one can easily note, that by increasing the polymer concentration the heat flux density maximum (q_{max}) becomes smaller and the time when q_{max} occurs ($t_{q\ max}$), becomes longer.

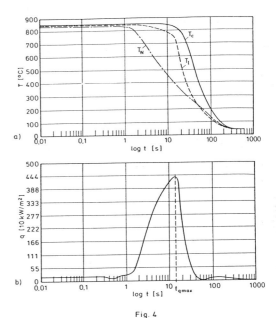

Fig. 4

Fig. 4 Measured temp. vs. time (a) and calculated heat flux density vs. time (b), when quenching the probe in polymer-solution (PAG) of 5% concentration, 40°C bath temp. and 0,8 m/s agitation rate.

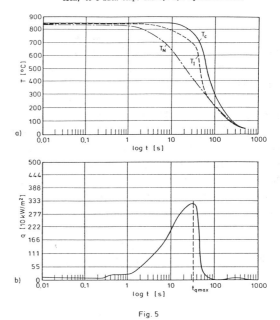

Fig. 5

Fig. 5 Measured temp. vs. time (a) and calculated heat flux density vs. time (b), when quenching the probe in polymer-solution (PAG) of 15% concentration, 40°C bath temp. and 0,8 m/s agitation rate.

The delay in the occurrence of the heat flux density maximum means that the first (vapour blanket) stage, of relative slow cooling, has been prolonged. Note also, that in case of oil quenching - see Fig. 3b, the heat flux density falls from its maximum to zero in about 86 seconds, while in the case of 25% concentration of particular polymer solution, (Fig. 6b) it falls in only 25 seconds, i.e. 3,4 times faster.

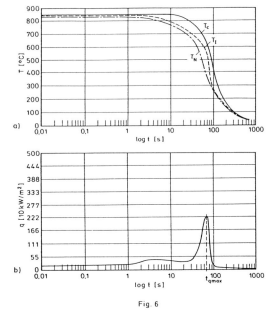

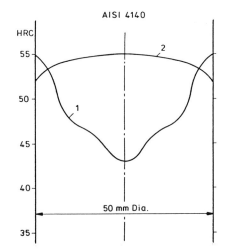

Fig. 6 Measured temp. vs. time (a) and calculated heat flux
density vs. time (b), when quenching the probe in
polymer-solution (PAG) of 25% concentration, 40°C
bath temp. and 0,8 m/s agitation rate.

Fig. 7a   Measured hardness after quenching a 50 mm Dia
x 200 mm bar made of AISI-4140 in:
1. Mineral oil of 20°C, without agitation
2. Polymer-solution (UCON-E) of 25% concentration,
40°C bath temperature and 0,8 m/s agitation rate.

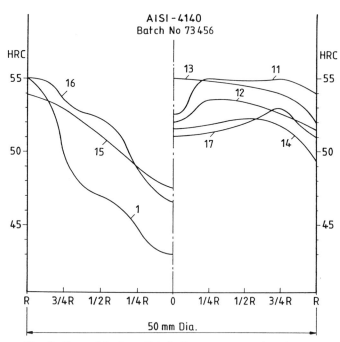

Fig. 7b Measured hardness distributions on cross-section of
50 mm Dia x 200 mm bars made of AISI-4140, quenched
according to quenching conditions in Table I.

This clearly distinguishes oil quenching from the particular polymer-solution quenching which is characterized by a very long period of slow cooling followed by a sudden change in the cooling rate. According to Shimizu and Tamura [4] [5] such pronounced discontinuous change in the cooling rate has a specific influence on transformation behaviour of steel concerned. It may, in some circumstances, (depending on steel hardenability and section size) result in "inverse" hardness distribution. This phenomenon is related to the incubation period (at different points of the cross-section) consumed before changing the cooling rate.

To find out whether or not this phenomenon will occur, two round bars of 50 mm Dia x 200 mm made of AISI-4140 were quenched under different quenching conditions:

- Bar 1 - in mineral oil of 20°C, without agitation,
- Bar 2 - in particular polymer-solution (PAG) of 25% concentration, 40°C bath temperature and 0,8 m/s agitation rate.

Fig. 7a shows the measured hardness distribution curves in the mid-length cross-section after quenching.

Further investigation has shown - see Fig. 7b, that a direct correlation exists between the time t_{qmax} and hardness distribution pattern, as it is shown in Tab. I. and Fig. 7b.
There is obviously a limit for t_{qmax} below which "normal" hardness distribution pattern exists - see left side in Fig. 7b, and above which "inverse" or "inverse-normal" hardness distribution exists - see right side in Fig. 7b.

Tab. I. Times $t_{q\,max}$ corresponding to Fig. 7b

Curve No	Quenching conditions	$t_{q\,max}$ s
1	Mineral oil 20°C, without agitation	14
11	Polymer-solution (PAG) 5%; 40°C; 0,8 m/s	16
12	Polymer-solution (PAG) 15%; 40°C; 0,8 m/s	33
13	Polymer-solution (PAG) 25%; 40°C; 0,8 m/s	70
14	Polymer-solution (PAG) 20%; 35°C; 1 m/s	30
15	Polymer-solution (PAG) 10%; 35°C; 1 m/s	12
16	Polymer-solution (PAG) 5%; 35°C; 1 m/s	13
17	Polymer-solution (PAG) 20%; 35°C; 1 m/s	47

In the case of AISI-4140 steel grade and 50 mm Dia size this limit for q_{max} to occur from immersion was 15 seconds ($t_{q\,max}$, crit. = 15 s).

CALCULATION OF THE HEAT TRANSFER

The software module II, named "PROBE" enables the calculation of the temperature dependent heat transfer coefficient vs. time or vs. surface temperature respectively; the cooling curves at any arbitrary point of the cross-section as well as temperature fields in mid-length cross-section of the probe for any desirable time of the actual quenching process.

Assuming that at mid length cross-section, only one dimensional, radially symmetrical heat conduction takes place, the temperature distribution along the radius was calculated by using one-dimensional Fourier equation in cylindrical coordinates:

$$\rho \cdot c \frac{\partial T}{\partial t} = \frac{1}{r} \frac{\partial}{\partial r} (r \cdot \lambda \cdot \frac{\partial T}{\partial r}) \quad (1)$$

ρ = density of the probe's material kg/m^3
c = spec.heat   J/kg K
λ = heat conductivity   W/mK
T = temperature   °C
t = time   s
r = radius   m
R = outer radius   m
q = heat flux density   W/m^2
α = heat transfer coefficient   W/m^{2}K
T_s = surface temperature of the body   °C
T_∞ = ambient temperature of the fluid °C

Integration of Eq 1 from r=0 to r=R leads to the heat flux on the surface:

$$q(t) = \frac{1}{R} \int_{r=0}^{r=R} \rho \cdot c \cdot \frac{\partial T}{\partial t} \cdot r \cdot dr \quad (2)$$

By taking as input the values of surface temperature measured by the probe the heat transfer coefficient from the heat flux on the surface can be calculated:

$$\alpha_{(t)} = \frac{q(t)}{T_{s\,(t)} - T_{\infty\,(t)}} \quad (3)$$

The problem was numerically solved - see [6] , by discretization of the cylinder into "n" control volumes giving a system of algebraic equations which can be solved by Three Diagonal Matrix Algorithm. The solution of this system gives the temperatures at every of the control volume points as a function of the surface temperature and the heat transfer coefficient in observed time.
Fig. 8 shows the calculated heat transfer coefficient vs. time and vs. surface temperature, respecitvely, when quenching the probe in mineral oil of 20°C without agitation. Fig. 9 shows the same for the case when the probe was quenched in a polymer-solution (PAG) of 25% concentration, 40°C bath temperature and 0,8 m/s agitation rate.
Fig. 10 shows calculated cooling curves for the center of a bar of 50 mm Dia, using heat transfer coefficients data from Fig.8 and Fig.9, respectively, -

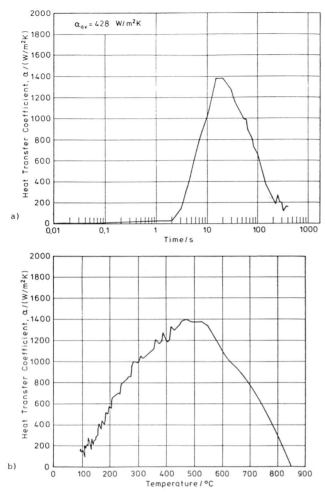

a)

b)

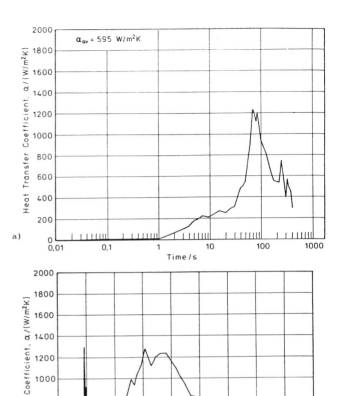

a)

b)

Fig. 8 Heat transfer coefficient vs. time (a) and vs. surface
temperature (b) when quenching from 850°C a bar of
50 mm Dia in mineral oil (KALENOL 25 S) of 20°C,
without agitation.

Fig. 9 Heat transfer coefficient vs. time (a) and vs. surface
temperature (b) when quenching from 850°C a bar of 50 mm
Dia in a polymer-solution (UCON-E) of 25% concentration,
40°C bath temperature and 0,8 m/s agitation rate.

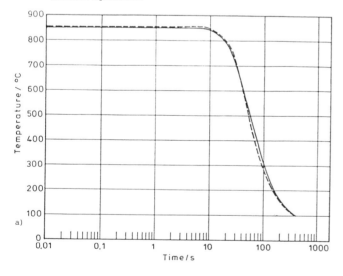

a)

b)

Fig. 10   Comparison of measured and simulated cooling curves for
the center of a 50 mm Dia x 200 mm bar, using variable α.
a) quenched in mineral oil of 20°C, without agitation
b) quenched in polymer-solution (UCON-E) of 25% concentr.;
40°C and 0,8 m/s

------- measured   ———— simulated

24

compared to cooling curves measured by the central thermocouple of the probe (T_C) having 50 mm Dia x 200 mm, for relevant quenching conditions.

If, instead of temperature dependent heat transfer coefficient data (variable α), a constant value of α is used for calculation of the cooling curve, the accuracy of such calculation is substantially decreased, as it can be seen from Fig. 11.

The error, as a function of time, compared to measured cooling curve in the case of oil quenching, is shown in Fig. 12, both for the calculation with a variable α and for calculation with a constant (average) value of α.
Fig.13 shows calculated temperature fields in the mid-length cross-section of the probe, when quenching it in oil of 20°C without agitation and Fig. 14 shows the same for quenching it in the above specified polymer-solution.

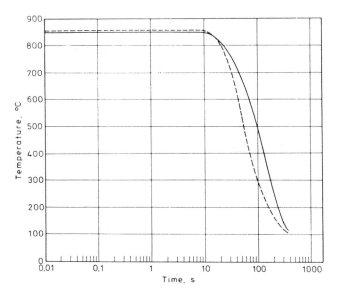

Fig.11 Comparison of measured and simulated cooling curves for the center of a 50 mm Dia x 200 mm bar, quenched in mineral oil of 20°C without agitation, using constant (average) value of α.

--------- measured   _____ simulated

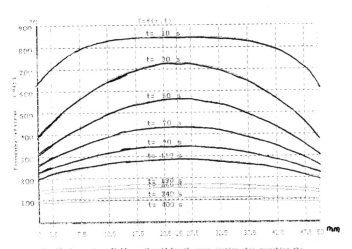

Fig. 13 Temperature fields on the mid-length cross-section when quenching the probe (50 mm Dia x 200 mm) in mineral oil of 20°C, without agitation.

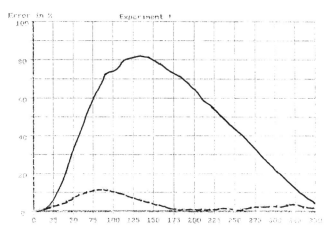

Fig.12 Error in percentage, based on measured cooling curves for center of a 50 mm Dia bar, quenched in mineral oil of 20°C, without agitation.
---------- calculated using variable α
_____ calculated using constant (average) value of α

The following software module developed will enable the verification of the measured surface temperature, $T_N = f(t)$, by calculating it from measured time-temperature data in the point 1.5 mm below the probe's surface, solving the "inverse heat conduction problem".
In order to increase further the accuracy of simulated cooling curves for specific steel grade, relevant thermal diffusivity (i.e. physical properties) should be used and latent heat of transformation should be taken into account.

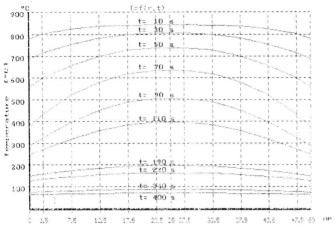

Fig. 14 Temperature fields on the mid-length cross-section when quenching the probe (50 mm Dia x 200 mm) in the polymer-solution (UCON-E) of 25% concentration, 40°C bath temp. and 0,8 m/s agitation rate.

CONCLUSION

The Workshop Designed System for Quenching Intensity Evaluation and Calculation of Heat Transfer Data, consisting of described hardware and software modules, aims to fulfill two main tasks:

I. To evaluate and compare the quenching intensity in real workshop conditions of different quenchants, quenching conditions and quenching techniques - regarding resulting depth of hardening, thermal stresses, sensitivity to cracking in martensite transformation range and possible pattern of the hardness-distribution curve.

II. To provide for, in every specified quenching process, real temperature -time data in order to enable accurate calculation of relevant temperature dependent heat transfer coefficients, cooling curves at arbitrary points of the round bar cross-section, and temperature fields during the whole quenching process. This task is the first step in every computer-aided modelling of a quenching process and resulting proper-

ties, and therefore needs to be performed as accuratelly as possible.

The used Temperature Gradient Method is applicable not only to cylindrical, but also to any other symmetrical shape of the probe (plate, sphere).

REFERENCES

[1] Liščić, B. and T.Filetin, Journal of Heat Treating, Vol 5, No 2, 115-124, (1988)

[2] Kobasko, N.I., Metallovedenie i termičeskaja obrabotka metallov, No 3 (1968)

[3] "HEAT TREATMENT AND SURFACE ENGINEERING" (Conference Proceedings), Edited by G.Krauss, p. 157-166, ASM-International (1988)

[4] N.Shimizu and I.Tamura, Transactions ISIJ, Vol 17, 469-476 (1977)

[5] N.Shimizu and I.Tamura, Transactions ISIJ, Vol 18, 445-450 (1978)

[6] S.Švaić: "Simulation of Cooling a Cylinder in the Surroundings of Arbitrary Chosen Temperature", Faculty of Mech.Eng., University of Zagreb (1991), - unpublished research.

Possibilities and Limits to Predict the Quench Hardening of Steel

H.M. Tensi and A. Stich
Technical University of Munich
Munich, Germany

Abstract

A definte prediction of the final hardness from the course of cooling is only possible by a quantification of the structural constituents formed during contineous cooling. In spite of this restriction fairly accurate hardness predictions can be made by assigning a hardness to a cooling curve. But, generally this is only possible for simple rotation symmetrical parts.

H-model and QTA-model are simplified methods to describe the heat transfer on the sample surface and are therefore not suited for a precise description of the mechanical properties. Especially, the wetting process can have a significant influence on the hardness distribution throughout a component volume.

The prediction of the metallurgical transformation during contineous cooling is a prerequisite for the calculation of residual stresses and distortion after quenching. But, in this very complicated field of quenching adequate research work is necessary in order to clarify many effects and correlations which are still uncertain or unknown.

THE HARDNESS OF A COMPONENT is determined by the microhardness of the different microstructures and microstructural combinations formed during cooling. Since different microstructural combinations can result in the same hardness, the relation (eq. 1)

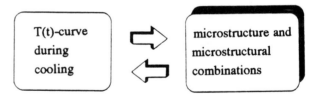

is non-reversible. A clear reversible relation only exists between the course of cooling and the microstructural formation (eq. 2):

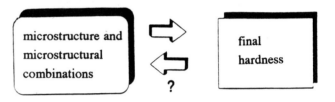

Therefore, a definite prediction of the final hardness from the course of cooling is only possible by using the following chain (eq. 3):

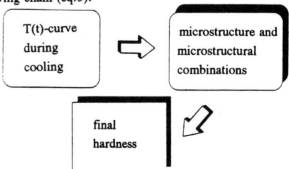

According to this relation a prediction of hardness always must result in a quantification of the transformation structure. In spite of this restriction, quite accurate hardness predictions can be made using the simplified way of assigning a hardness to a cooling curve (eq.4):

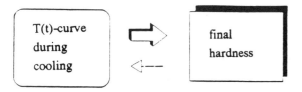

A prerequisite for the availability of this relation is that all metallurgical and thermal boundary conditions are kept as similiar as possible.

H-VALUE ACCORDING TO GROSSMANN

It is common known that during hardening of cylindrical bars the unhardened core zone D_u takes off with decreasing bar diameter. According to Grossmann the border areas between hardened and unhardened zone exhibit a structure of 50 % martensite. That means that all points at the border of depth hardening have followed the same or at least very similiar cooling.

To calculate the border of depth-hardening Grossmann introduced the H-value assuming NEWTONION-cooling. Points having undergone the same course of cooling are assumed to have the same H value. The "heat transfer equivalent" H comes from the non-steady-state heat conduction equation and is used instead of the Biot-number:

$$H \cdot D = Bi = (\alpha/\lambda) \cdot R \qquad (5)$$

hence it follows:

$$2H = \alpha/\lambda \qquad (6)$$

H : heat transfer equivalent in m^{-1};
Bi : Biot-number;
α : heat transfer coefficient in $Wm^{-2}K^{-1}$;
λ : heat conductivity in $Wm^{-1}K^{-1}$

The H value is independant on the sample dimension, but dependant on the type, composition, viscosity and temperature of the fluid, on the level of forced convection, and on the surface quality of the component. Some clues for H are listed in Table I.

Table I. Clues for the value H in $inch^{-1}$ [1].

agitation	air	oil	water
no agitation	0.02	0.25-0.30	0,9-1,0
moderate agit.	-	0.35-0.40	1.2-1.3
strong agitat.	0.05	0.50-0.80	1.6-2.0

The most serious drawback of the H-value is, that it is a constant value and cannot describe the heat transfer on the surface of a sample during quenching in an evaporable fluid.

Hardness predictions on the basis of the H-value can be easily done by using CRAFTS-LAMONT diagrams [1]. The CRAFTS-LAMONT diagram in Fig.1 describes the correlation between the distance x of the front-end in a Jominy sample and the diameter of the sample for constant heat equivalent values H and a radius ratio of r/R = 1. Lamont has developed these diagrams for different radius ratios in the range of $0 \leq r/R \leq 1$ in steps of 0.1.

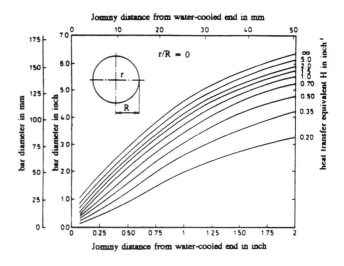

Fig.1 - CRAFTS-LAMONT diagram for a radius ratio of r/R = 1.

QTA-Method

In comparison to the H-model, the heat transfer on the surface is seperately defined for the different cooling phases film boiling, nucleate boiling and convective heat transfer [2].

The meaning of the model parameters is as follows:
- "Q" is the average heat flux density of the vapour blanket phase and is defined in the temperature range from the austenitizing temperature T_a down to 500°C. Note that the Leidenfrost temperature T_L is thus set to a constant temperature of 500°C. To determine the Q-value a standart steel probe with known hardenability is used.
- "T" is very near to the boiling temperature T_b of the fluid and does not depend on cooling conditions. Thus it can be determined from laboratory tests.
- "A" is the average heat transfer coefficient of the convection phase and can be calculated using known heat transfer laws.

The heat transfer on the surface within the QTA-model is illustrated in Fig.2. During the film boiling phase the variations in temperature are calculated by using the constant heat flux Q. In the nucleate boiling phase the heat transfer on the surface increases by up to two orders of magnitude so that the heat transfer can be calculated using the first boundary condition. In order to use this infinite high heat transfer coefficient the diameter of the sample is increased by $D' = D + 2r'$. According to Wünning $r' = 1$ mm for water and $r' = 2$ mm for oil quenching. The cooling in the convection phase is calculated by using the constant heat transfer coefficient A.

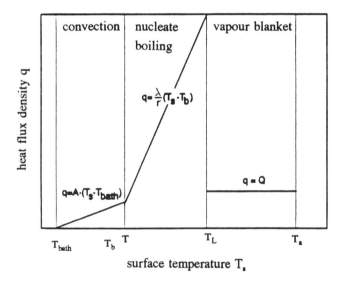

Fig.2 - Parameters of the QTA-model.

According to Wünning the value Q can be determined by a simple hardness measurement. A standardized cylindrical sample made of a low-alloyed steel is quench-hardened under workshop conditions and subsequently the hardness is measured at a radius ratio of $r/R = 0.7$. Comparison with a "calibration curve" yields the corresponding Q value.

The QTA-model was tested for qualification within the AWT [*] recently [3]. It was noted that
- the temperature curve at any point in a sample can be calculated with satisfactory accuracy, when the course of cooling down to 500 °C is not decisive,
- the Q-value cannot be rendered just like that to any given geometry,
- a single Q-value is not sufficient for the whole spectrum of quenching fluids, and
- the relation between hardness measured in the Q-probe at a radius ratio of $r/R = 0.7$ and corresponding Q value is not definite since different microstructures (resulting from different cooling rates) can have the same hardness.

Prediction of Hardness Distribution from Wetting Kinematics

During immersion cooling in an evaporable fluid the three different cooling phases can be simultaneously present on the sample's surface for a significant period of time. This causes great axial temperature differences and can result in completely different radial hardness distributions throughout the component volume. A method which takes into account this process of wetting is based on the following idea: The wetting time describes the thermal cycle on the surface of a component with adequate accuracy. Therefore, the same hardness is to be expected in locations having the same wetting time. The relation between surface hardness and wetting time can thus be used as a calibration curve for hardness prediction. It allows predetermination of the surface hardness resulting from different wetting processes [4].

Fig.3 illustrates the prediction of the surface hardness from measured wetting time t_b and calibration curve. When wetting proceeds slowly, the wetting edge arrives at location z^* (dot 1) after a wetting time of t_b^* (dot 2). In accordance with the calibration curve, this wetting time corresponds to a surface hardness of HV^* (dot 3). When this hardness value is transferred to the actual location of

---

* AWT: Arbeitsgemeinschaft Wärmebehandlung und Werkstofftechnik

the sample z^*, the surface hardness at this location is ascertained (dot 4).

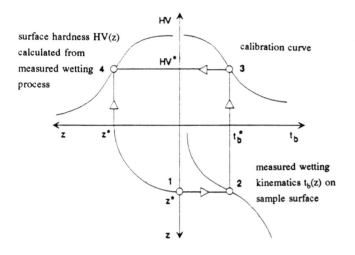

surface hardness HV(z) calculated from measured wetting process

calibration curve

measured wetting kinematics $t_b(z)$ on sample surface

Fig.3 - Prediction of surface hardness from measured wetting time t_b and calibration curve.

ne hardness distribution throughout the component v ne is determined by use of the CRAFTS-LAMONT diagrams and the relevant Jominy hardness curve. The H value for the cylindrical slice in question is determined from the calculated surface hardness value and the Crafts-Lamont diagram for the surface. When wetting proceeds slowly different H values must be calculated for the various front-end distances of the immersed sample [4].

As an example Fig. 4 depicts radial hardness profiles predicted (line) and measured as HV 10 values (dots) for two cross-sections of a cylindrical steel sample. Since wetting proceeds slowly (the total wetting time is about 32 s), the hardness distributions in different heights of the sample are completely different. The radial hardness profile at distance z = 10 mm is extraordinarily steep, while the hardness curve at distance z = 80 mm is nearly constant. In both cases the surface hardness was calculated from the calibration curve.

The main advantage of the described method is that it takes into account the process of wetting, the metallographic structure before austenitizing and the austenitizing conditions. However, the determination of the wetting process on a sample surface during immersion-cooling in a quenching tank under workshop conditions is very difficult. Moreover, the calibration curve cannot be rendered to other geometries without problems.

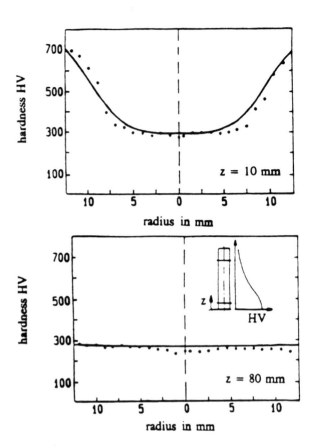

Fig.4 - Radial hardness distributions in a cylindrical part at two different distances z from the lower end; sample: dia 25 mm x 100 mm length, C 45; quenchant: non-agitated water at 35 °C.

Calculation of Hardness from Microstructure

The hardness distribution in components of any desired geometry can be predicted by using numerical methods such as finite difference or finite element methods provided the phase transformation of steels can be calculated with sufficient accuracy. Input are measured diagrams for isothermal transformation.

A frequently used method is as follows [5]: The thermal microstructural transformation of austenite at a constant temperature is calculated using a function according to the law of Avrami-Johnson-Mehl:

$$w = 1 - \exp\left(-b \cdot \left(\frac{t}{t_0}\right)^n\right) \qquad (7)$$

w : volume fraction of austenite transformed

b,n : coefficient and exponent of the austenite transformation kinetics,

t : time, t_0 : reference time.

The values of b and n are dependant on temperature and they are varying for ferrite, perlite and bainite transformation. For the determination of b and n the relevant isothermal TTT-diagramm is copied into a computer. The values of b and n are calculated from the curves of 1 % and 99 % volume fraction transformed according to equation (7).

Below the martensite start temperature M_s, it is assumed that the remaining austenite is transformed into martensite according to equations as:

$$w_M = 1 - \exp(c \cdot (M_s - T)^m) \qquad (8)$$

$$w_M = 1 - \left(\frac{T - M_f}{M_s - M_f}\right)^m \qquad (9)$$

w_M : transformed volume fraction of austenite into martensite,

M_s : martensite start temperature; M_f : martensite finish temperature,

T : transformation temperature, $T \le M_s$,

c,m : coefficient and exponent of austenite transformation.

If the transformation of austenite shall be calculated for any time temperature cycle, a contineous cooling curve is divided into constant temperature steps with appropriate times (Fig.5). It is assumed that the horizontal parts of this step function cause a transformation comparable to the transformation occuring at the individual temperatures in the isothermal TTT-diagram. By an iteration of the transformation steps the final microstructure is derived.

The final hardness can then be determined by applying an additivity rule if the hardness values of the different microstructures formed at defined temperatures are known:

$$HV = \sum \left(\sum(\Delta w_k(T_i) \cdot HV_k(T_i))\right) \qquad (10)$$

HV : final hardness at a defined location

w_k : volume fraction of microstructure k formed at temperature T_i

HV_k : microhardness of microstructure k formed at temperature T_i

The main advantage of this method is that it uses existing data about the transformation behaviour of austenite referring to isothermal TTT-diagrams. But the method also has some serious drawbacks: the isothermal TTT-diagram usually fails to give a sufficiently reliable information of the transformation behaviour of austenite due to deviations in the chemical composition, variations in the initial microstructure and different austenitizing conditions. In

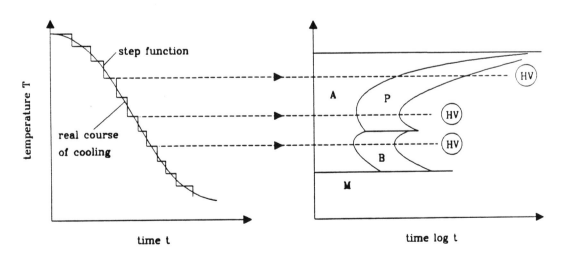

Fig.5 - Calculation of microstructure and hardness during contineous cooling by use of an isothermal TTT-diagram.

addition, the question appears whether it is exactly permissable to apply the isothermal transformation kinetics of austenite to a contineous cooling curve which is devided into a step function. Finally, equation (7) is only based on empirical data about the transformation behaviour.

References

1. Crafts, W. and J.L. Lamont, "Härtbarkeit und Auswahl von Stählen", Springer-Verlag, Berlin/Göttingen/Heidelberg (1954), 261 pages.
2. Wünning, J., Härterei Technische Mitteilungen 36 (1981) 5, p.231-241.
3. Lübben, Th., Bomas, H., Hougardy H.P. and P. Mayr, part 1: Härterei Technische Mitteilungen 46 (1991) 1, p.24-34; part 2: Härterei-Technische Mitteilungen 46 (1991) 3, p.155-171.
4. Tensi, H.M. and P. Stitzelberger-Jakob, in: Proceedings of the 6th Intern. Conf. on Heat Treatment of Metals, Chicago, 1988, p. 171-176.
5. Hougardy, H.P. and K. Yamazaki, steel research 57 (1986) 9, p.466-471.

Proceedings of the First International Conference on Quenching & Control of Distortion, Chicago, Illinois, USA, 22-25 September 1992

Application of Quench Factor Analysis to Predict Hardness Under Laboratory and Production Conditions

C.E. Bates
Southern Research Institute
Birmingham, Alabama

G.E. Totten
Union Carbide Chemicals and Plastics Co. Inc.
Tarrytown, New York

ABSTRACT

Cooling time-temperature profiles (cooling curves) can be readily obtained using stainless steel probes in the laboratory and under production quenching conditions. To optimally use this data, it is of value to interrelate the cooling curve shape with a property of interest such as hardness. One method of interrelating cooling curve shape and hardness is by quench factor analysis (QFA). This paper will focus on the use of QFA to successfully predict the as-quenched hardness of AISI 4130, 4140 and 1045 steel.

INTRODUCTION

Quenching refers to the rapid cooling of steel from the solution treating temperature, typically 845-870°C (1550-1600°F). Quenching is performed to prevent ferrite or pearlite formation and to facilitate bainite or martensite formation. After quenching, the martensitic steel is tempered to produce the optimum combination of strength, toughness and hardness.

Highly agitated cold water and brine solutions are excellent quenchants for maximizing the formation of as-quenched martensite. Unfortunately, these quenchants often produce large thermal gradients which increase the potential for deformation and cracking.

If distortion and cracking are to be minimized, the thermal gradients within the part must also be minimized during the quench. This often requires the use of alternative quenchant media such as: oil, aqueous polymers, salt baths, fluidized beds, etc. to mediate the relatively rapid heat transfer rates that accompany water and brine quenching.

The technical challenge of quenching is to select the quenchant medium that will minimize the various stresses that form within the part in order to reduce cracking and distortion while at the same time providing heat transfer rates sufficient to yield the desired as-quenched properties such as hardness.

To successfully predict the metallurgical consequences of quenching, it is necessary to determine the heat transfer properties produced by the quenchant medium during the cooling process. Cooling curve analysis has traditionally been considered to be the best method to obtain this information.

Recently, several methods for interrelating the cooling time-temperature behavior achievable with a quenchant with as-quenched hardness have been reported. These methods include empirical hardening power predictor (1,2), rewetting times (3), and rigorous analysis of the total cooling process (4).

Some of the advantages of QFA over previously published methods include:

- correlation with a time-temperature-property (TTP) curve which is calibrated to the property of interest,

- quench severity quantification using a single number (quench factor),

- no intermediate manual data interpretation since the calculation of quench factors can be automated using data files obtained from computerized data acquisition.

The objective of this paper is to describe an alternative method, QFA, to predict steel hardness under both laboratory and production conditions (5).

DISCUSSION

QFA Calculations

Quench factors are readily calculated from digital time-temperature (cooling curve) data and the C_T function describing the TTP curve for the alloy of interest using Equation I.

$$C_T = -K_1 \cdot K_2 \cdot \exp\left[\frac{K_3 \cdot K_4^2}{RT \cdot (K_4 - T)^2}\right] \cdot \exp\left[\frac{K_5}{RT}\right] \quad (\text{I.})$$

where:

C_T = critical time required to form a constant amount of new phase or reduce the hardness by a specified amount. (The locus of the critical time values as a function of temperature forms the TTP curve.),

K_1 = constant which equals the natural logarithm of the fraction untransformed during quenching, i.e., the fraction defined by the TTP curve,

K_2 = constant related to the reciprocal of the number of nucleation sites,

K_3 = constant related to the energy required to form a nucleus,

K_4 = constant related to the solvus temperature,

K_5 = constant related to the activation energy for diffusion,

R = 8.3143 J/(°K mole),

T = absolute temperature (°K).

It should be noted that the TTP curves are not synonymous with the well known TTT (time-temperature-transformation) or CCT (continuous-cooling-transformation) curves. Although TTP curves are related to metallurgical transformation behavior, they are empirically derived and calibrated with respect to the property of interest for each alloy and chemistry.

The constants K_1, K_2, K_3, K_4, and K_5 shown in equation I. define the shape of the TTP curve. The integration process is shown in Figure 1.

The incremental quench factor (q) for each time step in the cooling process is calculated according to equation II.

$$q = \frac{\Delta t}{C_T} \quad (\text{II.})$$

where:

q = incremental quench factor,

Δt = time step used in cooling curve data acquisition,

C_T = is defined by Equation (I).

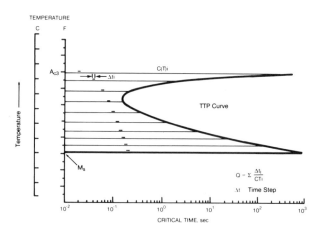

Fig. 1 - QFA Calculation Process

The incremental quench factor (q) represents the ratio of the amount of time an alloy was at a particular temperature divided by the time required for transformation to begin at that temperature.

The incremental quench factor is then summed over the entire transformation range to produce the cumulative quench factor (Q) as shown in Equation III:

$$Q = \sum q = \sum_{T_1 = Ar_3}^{T_2 = Ms} \frac{\Delta t}{C_T} \qquad \text{(III.)}$$

When calculating the quench factor (Q), the values of T_1 and T_2 are taken as the Ar_3 and M_s transformation temperatures respectively for the particular steel alloy. The cumulative quench factor (Q-Factor) is proportional to the heat removal characteristics of the quenchant as reflected by the cooling curve for the quenching process. Q-factors are single numbers that quantify quench severity which may be related to the metallurgical transformation behavior of steel.

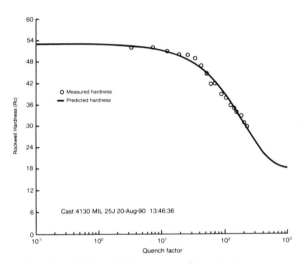

Fig. 2 - Q-Factor/Hardness Correlation

The cumulative quench factor, Q, can be used to predict the as-quenched hardness in steel by:

$$P_p = P_{min} + (P_{max} - P_{min}) \cdot \exp{(K_1 Q)} \qquad \text{(IV.)}$$

where:

P_p = predicted property (e.g., hardness),

P_{min} = minimum property for the alloy,

P_{max} = maximum property for the alloy,

exp = base of the natural logarithm,

$K_1 = \ln{(0.995)} = -0.00501$.

The correlation between quench factors and Rockwell hardness for one AISI 4130 alloy is shown in Figure 2. Equation IV describes the shape of the correlation curve.

Cooling Curve Data Acquisition and Analysis

The quench factor for a particular quenchant medium may be experimentally determined using cylindrical, sheet or plate probes or an actual part instrumented with thermocouples. The probes used for this work were constructed from AISI Type 304 stainless steel cylindrical steel bars with the length at least four times the diameter as shown in Figure 3. A Type K thermocouple was swedge-fitted into the geometric center of the bar.

The instrumented probe (or part) was heated to the austenitizing temperature of 840°C) and held for a sufficient length of time to assure temperature uniformity and a homogeneous structure and then quenched. Typically the holding time at temperature was one hour for one inch (25.4 mm) probes and two hours for 2 inch (50.8 mm) probes or parts.

Cooling curves were recorded as a function of temperature and time using a portable computer equipped with an A/D converter. A data collection rate of 5 Hz was used for this work.

Quench factors were then calculated from the cooling curve time-temperature data obtained using and the appropriate TTP Curve developed for the alloy of interest.

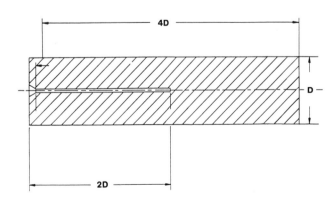

Fig . 3 - Schematic of Instrumented Bar Probe

QFA - Laboratory Application

QFA has been applied to successfully predict the as-quenched hardness of two relatively high hardenability steel alloys, AISI 4130 and AISI 4140 and a low hardenability plain carbon AISI 1045 steel after quenching in various oil and aqueous polymer media. Tables 1, 2 and 3 provide a comparison of the predicted and actual measured hardness data for these three alloys. The predicted hardness was based on QFA analysis of cooling curves obtained with Type 304 stainless steel probes. The "actual" hardness was measured on a steel bar of the alloy of interest with the same cross-section size. Both bars were heated and quenched in exactly the same manner.

Excellent correlations between measured and predicted hardness were obtained for the higher hardenability alloys, AISI 4130 and 4140. The predicted and measured hardness values for the low hardenability AISI 1045 agreed well until the martensite content fell below a Rc hardness of approximately 32. The reason for the inability of quench factor analysis to successfully predict hardness over the entire range of hardnesses is related to the difficulty in modeling the extremely rapid transitions from softer ferritic - pearlitic structures to martensite (6). For this reason, the predicted hardnesses, in the range of Rc = 35-45, must be verified experimentally in low hardenability steels such as AISI 1045.

TABLE 1

MEASURED AND PREDICTED AS-QUENCHED HARDNESS FOR AISI 4130[1]

Cooling Rate @ 1300°F (°F/S)	Quench Factor	Rockwell Hardness (Rc) Predicted	Measured
205.4	7.2	51.8	52
121.7	12.4	50.9	51
72.1	18.5	49.9	50
52.9	25.6	48.8	50
38.8	33.1	47.7	49
31.3	41.5	46.4	47
23.8	50.1	45.2	45
19.9	58.7	44.1	42
17.0	67.3	43.0	42
12.7	86.8	40.7	39
9.8	103.1	38.9	38
8.1	119.1	37.3	36
6.8	134.8	35.8	35
5.9	150.1	34.5	34
4.6	175.3	32.5	33
3.9	197.6	31.0	31
3.3	215.6	29.9	30

1. Chemistry of the Alloy was:

Carbon	0.315%
Manganese	0.576%
Silicon	0.379%
Chromium	1.140%
Nickel	0.103%
Molybdenum	0.244%
Copper	0.125%
Vanadium	0.014%

TABLE 2

COMPARISON OF PREDICTED VS ACTUAL ROCKWELL (Rc) HARDNESS VALUES FOR 4140 STEEL (MTL'S 31 AND 32)[1]

Material 31		Material 32	
Predicted Hardness (Rc)	Actual Hardness (Rc)	Predicted Hardness (Rc)	Actual Hardness (Rc)
57.3	56	56.5	57
57.1	56	56.3	57
56.7	55	56.0	56
56.3	54	55.6	56
56.2	52	55.6	54
55.2	53	54.8	54
51.6	53	51.6	53
53.8	50	53.6	54
49.7	49	50.0	54
48.6	49	49.0	46
44.6	42	45.5	38
25.6	31	27.5	34

1. The elemental compositions of these steels was:

Element	Composition (%) Material 31	Material 32
Carbon	0.439	0.394
Manganese	0.860	0.810
Silicon	0.259	0.268
Chromium	0.878	0.922
Nickel	0.076	0.138
Molybdenum	0.170	0.196
Copper	0.087	0.128
Vanadium	0.000	0.000
Boron	0.000	0.000

TABLE 3

COMPARISON OF PREDICTED VS ACTUAL ROCKWELL (Rc) HARDNESS VALUES FOR 1045 STEEL (MTL'S 27 AND 28)[1]

Material 27		Material 28	
Predicted Hardness (Rc)	Actual Hardness (Rc)	Predicted Hardness (Rc)	Actual Hardness (Rc)
59.8	59	58.0	60
59.0	58	56.9	53
58.4	59	56.2	59
55.7	57	52.7	56
53.4	56	30.2	28
45.7	54	49.8	52
48.0	55	40.4	46
35.4	23	43.8	43
38.2	26	29.0	24
35.8	34	31.0	26
25.3	27	29.3	23
19.1	22	19.7	24
16.9	18	14.6	16
20.0	19	13.1	18
12.3	18	15.2	18
		10.7	17

1. The elemental compositions of these steels was:

Element	Composition (%) Material 27	Material 28
Carbon	0.480	0.468
Manganese	0.834	0.749
Silicon	0.281	0.240
Chromium	0.035	0.054
Nickel	0.011	0.031
Molybdenum	0.170	0.016
Copper	0.091	0.015

QFA- Production Data

QFA has also been used with Type 304 stainless steel probes quenched under various production quenching conditions. In these cases, an instrumented Type 304 stainless steel probe was loaded into the furnace and soaked at the austenitizing temperature being used by the heat treater. The probe was then quenched in the production tank and a cooling curve was obtained. The cooling curves were analyzed by QFA and R_c hardnesses were predicted using the TTP curve parameters developed previously for the laboratory studies. In order to obtain comparative data, the AISI 4140 and 1045 bars used for hardness determination were also quenched immediately following the cooling curve data acquisition using an identical heating cycle and in the exact location and conditions used previously for the probe. Thus far, only AISI 4140 and 1045 hardness correlations under production quenching conditions have been studied.

The results that have been obtained thus far are summarized in Table 4. These data show that the predicted hardness gives the best correlation with measured values for the higher hardenability AISI 4140 alloy which is consistent with the data reported for AISI 4140 quenched under laboratory conditions (see Table 2). As expected, the reliability of the hardness prediction for the low hardenability AISI 1045 alloy was somewhat poorer since the hardnesses were < Rc = 32. This work verifies that intermediate predicted Rc hardness values for low hardenability AISI 1045 must be verified experimentally.

FUTURE WORK

The above results show that quench factor analysis has excellent potential as a physical property predictor for steel if appropriate cooling curve data is available. Additional work is in progress to:
- Develop quench factor solutions for a wide variety of steel alloys based on chemistry.
- Develop the ability to predict hardness at positions other than the geometric center of the test piece.
- Refine the predictive model to permit greater accuracy for low hardenability, plain carbon steel.

Table 4
COMPARISON OF PREDICTED AND MEASURED HARDNESS FOR PROBES QUENCHED UNDER PRODUCTION CONDITIONS

File No.	Cross Section Size (mm)	AISI Alloy	Rc Hardness Predicted	Actual
4140-2A1	25.4	4140	55.5	46.0
4140-2A2	50.8	4140	50.0	42.8
4140-4A1	25.4	4140	52.1	48.2
4140-4A2	50.8	4140	46.1	47.8
4140-1A1	25.4	4140	54.9	49.6
4140-1A2	50.8	4140	51.0	48.9
4140-3A1	25.4	4140	54.4	56.8
4140-3A2	50.8	4140	43.7	41.1
1045-2A1	25.4	1045	28.6	18.2
1045-2A2	50.8	1045	21.6	19.8
1045-1A1	25.4	1045	25.4	10.0
1045-1A2	50.8	1045	15.1	11.9
1045-3A2	50.8	1045	10.5	14.8

SUMMARY

It is becoming increasingly important to be able to successfully predict the metallurgical result, e.g. hardness, of a quenching process if significant quality improvements are to be gained. In this report, quench factor analysis was used to predict the as-quenched hardness of three steel alloys, AISI 4130, 4140 and 1045. (The accuracy of the QFA prediction of the low hardenability AISI 1045 steel was significantly poorer as the Rc hardness decreased below 35-45.) Work is continuing to expand the utility of this method to allow the accurate prediction of as-quenched hardness for other alloys as a function of chemistry. QFA can be readily applied to cooling curves obtained from production processes.

REFERENCES

1. S.O. Segerberg, *Heat Treat.*, **1988**, December, pp 30-33.

2. P. Damay and M. Deck, Int. Heat Treating Assoc. Meet., Lamans, September, 1990.

3. H.M. Tensi, G. Wetzel and Th. Kunzel, *Proceedings of Eigth International Heat Transfer Conference*, San Francisco, CA, pp 3031-3035, August 12 - 22, 1986.

4. B. Liscic and T. Filetin, *J. Heat Treat.*, **1988**, 57, pp 115-124.

5. C.E. Bates, G.E. Totten and K.B. Orszak, *Proceedings of 3rd International Federation for Heat Treatment and Surface Engineering - Quenching and Carburizing*, pp 46-64, Melbourne, Australia, September 1991.

6. J.S. Kirkaldy, Private Communication, 1988.

Proceedings of the First International Conference on Quenching & Control of Distortion, Chicago, Illinois, USA, 22-25 September 1992

The Mathematical Modeling of Temperature and Microstructure During Spray Cooling

P.D. Hodgson, K.M. Browne, R.K. Gibbs, T.T. Pham, and D.C. Collinson
BHP Research-Melbourne Laboratories
Mulgrave, Victoria, Australia

ABSTRACT

The post deformation cooling behaviour of hot worked steel has been modelled for a range of spray cooling systems. It has been proposed that the heat transfer coefficient is a function only of the water flux delivered to the steel surface and the temperature of the surface. Other constants are incorporated within this coefficient to reflect the complex variation with temperature, and the maximum in the heat transfer coefficient at a surface temperature of approximately 300°C. The model also includes submodels for the microstructural evolution during cooling and the relationships between final microstructure and mechanical properties. The model has been validated using both laboratory and mill data for cooling systems ranging from soft cooling, typical of that obtained during the accelerated cooling of plate, through to intense cooling required to produce quench and self tempered microstructures.

INTRODUCTION

There is now a worldwide trend to improve the properties of hot deformed steels, through the application of thermomechanical controlled processing. This has been applied to both hot rolling and hot forging [1]. Both phases of the forming route are controlled during thermomechanical processing: the deformation (rolling) phase, which determines the state of the austenite at the point of transformation, and the cooling phase, which determines the final microstructure and thereby the properties of the steel.

The design and optimisation of thermomechanical processing routes has been largely undertaken, to date, using laboratory and full scale mill investigations, where the composition, rolling history and cooling behaviour are investigated for their effects on the final microstructure and properties. Recently, though, there has been an attempt by many research groups to develop mathematical models which predict the final properties as a function of the rolling and cooling history. With such models a wide range of compositions, rolling mill types, cooling systems and processing scenarios can be quickly simulated. This allows even fictitious processes to be studied, which can be very useful when assessing possible hardware changes.

The models should incorporate all of the microstructural events that occur during reheating and rolling. At present, models for the microstructural changes during deformation and to predict the ferrite grain size have been developed for C-Mn [2-4], and Nb, V and Ti microalloyed steels [3-5]. The kinetics of the microstructure evolution during transformation have now been modelled by a number of groups for a range of steel types and cooling conditions [6-9]. The rate and extent of transformation affects both the cooling history and the final properties; the former through changes in the thermophysical properties of the steel and the generation of heat during transformation.

While the evolution of microstructure has received a certain amount of attention over the past decade, the modelling of the heat transfer process itself has received considerably less attention; particularly for total cooling systems where there may be interaction between sprays and the removal of water from the surface may be an issue. This is further complicated by the proprietary nature of much of this information. The general trend has been to collect data for a given cooling unit and to then regress empirical models based upon a certain expected temperature drop per unit time of cooling. In many cases this is adequate as the control of the cooling unit is essentially limited to varying the number of cooling sections that are operating for a given piece being cooled, with little scope for true flow rate control. Another reason why this works in current steel processing operations is that cooling has generally been limited to thin products, such as strip, and to situations where the cooling finishing temperature is high (>550°C). The latter means that the surface of the steel rarely cools into the temperature regime where there is an extremely strong and non-linear variation in the heat transfer coefficient with surface temperature.

However, the increasing efforts to extend the range of thermomechanical processes routes has now led to the identification of a number of desired cooling histories which now either involve cooling the total product to these temperatures, or at least cooling the surface of thicker product into this regime. Examples of this include the production of dual phase and TRIP strip steels where the coiling temperature may be between 500°C and room temperature, and quench and self-tempered bar, plate and structurals where the surface layers must be cooled below the martensite start temperature.

The aim of the work described here was to develop integrated thermal and metallurgical models for the cooling of steel after hot rolling. These models will then be used to develop new thermomechanical processing routes and may be applied on-line to improve the control of these cooling systems as more complex cooling histories are introduced. Previous papers [4,5,10-13] have dealt with the upstream thermal, mechanical and microstructure development models and the current work deals only with the post deformation cooling.

HEAT TRANSFER MODEL

The thermal response of a steel product to a particular cooling condition is described by the conduction equation with appropriate boundary conditions. For unidimensional heat transfer the conduction equation is :

$$\frac{1}{r^n}\frac{\delta}{\delta r}\left(r^n k \frac{\delta T}{\delta r}\right) + \dot{Q} = \rho c_p \frac{\delta T}{\delta t} \qquad (1)$$

where: $\dot{Q}$ is the rate of volumetric heat generation (W/m^3)

c_p is the specific heat (J/kg^{o}C)

k is the thermal conductivity (W/m^{o}C)

ρ is the density (kg/m^3)

r is the displacement (m)

For products with a circular cross section n = 1, while for plate or strip products n = 0. Equation (1) can be solved numerically using the implicit finite difference method. The initial temperature distribution within the product prior to cooling is obtained from upstream process models. During the cooling phase the product is subjected to a variable surface heat flux, q_s, which is based on the prevailing cooling condition.

The thermophysical properties (c_p, k and ρ) are determined separately for each node as a function of the temperature at that node and the volume fraction of each phase at that node. Therefore, when there is a mixture of phases present at a given time increment and nodal position, the average thermophysical properties are calculated for that node using the temperature variation in these properties for each phase and then applying a simple law of mixtures.

The accuracy of the finite difference solution generally improves with decreasing time and displacement increments, although reducing these too far imposes computational time penalties. To improve the accuracy of the finite difference solution, without increasing the processing time dramatically, a non-uniform nodal mesh is usually used, with the spacing between nodes decreasing near the surface. In the case of round products, a constant volume per node mesh is used. This provides the improved numerical accuracy where the thermal gradients are steepest.

Equation (1) also contains a term for the rate of heat generation during cooling. In the austenite temperature field, prior to any transformation, this is set to zero during the numerical solution. The evolution of heat is considered once the transformation reaction(s) commence and latent heat is evolved. The amount of heat released depends on the rate of transformation during a given section of the cooling history. In the current work,

this was calculated using the discretised form for the finite difference solution by:

$$\dot{Q} = \rho \cdot \Delta H \cdot \Delta X/\Delta t \qquad (Jm^{-3}s^{-1}) \quad (2)$$

where: ΔH is the latent heat of transformation (J/kg)

ΔX is the volume fraction of new phase formed during the time increment Δt

The rate of transformation is modelled using the method shown later, while the latent heat is obtained from standard data for each of the phases.

Heat Transfer In Water Cooling

The boundary conditions considered for (1) are natural air cooling and water cooling. The former is used for all periods where the stock is not directly in contact with water. Natural air cooling contains both radiative and convective components. For the implicit finite difference scheme these are combined into a single effective heat transfer coefficient:

$$h_{eff} = \varepsilon \sigma (T_s^2 + T_a^2)(T_s + T_a) + h_c \qquad (3)$$

where: T_s is the surface temperature of the steel (K)

T_a is the ambient temperature of the air (K)

h_c is the convective heat transfer coefficient(Wm^{-2}C^{-1})

ε is the emissivity of the steel surface

σ is Boltzmann's constant

This coefficient is then used for the boundary node to calculate the total heat flux of that node allowing for conduction from the interior node, which is then converted into a temperature change (using the finite difference formulation):

$$\rho c_p \frac{\Delta x}{2}\frac{\Delta T}{\Delta t} = -k\frac{\Delta T}{\Delta x} + h_{eff}(T_s - T_a) \qquad (4)$$

In this case, Δx is the internodal distance between the surface node and the adjacent node. For water cooling a similar set of equations is used, with h_{eff} in (4) being replaced by the heat transfer coefficient for water cooling. The exact mechanism by which the application of water extracts heat from the surface varies with temperature, as has been discussed in most standard heat transfer texts. At the same time the delivery of water to the surface by each of the cooling systems used commercially would be expected to change the efficiency of this heat extraction.

The current methods of post deformation water cooling of steel include immersion quenching into still, or stirred water, high velocity water immersion tubes, hydraulic sprays, air-water mist sprays, laminar flow nozzles and water curtains. The heat transfer coefficient for water cooling is a function of the method used, the steel surface temperature and the rate at which the water is applied or flowing over the surface. In general, though, the heat transfer coefficient has the same type of surface temperature dependence for all methods of water cooling. The heat transfer coefficient for quenching into still water is typical; measured values are shown in Figure 1.

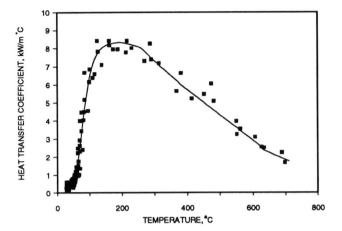

Figure 1. Measured heat transfer coefficient, as a function of steel surface temperature, for a 60 x 40 x 0.6mm stainless steel coupon heated to between 700°C and 950°C and vertically quenched into still water.

The heat transfer coefficient increases rapidly with increasing temperature until a steam barrier at the interface begins to dominate the heat flow. This reduces the heat transfer coefficient with increasing temperature until the film is fully established at about 700°C, although this is a function of the water flux. At higher temperatures the heat transfer coefficient tends towards a fairly constant value. These effects can be different in different types of cooling systems; for example, moving water has a very much greater convective component than still water. The variations in heat transfer coefficient with surface temperature, water flow and system characteristics, are most important in modelling the temperature history of the water cooled steel and hence its microstructure.

Hydraulic Spray Cooling

The heat transfer coefficient associated with hydraulic water sprays has a similar type of dependence on surface temperature as immersion quenching. The other important variable in this case is the water flux rate. While there are some data in the literature for water spray cooling, none cover the full temperature and water flux rates used in steel-making and fabrication processes. Accordingly, the convective heat transfer coefficient, h_c, associated with hydraulic spray cooling of steel, has been determined from experimental data measured in these laboratories and combined with published data.

The most reliable published heat transfer coefficient for water cooling surfaces in the range 400 to 800°C is that of Mitsutsuka [14], which is based on seven independent data sources and is also in excellent agreement with more recent work by Ohnishi et al. [15]. Mitsutsuka's function for the convective component of the heat transfer coefficient is:

$$h_c = 3.31 \times 10^6 \ \dot{w}^{0.616} \ T_s^{-2.445} \qquad (5)$$

where $\dot{w}$ is the water flux rate and T_s is the surface temperature.

In this temperature range the heat transfer coefficient decreases with increasing temperature. At the higher temperatures, somewhere above 600°C, it is well established [14, 16-19] that the heat transfer coefficient becomes less temperature dependent although there is no universal agreement on its values at high temperatures. On the other hand, at low temperatures, it has been shown [16, 18, 20, 21] that the heat transfer coefficient has a maximum somewhere between 200 and 300°C. Using the experimental method described elsewhere [] a modified equation was determined for a laboratory hydraulic cooling system for water fluxes from 0.16 to 62 L s^{-1} m^{-1}. For a vertical surface exposed to these sprays the equation is:

$$h_c = 3.15 \times 10^6 \ \dot{w}^{0.616} \left[700 + \frac{T_s - 700}{\exp\{(T_s - 700)/10\} + 1} \right]^{-2.455}$$

$$\cdot \left[1 - \frac{1}{\exp\{(T_s - 250)/40\} + 1} \right] \qquad (6)$$

The correlation between the calculated temperature drops during experimental cooling using the new equation and those actually measured is shown in Figure 2, while Figure 3 shows some examples of the predicted form of the heat transfer coefficient as a function of surface temperature at a range of water fluxes using (6) and Mitsutsuka's equation. It can be seen that the two equations give similar values between 400 and 700°C, but that equation (5) leads to extremely high predictions of the heat transfer coefficient at 300°C and below. At higher temperatures equation (5) continues to decrease, whereas most workers accept that the heat transfer coefficient should become independent of temperature.

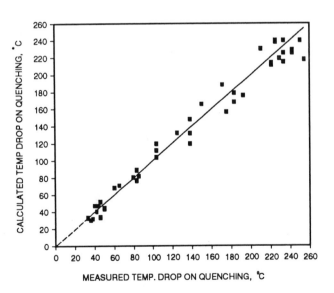

Figure 2. Comparison of the calculated temperature drop of a steel plate using equation (6) with the experimentally measured temperature.

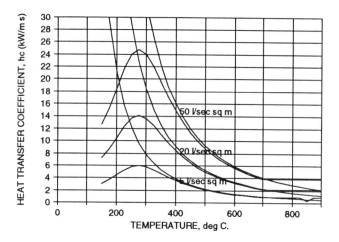

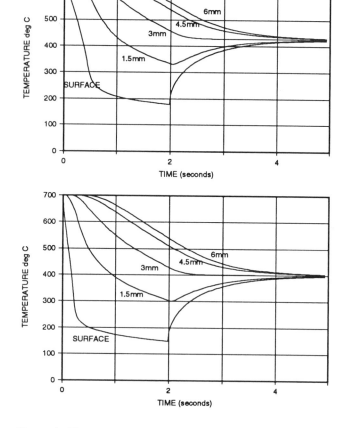

Figure 3. The variation of the heat transfer coefficient for water spray cooling of steel as a function of steel surface temperature and water flux, using equations (5) and (6).

Cooling At High Water Flux

Because of the characteristic temperature dependence of the heat transfer coefficient, high water flux rates at the surface of a block of steel quench the surface rapidly to a fairly low temperature around 200°C and this then remains fairly constant during the quenching process. The internal cooling of the steel is then dependent on its thermal diffusivity rather than the water flux. The calculated cooling of a 12mm thick slab by water fluxes of 50 and 100 $Ls^{-1}m^{-2}$, shown in Figure 4, illustrate the typical quenching behaviour at high water flux rates and indicates the small effect of doubling the already high water flux.

This characteristic of water cooling has two important practical effects. Firstly, the water quenching of steel at very high water flux rates is essentially regulated by quenching time and, hence, is greatly affected by processing speed. Secondly, very efficient cooling is possible (in terms of water use) by rapidly quenching the steel surface to about 200°C with high water flux, then decreasing the water flux while maintaining the surface at a temperature where the heat transfer is a maximum.

TRANSFORMATION MODEL

The transformation model is currently restricted to the formation of ferrite, pearlite and martensite with no account taken of the various bainites that can form during high rate cooling of certain steels. The approach taken is similar to that proposed by a number of authors [7-9], where the progress of transformation for the diffusional reactions is followed by an Avrami type equation, assuming additivity, while the fraction of martensite formed is modelled as a function of holding temperature below the martensite start temperature.

For the diffusional phase transformations of austenite to ferrite and pearlite, the formation of a new phase on cooling is only possible once the temperature is below the equilibrium transformation temperature. This temperature is dependant on the alloy content of the steel.

The calculation of the equilibrium transformation temperature for austenite to ferrite, the Ae_3, is based on the method developed by Kirkaldy and Baganis [22]

Figure 4. The temperature distribution in a 12mm thick slab, cooled from 800°C on both sides by a water spray flux of $50Ls^{-1}m^{-2}$ (upper) and $100Ls^{-1}m^{-2}$(lower). Note that doubling the water flux only reduces the final temperature by a further 25°C.

and has been extended to include vanadium and niobium. This technique calculates and compares the chemical potential of the austenite and ferrite phases at a given temperature for all of the elements present in the steel. The equilibrium transformation temperature is obtained when the two chemical potentials are equal. While this method is complex mathematically, it offers the advantage of being able to be handle a wide range of steels. Other empirical approaches to obtain the Ae_3 temperature tend to be limited to a particular steel or group of steels. However, for the austenite to pearlite reaction, the equilibrium transformation temperature (Ae_1) is an empirical equation by Andrews [23]. The use of the empirical formulation for the Ae_1 has been justified by the fact that the Ae_1 temperature does not vary greatly with composition in contrast to the Ae_3.

As the temperature is reduced below the equilibrium transformation temperature for the steel, a finite amount of time is needed before an observable amount of new phase forms. This incubation time is dependant on the thermal history of the steel below the equilibrium transformation temperature. The method of Hawbolt and coworkers [6-8] is currently used to estimate the reaction start time and temperature, however, this method is limited to smooth continuous cooling conditions. There still appears to be a need to develop more complex models for the initiation of the diffusional reactions, based upon and additivity approach. This is currently being pursued.

The kinetics of the growth of ferrite and pearlite are described using the Avrami equation [24-26]:

$$X = 1 - \exp(-b(T) \cdot t^n) \qquad (7)$$

where: X is the volume fraction transformed
t is time
b, n are constants for a particular steel at a given temperature.

True additivity is assumed to hold for the kinetics, a feature which has been confirmed by other workers [7,8]. According to the Avrami equation the amount of new phase formed will eventually reach unity at long holding times. However, in the case of ferrite formation there is an equilibrium volume fraction of ferrite that cannot be exceeded at a given temperature. For this reason the equilibrium amount of ferrite which could be formed is calculated at each time step and then used to normalise the Avrami equation. This follows closely the method developed by Hawbolt and coworkers [6-8].

The formation of martensite is modelled by first calculating the martensite start temperature from the chemistry of the steel [27]:

$$M_s = 512 - 453C - 16.9\,Ni + 15\,Cr - 9.5\,Mo + 217\,C^2 - 71.5$$
$$C\,Mn - 67.6\,C\,Cr \qquad (8)$$

The volume fraction of martensite is then modelled using an equation proposed by Koistinen and Marburger [32]:

$$X_M = 1 - \exp -[0.011 \cdot (M_s - T)] \qquad (9)$$

where: X_M is the volume fraction martensite
M_s is the martensite start temperature given in (15) above
T is the current temperature

The latent heat of transformation for any new phase formed is calculated at the end of each time step for every node in the finite difference mesh. This is done by calculating the latent heat of transformation at the current temperature and then applying equation (2) to obtain the heat generated per unit volume.

From the heat generated, the temperature rise in the steel is calculated using the thermal properties of the current phases present at that node in the steel. This temperature rise is then added to the temperature of the steel. Provided that the time step is small this method gives excellent predictions of the steel thermal history and microstructure evolution.

MODEL VALIDATION AND PERFORMANCE

The range of thermomechanical processing routes covered by this work extended from relatively soft accelerated cooling, through to the intense cooling required to produce quench and self tempered microstructures. The effect of the transformation on the actual cooling curve varies for each of these scenarios and the composition of steel.

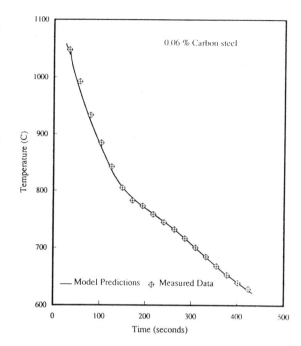

a)

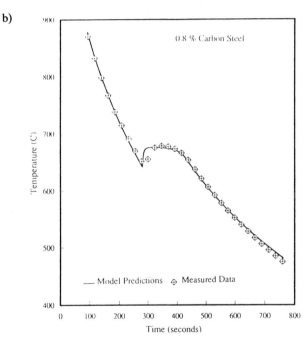

b)

Figure 5. Measured and predicted centre-line temperature for the air cooling of a) 0.06 C and b) 0.8 C, 20mm plate.

Examples of modelled and measured cooling curves for low (0.06) and high (0.8) %C - 1 % Mn steels under slow cooling conditions (Figure 5) show that the evolving microstructure significantly alters the cooling curve of the steel, and that the integrated microstructural-thermal models are able to reproduce the cooling behaviour very accurately under these conditions. For the low carbon steel there is relatively little heat of transformation and the distinct change in cooling rate is due to the change in thermophysical properties; especially the specific heat. The ferromagnetic-paramagnetic transformation effects on the specific heat must also be included for this steel. For the eutectoid steel the change in cooling behaviour is almost entirely due to the heat of transformation, with a smaller effect from the difference in thermophysical properties between the austenite and pearlite.

For low C steels Kumar et al [6] have also shown the importance of modelling the phase change to enable the accurate prediction of the cooling history on the run-out table of a hot strip mill. There, the difference between the model predictions with and without a transformation model was approximately 40°C during cooling from 900 to 550°C. However, recent simulations using the model have also shown that when processing low C steels it is not necessary to have *detailed* transformation models integrated within the thermal model.

Figure 6 shows that the cooling history can be predicted for a low C steel by simply assuming that transformation goes to completion at the transformation start temperature (Ar3), and using the specific heat of the ferrite for all calculations below this temperature. The heat of transformation from this phase is very small, particularly under the high heat flux cooling used for strip. The main source of error will be in neglecting any heat generation due to the formation of pearlite or other carbide phases. Therefore, this simplified approach will lead to increasing error as the C and Mn contents of the steel increase. The heat transfer coefficient predicted by (6) was used throughout this work without any modification.

The results show a good agreement, both over the temperature range where transformation is occurring and at lower temperatures, where it is possible to assess the performance of the hydraulic spray cooling model.

should be noted that it is very difficult to accurately measure the contact lengths of the sprays on the strip surface for all sprays. Moreover, in strip cooling there is a very large, rapidly moving horizontal surface. It is expected that water from upstream sprays will be dragged under subsequent sprays, thereby interfering with water delivery to the surface. This would lead to an overall reduction in the *useful* water flux reaching a point on the surface. In the laboratory simulation the sample was much smaller and the velocity lower, so that the water should easily leave the plate surface without interfering with subsequent water delivery. Importantly, though, the relatively simple approach taken here does provide a reasonable representation of a complex cooling system.

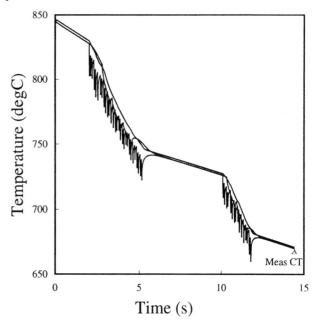

Figure 7. Predicted cooling curves and pyrometer measurement at coiler for 3 mm 0.06C strip.

The ability of the model to follow the dynamics of the cooling process is illustrated in Figure 8, where the thermal profile along the strip was modelled and compared with the surface temperature measurements at the coiler.

Air-mist cooling has also been assessed using the current model in the laboratory cooling rig [28]. Again for a low C steel it is possible to ignore the details of the tranformation reaction in the way described above. The results in Figure 9 also include the rolling phase, where a thermal model described previously was used. In this case the cooling was followed to much lower temperatures and the details of the cooling behaviour were very closely followed by the model. In these simulations the heat transfer coefficient predicted by (6) was scaled by a factor of 2 to match the measurements.

To produce quench and self tempered structures requires the surface layers for the product to be cooled below the M_s, while leaving the core of the steel at a sufficiently high temperature to temper the martensite by conduction of heat from the centre to the surface, once the material has left the quenching zone. This requires the intense cooling previously discussed. It has been found that the model provides a very good estimate of the exceptionally steep rate of recovery of the surface. The model

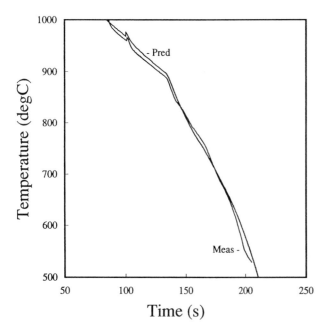

Figure 6. Predicted and measured cooling curves for the laboratory hydraulic spray cooling of a low C, 20mm plate after hot rolling.

This model was then applied to hydraulic cooling on a hot strip mill (Figure 7). Here, there are periods where the steel is under hydraulic sprays. The heat transfer coefficient for the strip simulations was calculated using equation (6), although some reduced scaling of the heat transfer coefficient was required. This could indicate that the model has certain deficiencies, although it was also able to predict the depth of the martensite layer by combining the temperature profile at the end of the quench operation with equations (8) and (9).

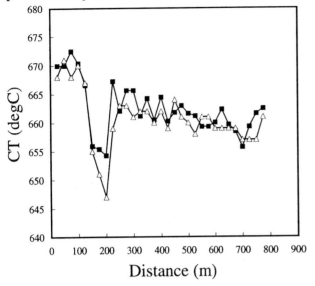

Figure 8. Predicted and measured centre temperatures for laboratory hydraulic cooling of a 20mm low C steel plate.

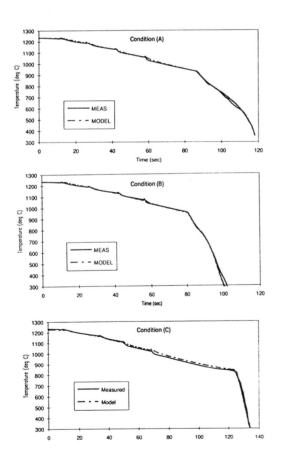

Figure 9. Predicted and measured centre temperatures during hot rolling and air-mist cooling. Conditions in [28].

The accurate prediction of the tempering temperature, which is the maximum temperature that the surface achieves after quenching, again depends upon the incorporation of a transformation model for the ferrite and pearlite formed in the core region. The modelling of transformation is very difficult under these conditions. The central region of the product will behave as if experiencing continuous cooling conditions. However, other sections of the product will have initially experienced a high cooling rate, then followed by a lower cooling rate. The modelling of transformation under these conditions of *discontinuous* cooling is an area which has received very little attention, although it is of significant importance in many practical cooling conditions. For low C and Mn contents it is again possible to take a simple approach to this transformation reaction, while for higher C steels the significant contribution from the heat of transformation requires a complete model.

During the course of the laboratory investigations to evaluate the model performance, it was noted that quite variable cooling curves could be obtained for identical set-up conditions. Further investigation suggested that the scale layer on the steel surface was affecting the cooling conditions, with patches of coarse scale actually *increasing* the cooling rate. This was initially suprising given that scale is generally regarded as a thermal insulator. At the same time it can be imagined that the scale will cool very quickly with the surface temperature in the scaled region reaching the high heat transfer coefficient values in (6) more quickly than the unscaled region. The thermal model was, therefore, used to examine the competing role of scale as a thermal insulator and as a low temperature surface. The results for a range of water fluxes (Figure 10) show that at low fluxes coarse, primary scale formed during reheating and not removed by the descaling and rolling practices used, can markedly increase the cooling rate in the centre of a 20mm plate. The resultant predicted cooling curve for one set of scale and cooling conditions (Figure 11) shows similar differences in the final temperature to those observed. No control experiments were performed in the current work with different scale layers.

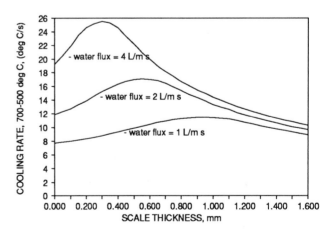

Figure 10. Predicted effect of scale layer on the average cooling rate between 700 and 500°C of a 20mm plate for various water fluxes under hydraulic cooling.

In most full scale rolling mills the steel is heavily worked and various descaling practices employed to minimise the amount of rolled-in high temperature scale. Under these conditions the steel will have a very thin scale layer which will not markedly alter the heat transfer coefficient predicted from laboratory measurements. For forgers the situation may be more critical if accelerated cooling, rather than direct quenching is to be used to produce new steels.

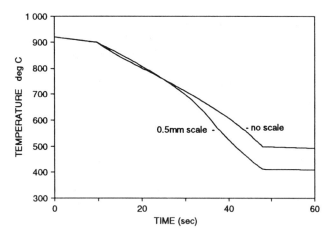

Figure 11. Average temperatures during cooling of a 20mm thick block of steel without scale (slower cooling case) and with 0.5mm of scale. The scale conductivity is $2Wm^{-1}°C^{-1}$.

FUTURE DEVELOPMENTS

The current work has shown that it is both necessary and possible to integrate thermal and metallurgical models for the post deformation cooling of steel. This has, of course, been known for some time and there are many groups taking this approach for both conventional heat treatment and thermomechanical processing. There are, though, still a number of areas that require further development before this method can be more widely applied.

As mentioned in the previous section, the problem of how to handle discontinuous cooling in the transformation model has received relatively little attention. This relies upon the application of the additivity principle which has been successfully applied by some groups, but not by others to predict the start of the transformation reaction. Even when it has been shown to apply, it has never been under the extreme conditions described above, where there may be intense quenching followed by heating, due to conduction from the central layers, and then finally a period of slow cooling. It appears doubtful that additivity will apply for the transformation start reaction under these conditions, as the shape of the TTT curve at these temperatures is not generally known for the steel types being considered.

Another feature which is still to be adequately modelled is the bainite/acicular ferrite phases. These are increasingly being utilised in thermomechanically processed steels to provide unique property packages. The formation of these phases in ideal alloy systems has been reasonably well quantified. However, the formation of these phases in competition with ferrite and pearlite to produce a mixed final microstructure is still in the early stages of development. One major obstacle still lies in determining which phases are actually forming.

CONCLUSIONS

The work described here has shown the a generalised equation has been developed which satisfactorily relates the heat transfer coefficient during spray cooling to the water flux and the surface temperature. The application of the model suggests that this coefficient does requires scaling when it is applied to the cooling of large components, or when air-mist rather than hydraulic cooling is used. This suggests that need for a further geometric function within the heat transfer coefficient.

The model also incorporates a basic model for the transformation kinetics for the ferrite, pearlite and martensite reactions. This affects the thermal model through changes to the thermophysical properties and the heat of transformation. However, it has been demonstrated that for low C steels a much simpler approach to the transformation reaction can be considered for most cooling scenarios.

Finally, the role of scale has been highlighted. In some cases a thick scale layer formed during high temperature reheating and not adequately removed during the forming process can lead to marked increases in the cooling rate. This is of particular importance for experimentalists who use laboratory equipment to develop new thermomechanical processing routes, as it may lead to poor reproducability for the same experimental conditions.

ACKNOWLEDGEMENTS

The authors would like to thank The Broken Hill Proprietary Company Limited for permission to publish this work. They also acknowledge the support and assistance provided by many people both at these laboratories and elsewhere throughout the Company; particularly Ron Gloss.

REFERENCES

1. C.I. Garcia, A.K. Lis and A.J. DeArdo, Proc. Int. Symp. on Microalloyed Bar and Forging Steels, ed. M. Finn, C.I.M., 1990, p25.
2. P.J. Campbell, P.D. Hodgson, M. Lee and R.K. Gibbs, Proc. Int. Conf. on Phys. Metallurgy of Thermomechanical Processing of Steels and Other Metals, 'Thermec 88', Ed. I. Tamura, Tokyo, Japan, ISIJ, (1988), p.761.
3. C. M. Sellars, 'Hot Working and Forming Processes', ed. C.M. Sellars and G.J. Davies, Metals Society, London, 1980, p.3.
4. P.D. Hodgson and R.K. Gibbs, Int. Symp. "Mathematical Modelling of Hot Rolling of Steel", Ed. S. Yue, C.I.M., 1990, p.76.
5. R.K. Gibbs, P.D. Hodgson and B.A. Parker, Morris E. Fine Symposium, Ed. P.K. Liaw, I.R. Weertman, H.L. Marcus and J.S. Santer, TMS, 1991, p73.
6. A. Kumar, C. McCulloch, E.B. Hawbolt and I.V. Samasekera, Mat. Sci and Tech., 7, 1991, p.360.
7. E.B. Hawbolt, B. Chau and J.K. Brimacombe, Metall. Trans., 14A, 1983, p1083.
8. E.B. Hawbolt, B. Chau and J.K. Brimacombe, Metall. Trans., 16A, 1985, p565.

9. M. Umemoto, N. Nishioka and I. Tamura, J. Heat Treating, 2, 1981, p130.

10. P.D. Hodgson, K.M. Browne, D.C. Collinson, T.T. Pham and D.C. Collinson, 3rd IFHT International Seminar "Quenching and Carburising", Melbourne, Australia, 1991, to be published.

11. P.D. Hodgson, J.A. Szalla and P.J. Campbell, 4th Int. Steel Rolling Conference, Deauville, France, (1987), paper C8.

12. L. D. McKewen, R. K. Gibbs and B. Gore, 32nd Mechanical Working and Steel Processing Conference Proceedings, ISS-AIME, 1991, p.87.

13. J. A. Szalla, G. Glover and P. M. Stone, 29th Mechanical Working and Steel Processing Conference Proceedings, ISS-AIME, 1988, p87.

14. M. Mitsutsuka, Tetsu-to-Hagane, 69, 1983, p268.

15. A. Ohnishi, H. Takashima and M. Hariki, Trans. ISIJ, 27, 1987, B299.

16. M. Mitsutsuka, Tetsu-to-Hagane, 54, 1968, p1457

17. H. Muller and R. Jeschar, Arch. Eisenhuttenwes., 44, 1973, p589.

18. E.A. Mizikar, Iron and Steel Engineer, June 1970, p53.

19. K. Sasaki, Y. Sugatani and M. Kawasaki, Tetsu-to-Hagane, 65, 1979, p90.

20. H. Kamio, K. Kunioka and S. Sugiyama, Tetsu-to-Hagane, 63, 1977, S184.

21. M. Mitsutsuka and K. Fukuda, Tetsu-to-Hagane, 69, 1983, p262.

22. J.S. Kirkaldy and B.A. Baganis, Metall. Trans., 9A, 1978, p495.

23. K. W. Andrews, JISI, 203, 1965, p721.

24. M. Avrami, J. Chem. Physics, 7, 1939, p1103.

25. M. Avrami, J. Chem. Physics, 8, 1940, p212.

26. M. Avrami, J. Chem. Physics, 9, 1941, p177.

27. D.P.Koistinen and R.E. Marburger, Acta Met., 7, 1959, p59.

28. R. E. Gloss, R. K. Gibbs and P. D. Hodgson, 3rd IFHT International Seminar "Quenching and Carburising", Melbourne, Australia, 1991, to be published.

Proceedings of the First International Conference on Quenching & Control of Distortion, Chicago, Illinois, USA, 22-25 September 1992

The M-7 Steel: Its Sub-Zero Treatments, Mechanical Properties, and Temperatures

D.M.K. Grinberg and A. Grinberg
Universidad Nacional Autonoma de Mexico
University City, Mexico

I.V. Carrasco
Universidad Autonoma Metropolitana-
Azcapotzalco
Azcapotzalco, Mexico

ABSTRACT

Steels for high speed tools are usually submitted to quenching followed by at least two tempering treatments to eliminate the retained austenite, to decompose martensite in less fragile componentes, and to enhance the precipitation of complex carbides which produce desirable hardening. Alternatively, a sub-cero treatment at temperatures between -75 C and -100 C is introduced between the tempering treatments.

We present here a comparison of the results obtained with sub-zero treatments (after the first tempering) at temperatures of -20, -40, -60, -80, -100 and -196 C, followed by the second tempering, with those obtained after two temperings.

The samples were tested under torsion, impact and hardness. Distortion was measured on Navy "C" specimens.

The treatments at -40 C and -60 C give similar or better results than the double tempering tretment, which we took as a reference, but beyond this temperature interval distortion is higher, mainly between -80 C and -100 C (the interval suggested in the literature). A similar behavior is found in all the mechanical properties tested.

INTRODUCTION

Tool steels such as the M7, with high contents of carbon and alloy elements, are typical examples of steels which can retain high amounts of austenite after quenching. An austenitizing temperature higher than the usual one and capable of dissolving additional amounts of complex carbides, or an insufficient cooling during quenching, without exceding the M_f temperature, will increase the amount of retained austenite expected for the type of the heat treated steel [1].

When large amounts of retained austenite, between 15 and 30%, are formed in high-speed steels, they can be eliminated by using multiple tempering treatments. During the first tempering internal stresses are relaxed and martensite loses carbon and alloy elements which precipitate as complex carbides, whereas the tetragonality of martensite diminishis; part of the retained austenite is converted into lower bainite and, due to the diminution of dissolved alloy elements, another part remains ready for subsequent transformation into martensite in the following cooling down. There are now two possible options: to perform a second tempering and quenching, repeating the previous procedure, or to submit the steel to a sub-zero treatment which allows it to reach the M_f temperature and the total or partial convertion of retained austenite into martensite. Obviously, this last option must be followed by a new tempering to avoid the spontaneous tempering of this last martensite during service conditions. In this way the dimensional stability of the tool is assured.

EXPERIMENTAL PROCEDURE

In this research a Mexican M7 steel was used; its chemical composition is as follows:

C=1.00w%   Mn=0.31w%   Si=0.50w%
W=1.92w%   Cr=4.00w%   Mo=8.40w%
V=2.04w%   S=0.03w%

These values are between the average values specified for this steel, which was chosen taking into account considerations such as internal demand and costs.

A high speed tool steel must exhibit simultaneously the following properties: 1) high hardness even at high temperatures as well as a high wear resistance; 2) reasonable low fragility; 3) small dimensional distorsion to avoid changes in size and shape of the tools due to a high internal stresses. (2).

Several tests were performed for double tempering and different subzero treatments: torsion, impact, hardness and distortion.

The heat treatment conditions for double tempering were as follows:

Preheating: 1st. T=780 C,   t=5 min
                 2nd. T=850 C,   t=5 min
Austenitizing: T=1,190 C,   t=5 min
Hot quenching: T=550 C,   t=5 min
Oil quenching: T=65 C,   t=10 min
Tempering: 1st. T=560 C,   t=120 min
                 2nd. T=540 C,   t=120 min

Pre-heating, austenitizing, hot quenching and tempering were all performed in salt baths installed in them tools heat treatment plant of the factory that produces this steel, by using their usual procedures, with a previous control of temperatures. The permanence time in the baths were long enough as to assure the temperature homogeneity.

The double tempering treatment was chosen as the reference state for the sub-zero treatments, which were performed in between the first and second tempering, at -20, -40, -60, -80, -100 and -196 C.

Since an industrial cooling chamber with small thermal gradientes was not available, the chamber shown in Fig. 1 was made. It consists in a cylindrical stainless steel container isolated with fiber-glass and epoxi-resin, with a metallic shield, to contain liquid nitrogen. The samples were put inside a cooper box, with a copper lid, thick and big enough as to contain a Navy "C" specimen and samples for mechanical tests.

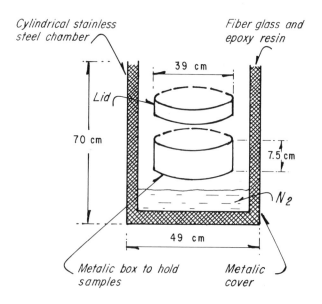

Cylindrical stainless steel chamber

Fiber glass and epoxy resin

Lid

70 cm

39 cm

7.5 cm

N_2

49 cm

Metalic box to hold samples

Metalic cover

Fig. 1

To determine the magnitud of thermal gradients the following procedure was used: liquid nitrogen was introduced in the chamber until a certain level, the box with samples was hanged above liquid nitrogen level, whose gasification cools down the box. To find out the best conditions for an adequate temperature distribution each sample was connected to a thermocouple and the cooling curves were recorded. By filling the chamber up to 5 cm in height of coolant and the box suspended at different heights, the maximum temperature diferences were ±2.5 C. With 10 cm of coolant and the box placed at 20 cm over the liquid level the temperature differences were ±1.5 C. The following step was the calibration of the box as a function of the permanence time in the atmosphere of nitrogen. The chamber was placed near the salt baths to minimize the stay of samples at room temperature.

Mechanical tests

The results of Rockwell C hardness tests are shown in Table 1; each value is an average of five measurements. A comparison between results obtained in sub-zero treatment and double tempering can be made.

In the same Table values of impact tests (average of four determinations), as well of maximum load, maximum angle of rupture, modulus of rupture, modulus of rigidity and toughness modulus obtained in torsion tests, are included.

To calculate these values the formulae presented in the Apendix were used.

The effects of the heat treatments on hardness and Charpy tests are presented in Fig. 2, whereas those corresponding to torsion tests are shown in Fig. 3.

Distortion tests

Distortion tests were performed on Navy "C" specimens, which were measured in a national metrology laboratory. Fig. 4 illustrates its form.

Table 2 shows the dimensions of these samples before and after the heat treatment, the differences between these measurements and the porcentual deviation.

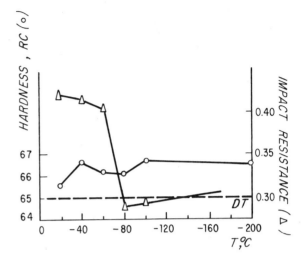

Fig. 2

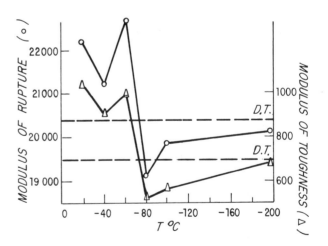

Fig. 3

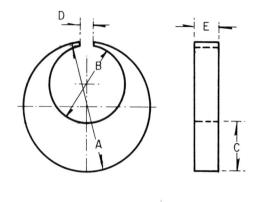

Fig. 4

Fig. 5 summarizes the differential values.

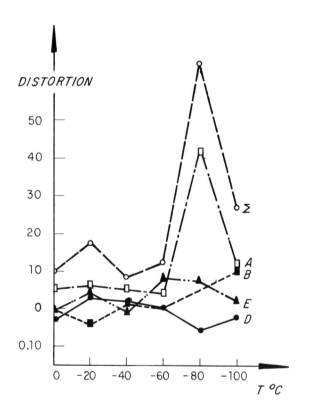

Fig. 5

DISCUSION

In the literature (2) (3) the interval between -75 and 100 C is mentioned as the very best for sub-zero treatments. This is based on the idea that to obtain the total transformation of retained austenite to martensite it is only required to reach the Mf value, which is below room temperature. However, in this temperature interval a minimum value in the mechanical properties of the steel is found, which are worse than those obtained with double tempering.

By taking as a reference for distortion the sample submited to double tempering, the sub-zero treatment at -40 C generates the best results in this sense. It is followed by the -60 C and afterwards the -20 C treatments. The results at -80 and -100 C are unfavorable in nearly all the dimensions. (Fig. 5).

We think that it would be importante to explore the possibilities for industrial applications of the interval around -40 C. Even though the hardness increses only in 2.5%, the impact resistance increases in 43.6% and the rupture and toughness moduli increase in 3.4 and 22.38%, respectively. On the other hand, the distortion behavior is optimum. It is easy to reach this temperature by simple means with low costs because the duration of the treatment is short.

CONCLUSIONS

1   The results of hardness and impact tests, in spite of their wide use in the industry, do not provide enough information to evaluate the heat treatments.

2   Concerning hardness values, all the sub-zero treatments are better than double tempering.

3   Sub-zero treatments improve the impact resistance of M7 steel with exception of those made at -80 C and -100 C.

4   Sub-zero treatments between -40 and -60 C produce better results in torsion tests of the M7 steel than those at lower temperatures.

APENDIX

Modulus of rupture (in torsion) (4).

$$S_{su} = M_{tu} \; r/J$$

where:

M_{tu} = maximum torque.

r = radius of the sample.
  = 0.63 cm

J = polar moment of inercia of the transversal section.

Modulus of rupture (in torsion) (4).

$$E_s = M_t \; L \, / \, \theta \, J$$

where:

M_t = torque.

L = gauge length.

θ = rotation angle (in radians)

Toughness modulus (in torsion) (4).

$$T_s = m_{tu} \; \theta_f \, / \, A \, L$$

where:

M_{tu} = maximum torque.

θ = maximum angle at rupture.

A L = volume of the cylindrical sample.

REFERENCES

(1) Roberts, G.A., J.C. Hamaker and A.R. Johnson. "Tool Steels". American Society for Metals, 1971.

(2) ASM Handbook, Vol 4, 1991, p.752.

(3) Becherer, B.A. and L. Ryan. "Control of Distortion in Tool Steels". ASM Handbook, Vol 4, 1991, p. 762.

(4) Marin, J. "Mechanical Behavior of Engineering Materials". Prentice-Hall , USA, 1965, p. 8, 66, 69.

Heat Treatment	Double Tempering	-20 C	-40 C	-60 C	-80 C	-100 C	-196 C
Hardness Rc	65.0	65.6	66.6	66.1	66.1	66.7	66.5
ΔRc %	---	+0.9%	+2.5%	+1.5%	+1.7%	+2.6%	+2.3%
Impact Kg/m	0.30	0.44	0.43	0.41	0.28	0.29	0.32
ΔR_{imp} %	---	+46.7%	+43.3%	+36.6%	-6.7%	-3.3%	+6.6%
M_{su} Kg/cm	8,016	8,742	8,358	8,941	7,525	7,801	7,900
ΔM_{su} %	---	+23.0%	+3.4%	+10.6%	-5.6%	-3.3%	-3.1%
θ (radians)	1.099	1.501	1.361	1.396	0.855	0.890	1.065
$\Delta\theta$ %	---	+36.6%	+23.8%	+27.0%	-22.2%	-19.0%	-3.1%
S_{su} Kg/cm^2	20,406	22,255	21,277	22,761	19,156	19,859	20,111
ΔS_{su} %	---	+9.1%	+4.3%	+11.5%	-6.1%	2.7%	-1.4%
T_s $\frac{Kg\ cm}{cm^3}$	707	1,052	912	1,001	516	563	675
ΔT_s %	---	+48.8%	+22.0%	+41.6%	-27.1%	-20.4%	-4.5%

TABLE 1-COMPARATIVE EFFECTS OF DIFFERENT HEAT TREATMENTS ON THE MECHANICAL PROPERTIES OF THE M7 STEEL.

	A	B	C	D	E
Double Tempering Initial length Final length Δl %	126.58 mm 126.63 mm +0.05 +0.04%	73.80 mm 73.79 mm -0.01 -0.01%	--- --- --- ---	12.75 mm 12.72 mm -0.03 -0.24%	25.35 mm 25.343 mm -0.007 -0.03%
-20 C Initial length Final length Δl %	126.50 mm 126.56 mm +0.06 +0.05%	73.77 mm 73.81 mm -0.04 -0.05%	--- --- --- ---	12.75 mm 12.78 mm +0.03 +0.24%	25.38 mm 25.423 mm +0.043 +0.17%
-40 C Initial length Final length Δl %	126.60 mm 126.65 mm +0.05 +0.04%	73.66 mm 73.67 mm +0.01 +0.01%	--- --- --- ---	12.74 mm 12.76 mm +0.02 +0.16%	25.40 mm 25.396 mm -0.004 +0.02%
-60 C Initial length Final length Δl %	126.52 mm 126.56 mm +0.04 +0.03%	73.74 mm 73.74 mm 0.00 0.00%	--- --- --- ---	12.70 mm 12.70 mm 0.00 0.00%	25.35 mm 25.432 mm +0.082 +0.32%
-80 C Initial length Final length Δl %	126.26 mm 126.38 mm +0.42 +0.10%	73.73 mm 73.78 mm +0.05 +0.07%	--- --- --- ---	12.78 mm 12.72 mm -0.06 -0.47%	25.33 mm 25.407 mm +0.077 +0.30%
-100 C Initial length Final length Δl %	126.62 mm 126.74 mm +0.12 +0.94%	73.79 mm 73.89 mm +0.10 +0.14%	--- --- --- ---	12.74 mm 12.72 mm -0.02 -0.16%	25.32 mm 25.397 mm +0.027 +0.30%

TABLE 2-COMPARATIVE EFFECTS OF DIFFERENT HEAT TREATMENTS ON THE DIMENSIONS OF THE NAVY "C" SPECIMENS.

Eleven Years Quenching With P.A.G. (Polyalkylene Glycsol)

E. Varela and P. Córdoba
Acindar S.A.
Buenos Aires, Argentina

ABSTRACT

Today, ACINDAR S.A. is the most important high-quality specialty steel manufacturer in Argentina. Entirely owned by argentinian capitals, it has two plants. Plant #2, located at Villa Constitución, produces 750,000 tons/year of carbon steels. Plant #1, located at Buenos Aires, produces 200,000 tons/year of high grade and special steels.
Coming from Plant #1, customers receive about 2,000 tons/year of alloyed steel products used in different fields of the industry: automotive (gears, pinions, crank shafts, etc.), oil (sucker rods, drill collars, shaft spindles, etc.) and others. These products may be delivered under different heat treatment conditions, such as: homogenized, normalized, soft annealed, quenched and tempered, stress relieved, etc.
Since 1980, we have been opereating with an installation which has a tilt-up type integral quenching furnace with a quenching manipulator and a 160,000 lts. quench-tank with a solution 20% (V/V) of a Polyalkylene Glycol (P.A.G.). Prior to 1991 we have processed about 20,000 tons. of different steel grades with good results and without any problem.
However, at this time some problems in metallurgical properties (quench cracks, low toughness) in quenched products began to occur. It was decided to contact then the P.A.G. supplier, who considered that the bath life was coming to an end.
Different alternatives were then evaluated to select the new quenchant media. Several P.A.G. products available in domestic market were tested in laboratory and industrial tests. In this report, the results of these experiences are detailed.

FROM 1980 TO 1988, A TOTAL OF 20,000 tons of different materials were quenched with excellent results and without any operative problems. At that moment, some irritating vapors appeared to be emanating from the bath during quench that lead us to contact our dealer in order to test a sample of the used quenchant. From this sample, chemical analysis and cooling curves data were obtained.

Gel Permeation Chromatography (G.P.C.) data showed that substantial polymer degradation has occurred (Fig.1). Thus, the irritating vapors were probably due to the evolution of the volatile low molecular weight degradation by-products.

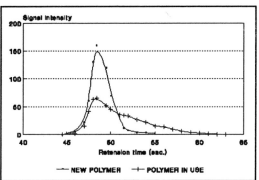

Figure 1. Comparison of fresh and in use polymer samples

Cooling curves analyses confirmed that degradation had occurred, since much faster cooling rates were obtained for the used quenchant versus a fresh one (Fig.2). On the basis of these results, the supplier recommended quenchant bath replacement.

This fact, together with the appearance of some metallurgical problems in quenched materials, lead to the evaluation of different polymer quenchant alternatives for the bath replacement. With this purpose, laboratory and industrial tests were simultaneously made with P.A.G. products available in domestic market.

EXPERIMENTAL PROCEDURE

Laboratory tests consisted of comparative analysis of samples provided by local suppliers. In each case the following data were evaluated:
- Density at 20°C and 40°C.
- Flash point.
- Cloud point 16% V/V
- pH 1% and 5%

Quenched Solutions at 10,15 and 20% were also prepared, and tested to obtain the cooling curves at different temperatures. These tests were performed with a $1cm^3$ of volume probe (Type 304 Stainless Steel), minimum agitation, 40 cm/min recorder, and 750 cm^3 of quenchant bath. The curves obtained were compared with those of the used polymer at 20% and water.

To verify the quenching behavior of the alternative polymers under industrial tests, bars of the same lenght, diameter, grade and heat were used, in order to ensure conditions equal to the productive process.

The steel grades analyzed were:
- SAE 4145 (ϕ 131 mm)
- SAE 4340 (ϕ 130 mm)
- SAE 4140 (ϕ 69 mm)

After quenching and tempering, the mechanical characteristics obtained were measured.

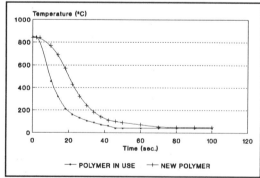

Figure 2. Comparison of fresh and in use polymer (B.Temp: 40°C)

RESULTS AND DISCUSSION

Generally routine chemical analyses are performed to provide assurance that the bath composition is correct and that it had not undergone any degradation. The parameters usually controlled are: Viscosity, which depends on the molecular weight of the polymer used to formulate the quenchant. The viscosity of the quenchant bath can be compared to a standard viscosity/concentration chart to determine if molecular weight degradation took place.

Cloud point (of a new polymer) is indicative of the polymer composition. This parameter can be affected by many variables, for that reason it is considered as an emperical indicator of the bath performance variation.

Refractive Index is primarily a function of concentration. This index is not sensitive to molecular weight variation, but it is sensitive to substantial polymer degrations, polymer composition, impurities, and temperature. Since it is a rapid method, we have been using it in our plant to determine the bath concentration.

We have been also using the delta factor as a measure of polymer degradation. This factor shows the relationship between the concentration data obtained through the viscosity and refractive index measurements. The difference between both values is reflective of the degradation. When the anlysis was carried out the delta factor of the polymer in use was 8-10.

Typical properties of used and alternative polymers are shown in TABLE I.

No important variations between the studied products were registered, regarding density.

Polymer #2 reported a flash point of 180 °C. The other ones didn't.

As this test is made after water evaporation, it is considered that there is no flammability risk.

The difference in viscosity among the polymers is quite important. For example, 1,500 cst was measured in polymer # 2, while 500 cst (aprox.) was the value for the #1. Remember that the water viscosity is 1.0 cst.

If we take the pure product and dilute it 20%, the differences in viscosity are reduced.

Polymer #3 reported a very low cloud point (74°C), compared with the others (84-92 °C). As with the bath after continuous use, its temperature increases aprox. 6 °C (in our case). This could restrict the possibility of use at higher temperatures of this product, for example, to lower the quenchant severity.

TABLE I. Comparison of characteristics

CHARACTERISTICS	POL. IN USE	POLYM. #1	POLYM. #2	POLYM. #3
Viscosity at 40°c (cst)	646	582	1500	512
Firing point of the remainder	NONE	NONE	180 °C	NONE
Cloud point 16% V/V	90 °C	90 °C	84 °C	74 °C
PH 1 %	9	9-11	8	9
PH 5 %	9	9-11	8,3	9
Density at 20°c (g/cm³)	1,140	1,101	1,108	1,138

LABORATORY TESTS

With each supplier's sample, solution concentrations of 10, 15 and 20% were prepared. Using our laboratory quenchant equipment, the cooling curves at 60 and 70 °C were drawn, and they were compared those of the polymer in use and water (Fig. 3, 4 and 5).

In Fig.3a, it can be clearly seen that the curves from the polymer in use at 20% and the fresh polymer are quite similar. The test at 60 °C showed an approach of the polymer in use curve to that from the water, while the solutions at 15% did not show substantial behaviour differences.

From the cooling curves analyses, at different conditions, it can be concluded that as the tested solutions concentrations approach that of the polymer in use, behaviors difter more and more, specially during second and third cooling stages.

In tests performed at 70 °C the cooling curves were similar, in every case, and the differences decrease as the quenchant severity lowers.

INDUSTRIAL TESTS

In order to verify the alternative polymers behavior, quenchant tests were carried out with commonly used steel grades, and under the production process conditions. The quench-tank used for these tests has a 2,500 lts capacity.

Samples were sent to the Metallurgical Laboratory after the treatment (quenching and tempering), in order to evaluate the mechanical characteristics obtained.

Figs. 6,7 and 8 show no remarkable hardness differencies (HRc) among the samples quenched with the different polymers. The values obtained are located in the expected range, considering the operative conditions. The mechanical characteristics showed identical behaviour.

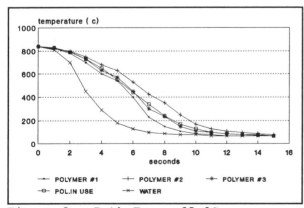

Figure 3a. Bath Temp.:60 °C
Concentration:10%

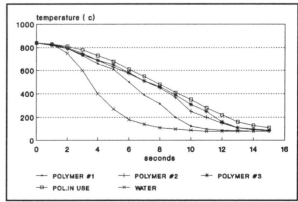

Figure 3b. Bath Temp.:70 °C
Concentration:10%

TABLE II. Mechanical characteristics

		POL.IN USE	POL. # 1	POL. # 2	POL. # 3
SAE 4140 Ø 69mm	σR	115,5	115	112,1	115,8
	σ0,2	105,8	103,9	100,5	106,3
	Δl(%)	17	14,8	16,6	16
	Ψ	51,6	52,2	51,2	52,9
SAE 4340 Ø 180mm	σR	105,4	100,6	102,2	100,8
	σ0,2	92,6	87,7	89,2	86,4
	Δl(%)	17,8	18,2	18,6	16,8
	Ψ	57,9	59,7	61,3	60,4
SAE 4145 Ø 131mm	σR	112,7	115,6	106,6	112,7
	σ0,2	91,3	102,7	95,1	98,8
	Δl(%)	15,2	13,6	16	15,2
	Ψ	51,7	42,3	52	59,9

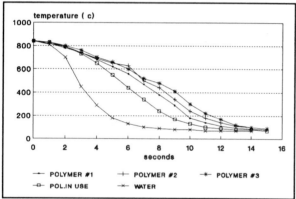

Figure 4a. Bath temp.: 60°C
Concentration: 15%

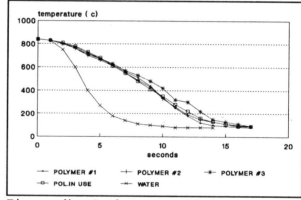

Figure 4b. Bath temp.: 70 °C
Concentration: 15%

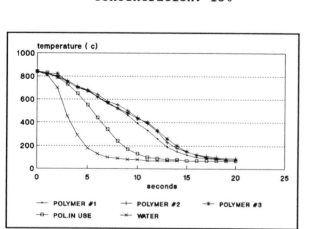

Figure 5a.Bath temp. 60 °C
Concentration: 20%

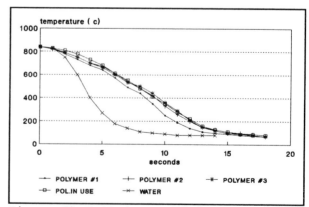

Figure 5b.Bath temp.:70 °C
Concentration:20%

In Table II the measured values of σ_R (Kg/mm2) , $\sigma_{0.2}$ (Kg/mm2), Δl (%) and Ψ (%) are indicated.

CONCLUSION

* The quenchant polymer #2 reports higher viscosity than the #1. In cooling curves both polymers behaviour is similar, being the #1 slightly slower than the #2.
* The quenchant polymer #3 has a cloud point lower than the others. This fact prevents the use of an inmediate action tool to modify the quenching severity.
* The polymer #2 is cheaper than #1, but no experimental data is known.
* During industrial tests all the polymers behaved satisfactorily.
* As a result of the studies reported here, it was decided that there was no reason to convert from polymer #1 which is currently in use. This allows continuity in the plant since the operators do not have to be retrained to work with another quenchant. In addition to this fact and the excellent lifetime achieved previously, the excellent support provided by the supplier was especially valuable.

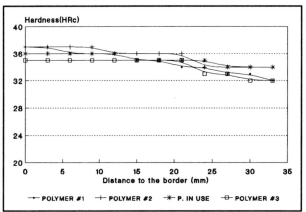

Figure 6-SAE 4140-Diam. 69mm
Conc.:12,5%-Temp.:60 °C

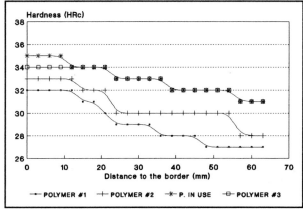

Figure 7. SAE 4340-Diam. 130mm
Conc.:12,5%-Temp.:60 °C

REFERENCES

1-G.E.Totten. "Analysis of used UCON Quenchant HT from ACINDAR S.A."-Technical Service Report-UNION CARBIDE Corp.,May 1988.
2-Bob Brennan. "Advanced Polymer Extends Quenching Technology".-Industrial Heating,pag.34-36, September 1988.
3-A.Beck. "Some factors to consider in Quench Design using Polymer Quenchants".-Heat Treating,pag.55-60, May 1977.
4-Research and Development. "Evaluation of quenched samples"-Internal report, ACINDAR S.A., August 1989.

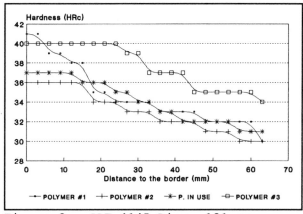

Figure 8 - SAE 4145-Diam.:131 mm
Conc.:12,5%-Temp.:60 °C

Conversion and Use of Polymer Quenchants in Integral Quench Furnace Applications

D. Diaz
Union Carbide Quimicos Y Plasticos
Mexico City, Mexico

H. Garcia
Spicer Group-Mexico
Queretaro, Mexico

B. Bautista
Spicer Group-Mexico
Celeya, Mexico

ABSTRACT

Although they are excellent quenchants, the use of oils in integral (sealed) quench furnaces present an often unacceptable fire risk. To address this potential problem, test-quenching work was recently completed at the Spicer Group - Mexico, to evaluate the potential use of aqueous polymer quenchants in various integral quench (IQ) furnace quenching processes. The conversion procedure (oil to polymer) and subsequent test-quench results of this work will be described here.

INTRODUCTION

Heat treaters have used various aqueous polymer quenchants for over thirty years. These quenchants are often used as fire-resistant alternatives to medium-fast speed quench oils or as additives to water, when used in low concentration, to provide a more uniform quench.

The polymers that have been used as quenchants include: poly(vinyl alcohol), poly(sodium acrylate), poly(vinyl pyrrolidone), poly(ethyl oxazoline) and poly(alkylene glycol) (1). Of these polymers, various poly(alkylene glycol)'s - (PAG's) are most often used in quenching applications today. Recently, a paper was published providing details on the selection and use of PAG quenchants.(2).

Aqueous polymer quenchants are being used increasingly for quenching forgings (3-6), open-tank quenching of high-hardenability steels (7), patenting of high-carbon steel wire and rod (8), induction hardening (9), IQ furnace applications (10), railroad rails (11), etc. Of these applications, perhaps the use of aqueous polymer quenchants in IQ furnaces is both the least known and most poorly understood. Some of the concerns that have prevented broader use of aqueous polymers in IQ furnace quenching processes include: potential water vapor contamination (from the quenchant) of carburizing atmospheres, corrosion problems, and poor quenching performance, e.g. distortion and cracking, due to insufficient quench severity reduction and non-uniformity.

The objective of this paper is to describe the use of a PAG quenchant in two IQ furnace quenching processes including conversion, case histories and description of the periodic maintenance procedure currently in use. It will be shown that PAG quenchants effectively control distortion and prevent cracking while providing the desired hardness. It will be shown that PAG quenchants are more cost-effective and versatile than many accelerated quench oils. It will also be shown that quench uniformity superior to that previously attainable with a medium-speed quench oil (Houghtoquench G) was achieved.

DISCUSSION

1. Quenchant Selection

The PAG polymer quenchant, in this case UCON® Quenchant A, was selected after conducting a test-quench program in the IQ furnace located in Waukesha, WI and made available by Tenaxol Inc. For these tests, carburized AISI 8620 and 8617 universal joints were racked and quenched in different UCON® Quenchants at varying concentrations and bath temperatures. This work demonstrated the feasibility of the potential use of a PAG quenchant for the production of these parts.

2. Furnace Descriptions

The next step was to conduct a rigorous trial under actual production conditions using a Surface Combustion All-Case IQ furnace. This furnace is typical of many IQ furnaces commonly used in the heat treating industry which is particularly suited to the use polymer quenchants since: the quench tank is adequately sized (4,000 liters), it has excellent agitation (single propeller) to facilitate a uniform quench and it employs an adequate heat exchanger system to control the bath temperature rise during the quench ($\approx$10°C maximum).

A second trial was conducted with an AFC (Atmosphere Furnace Company) furnace which is also ideal for use with PAG quenchants. This furnace utilizes a 12,000 liter quench tank with excellent agitation provided by a dual propeller system.

3. Oil Conversion Procedure

Earlier it was shown that PAG polymers mediate heat transfer during quenching by film formation (2). It has also been shown that if quench cracking and distortion are to be minimized it is critically important to minimize any potential source of non-uniform heat transfer (12).

The presence of a sufficient quantity of quench oil in the system may produce a non-uniform film around the part during the quench (since oils are generally incompatible with water soluble polymers). The polymer film heterogeneity will produce varying heat transfer rates across the surface which may result in surface thermal gradients during the quench which are sufficiently high to yield localized stresses. This is a potential source of distortion or cracking. Thus it is critically important to minimize any residual quench oil if optimal as-quenched results are to be obtained.

The following conversion procedure was followed to minimize the presence of residual quench oil:

1. The quench oil was drained (or pumped) from the quench tank.

2. The quench tank was then washed with an aqueous detergent solution to emulsify any residual oil to facilitate removal.

3. The tank was then washed repeatedly with water to assure removal of any residual detergent. (This is essential since excessive foaming may also result in a non-uniform quench.)

4. After the system has been thoroughly cleaned, the PAG quenchant concentrate is added and diluted to the desired

concentration. The concentration of the quenchant is determined with the use of a hand-held refractometer and a standard calibration chart available from the quenchant supplier.

4. Surface Combustion All-Case Furnace Test Results

The control quench process used in the Surface Combustion All-Case furnace involved carburizing (7 hour cycle time) AISI 8620 or 8617 universal joint crosses at 960°C. A typical load weight varies from 200-350 kg. The final surface carbon content was 1.6-1.8% with a surface hardness of 59-64 Rc and core hardness of 45 Rc after quenching. The surface martensite content was 56% (Figure 1). These properties were easily obtained with the quench oil.

The conversion from the quench oil to 11.5% of UCON® Quenchant A was performed following the procedure outlined above. The initial quenchant temperature was 40°C and was not allowed to exceed 50°C during the quench. The carburizing cycle used for this work was identical to that used for the oil-quenching process.

Metallurgical analysis (see Table 1) showed that the surface martensite content had increased from 56% for the oil quench to 86% for the PAG quench as shown in Figures 1 and 2 respectively. This quenching condition produced a surface hardness of 63-64 Rc. The core hardness was 44-45 Rc.

TABLE 1
Comparison of AISI Carburized 8620 Metallurgical Properties of an Oil and Polymer Quench

	Oil Quench	Polymer Quench
% Martensite (at surface)	56	86
% Carbon (at surface)	1.6 - 1.8	1.6 - 1.8
Hardness at Surface (Rc)	59 - 64	63 - 64
Hardness at Core (Rc)	45	44 - 45

Fig. 1 - Martensite Content of Oil Quenched Carburized AISI 8620 (56%)

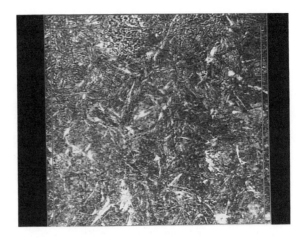

Fig. 2 - Martensite Content of PAG Quenched Carburized AISI 8620 (86%)

The PAG quenchant was a particularly cost-effective alternative to oil. In this case, the annualized cost of the PAG quenchant was 28% less than the oil. This was due to a combination of both a lower cost/unit volume on a diluted basis and substantially lower drag-out rates for the PAG quenchant.

5. AFC Furnace Test Results

The AFC furnace is used to heat treat constant speed joints made from AISI 8617 which is carburized at 945°C to a case depth of 1.5 mm with a carbon content of 1.2-1.4%. These parts are particularly prone to cracking and distortion due to their dimensional complexity. The metallurgical specifications are summarized in Table 2.

TABLE 2
Summary of Metallurgical Properties Obtained in the AFC Furnace Test-Quench Program

Property	Polymer Quench
Case Depth (mm)	1.5
% Carbon at Surface	1.2 - 1.4
Surface Hardness (Rc)	58 - 62
Core Hardness (Rc)	35 - 45
Max. Distortion (mm)	0.2

In this case, the quenchant concentration and optimal bath temperature were established by cooling curve analysis using an IVF Quenchotest instrument (13) and 12.5 x 60 mm INCONEL 600 probe. The cooling curves for various PAG quenchant concentrations and quenching conditions were obtained and compared to the quench oil (Houghtoquench G) currently in use. Although the cooling temperature-time profiles for the PAG quenchant were not identical to the oil, this study showed that approximately the same thermal traces in the most critical heat transfer regions could be obtained. It is recognized that substantially different agitation conditions are present in the quench tank itself. Nevertheless, these studies showed that cooling behavior similar to a medium-speed quench oil was possible. On the basis of this work, it was assumed that comparable quench severities could be obtained in the AFC furnace quench system, although this was not confirmed by cooling curve analysis under production quenching conditions.

This furnace (which was actually designed for an oil quenchant) was purchased new. However, the only quenchant that has been used thus far is UCON® Quenchant A (24%). This concentration was established from the cooling curve analyses reported above. The starting quenching temperature was 50°C and was not allowed to exceed 60°C.

The results of this test-quench program showed that the hardness specifications were easily obtained. The distortion was comparable to that obtained for the same type of parts quenched in oil in other furnaces performing the same heat treatment operation.

Parts quenched in oil typically produce an average hardness of 61.26 ± 1.23 Rc. The PAG quench produced a higher average hardness with a lower standard deviation (62.87 ± 0.11 Rc). Therefore, another advantage of the use of the PAG quenchant for this process was a substantial improvement in process uniformity.

It is important to note that no corrosion problems have been encountered. Also, no water contamination of the endo atmosphere has occurred.

6. Quenchant Maintenance

For optimal quenching performance, it is important that precautions be taken to assure that the PAG quenchant remain uncontaminated during use. This includes solid contaminants such as scale, etc. Therefore, a filter, preferably approximately 5 micron, should be used in the quenchant recirculation system. The only other maintenance procedure performed by the engineers at both of these shops is periodic determination of quenchant concentration using a hand-held refractometer. The quenchant concentration should be maintained within approximately ± 1.0%.

Periodic quenchant evaluations are also performed by the quenchant supplier as a free service. One of the objectives of this level of maintenance is to assure that the quenchant concentration measurement by hand-held refractometer is properly calibrated. Another objective is to determine if additional corrosion inhibitor must be added to assure continued corrosion protection of the system. Finally, tests are performed to determine if any unexpected contamination has occurred.

7. Application Notes

PAG quenchants perform well if the following precautions are taken:

- Excessive bath loadings are avoided.

- Proper racking procedures should be used to permit adequate flow of the quenchant through the load during the quench.

- The agitation system must **always** be on before and during the transfer of the load from the furnace vestibule to the quench chamber to minimize the potential for water contamination of the furnace atmosphere.

- The bath temperature rise should not exceed 10°C.

- The concentration of the quenchant should be maintained to ± 1.0%.

- Periodic replenishment of the corrosion inhibitor is required to assure long-term furnace protection.

CONCLUSION

Aqueous PAG quenchants are excellent fire-resistant alternatives to quench oil for many IQ furnace quenching applications. On the basis of this work, it is clear that significant physical property improvements relative to oil are possible with the use of PAG quenchants. In addition, substantial improvements in process quality and cost reduction are possible, as long as proper maintenance procedures are followed.

REFERENCES

1. G.E. Totten, *Adv. Mater. and Process.*, **1990**, 137(3), p 51-53.

2. G.E. Totten, K.B. Orszak, L.M. Jarvis and R.R. Blackwood, *Ind. Heat.*, **1991**, October, p 37-41.

3. W.G. Patton, *Iron Age*, **1973**, 212, p 46.

4. H. Yamamoto, Y. Odagiri, M. Miyashita and F. Nishiyama, *Netsu Shori*, **1989**, 29 (1), p 14-18.

5. E. Sprow, *Tooling and Production*, **1983**, 49(9), p 22.

6. R.S. Krysiak, *Mech. Working and Steel Process. Proceed.*, **1988**, p 255-259.

7. R. Creal, *Heat Treat.*, September, **1982**, p 21-23.

8. K.J. Mason and T. Griffin, *Heat Treat. of Metals*, **1982**, 3, p 77-83.

9. O. Sparks, *Heat Treat.*, March, **1978**, p 18-22.

10. A.J. Beck, *Heat Treat.*, May, **1977**, p 55-60.

11. E.L. Kolosova, L.P. Shcherbakova, Yu.G. Ersmondt and A.V. Kharov, *Steel in the USSR*, **1988**, 18, p 557-559.

12. R.K. Zhelokhovtseva, *Steel in the USSR*, **1985**, 15, p 238-239.

13. The Quenchotest is a portable cooling curve data acquisition instrument which is manufactured by: The Swedish Institute for Production Research (IVF), Molndalsvagen 85, S-412 85 Gothenberg, Sweden.

Proceedings of the First International Conference on Quenching & Control of Distortion, Chicago, Illinois, USA, 22-25 September 1992

Advances in Fluidized Bed Quenching

A. Dinunzi
Procedyne Corp.
New Brunswick, New Jersey

ABSTRACT

Fluid Bed Heat Treating has been in use for about 10 years. One of the fastest emerging areas of interest has been in the use of fluidized beds for quenching of various grades of steel.

This paper discusses recent process advances in fluid bed quenching and outlines the various grades of steel that can be successfully fluid bed quenched.

I.   Introduction

The portion of any metal Heat Treatment cycle occupied by the quench is minor. Yet it is the most critical factor in determining whether the required mechanical, dimensional and surface properties are met.

Significant efforts have been made in the last 20 years to obtain more control and variability of quench rates. This has included advances in water based synthetics and high pressure gas quenching.

Added to this list have been a series of advances in the use and application of fluidized beds as a quenching medium. Fluidized bed quenching technology has been shown to help alleviate many of the control problems associated with liquid quenchants. Liquid quenchants exhibit complex cooling behavior (vapor phase, nucleate boiling, convection). The heat transfer mechanism in fluid beds is the same throughout the entire temperature range of quenching and is dominated by the properties of the gas phase. This leads to a quench rate that is reproducible, does not degrade with time and can be adjusted within wide limits.*

In addition to this flexibility, fluid beds can operate over a wide temperature range and do not have the environmental drawbacks of oils or salt quenchants.

The enhanced flexibility and control offered by fluid beds have led to a variety of unique quenching applications. These will be the subject of this paper.

II. Fluidized Bed
    Technical
    Characteristics

Delano and Van Den Sype[1] investigated the operating properties of Fluid Beds and what factors affect the quench speed. It was found that Fluid Beds, operating at ambient with twice minimum fluidization under N_2, were intermediate between gas quenching and oil quenching, but did not have a film co-efficient sufficient to through harden most medium alloy or case hardened materials.

However, the rate of quenching or heat

transfer can be adjusted by altering the operating conditions of the fluid bed. Figure 1** below lists the 4 variables that were investigated for their effect on the heat transfer rate.

FIGURE I
Parameters Affecting Heat
Transfer in Fluid Beds

Parameters	Effect	Comment
A) Particle size (d)	As dia.↘ h↗	
B) Vol. Heat Cap. (Cv)	As Cv↗ h↗	Al_2O_3 is preferred
C) Fluidizing Conducting (k)	As K↗ h↗	H_2 or He is highest
D) Fluidizing Gas Flow (Ug)	h becomes max. at Ug≈5 x U_{MFF} (Min. Fluidization Flow)	

The two variables most strongly effecting heat transfer rates were the rate of fluidizing gas flow and the thermal conductivity of the gas. In Section V we will review how adjustments to these two variables can yield enhanced quench results.

**M.A. Delano and J.Van Den Sype, "Fluid Bed Quenching of Steels:Applications are widening," Heat Treating, December 1988.

III. Use of Fluid Bed for
     Quenching of Air
     Hardening Steels

Fluidized Beds have
been used for a number of
years as a quenching
medium with air harden-
ing tools steels. There
are two critical issues
in the quenching of tool
steels. The quench rate
must be severe enough to
effect full metallurgical
transformation of thick
sections while at the
same time not causing
severe distortion or
cracking.

Vacuum furnace tech-
nology has made signifi-
cant advances in the area
of high pressure gas
quenching. These ad-
vances have come,
however, with a high
initial equipment
purchase price.

Figure 2 shows a
comparison of quenching
rates (to 1000°F, from
2200°F, for M-2 grade
High Speed Steel) between
an ambient fluid bed
operating on nitrogen and
vacuum furnaces operating
at 2 Bar and 6 Bar quench
pressures.

The fluid bed has
slightly higher quench
rates than the 6 Bar
unit. The initial
capital cost for a
fluidized bed system is
significantly less than
the vacuum.

A second critical
consideration in the
quenching of air
hardening tool steels is
the ability to control the
quench process to minimize
distortion and prevent
cracking.

The two major contri-
butors to cracking and
distortion are too severe a
quench rate and non uniform
cooling of all part surfaces.
In vacuum systems cooling gas
is injected into the work
chamber by a high horsepower
blower through an internal gas
distribution system. It is
then drawn out of the chamber
through a heat exchanger and
reinjected into the work
chamber. An accelerated
cooling rate is obtained by
injecting the cooling gas at
increased pressure.

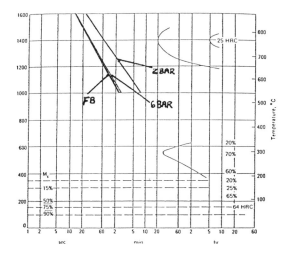

FIGURE 2
Quenching Rates to 1000°F
Comparison of Fluid Bed Quench
operating on N_2 and Vacuum
Furnaces operating at 2 Bar
and 6 Bar Quench Pressures.

If the work load in the vacuum chamber is dense or the parts are intricate in shape the flow of quench gas can be non-uniform or short circuited. This can cause either slack hardness or distortion due to differential cooling of part surfaces.

The fluid bed is much less sensitive to load density or part geometry. Because of the liquid like characteristics of the fluid bed the parts are fully surrounded by aluminum oxide. The high heat capacity of the aluminum oxide does not require complete gas flow over all surfaces since heat is removed via conduction.

Additional control over quenching speeds and hence distortion can be obtained by use of multiple fluidized beds operating at various temperatures.

A Procedyne customer, routinely processing 1000-2000# gross loads of H-11 and H-13 forge dies and hot work tooling which has section thicknesses of up to 8" (200 mm) and critical distortion tolerances, uses the furnace arrangement shown in Figure 3 to produce work with full hardness and minimal distortion.

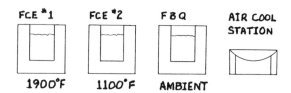

FIGURE 3
FLUIDIZED BED FURNACE
ARRANGEMENT FOR PROCESSING
H-11 & H-13 FORGE DIES AND
HOT WORK TOOLING

The heat treating procedure for the dies is as follows:

1) Preheat work to 1100°F

2) Austenitize at 1900°F

3) Step quench at 1100°F for 15 minutes

4) Fluid Bed Quench at ambient for 5-7 minutes to approx. 550°F

5) Air Cool to room temperature.

Figure 4 shows the cooling rate superimposed on the iso-thermal transformation diagram for H-11 material.

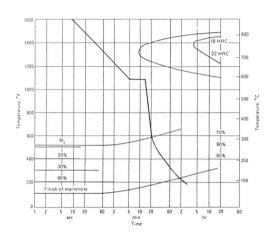

IV. Wire Patenting

One of the oldest applications for the use of fluidized bed quenching is in the patenting of steel wire. Patenting involves heating steel wire, with varying carbon contents, to austenitizing temperature and cooling the wire at such a rate as to prevent the formation of transformation products.

A typical patenting line consists of an austenitizing furnace followed by a holding or quench furnace. The austenitizing furnace could be either a high temperature lead or salt bath, an open gas fired unit or a protective atmosphere type using alloy tubes filled with a reducing gas. The quenches used are generally liquid salt or lead.

Because of environmental concerns fluidized beds have been viewed as a desirable alternative to both lead or salt quenches. The major drawback to their use has been twofold. First the comparative heat transfer rate between salt/lead and fluidized bed is very poor. Secondly, fluidized bed quenches are not compatible with salt and lead austenitizing lines because of dragout contamination from the high temperature furnace into the quench.

For atmosphere type austenitizing units, Procedyne has developed a successful solution. It has compensated for the lower relative heat transfer rate of fluid beds, compared to salt /lead, by implementing a two stage

Figure 4
Cooling Rate of H-11
Superimposed on
Isothermal Trans-
formation Diagram

This procedure allows the large tooling sections to equilibrate at a temperature below the "nose" of the cooling curve prior to being cooled to a point just above the M_s line. The parts are then allowed to slowly cool as they begin to transform. This procedure dramatically reduces distortion without compromising hardness. Thorough analysis of metallurgical structure has shown this procedure prevents carbide precipitation at the grain boundaries.

quench.

The wire exists the austenitizing unit at 1600°F and enters the first stage quench which is operating at 250-400°F. This unit is generally 2-4' in length. The high ▲T between the wire and bed temperature compensates for the lower heat transfer rates found in fluid beds. The wire exits this initial quench zone at 950-1100°F and enters the 2nd quench zone which is operated at the expected wire exit temperature from the first quench zone. The net effect of this arrangement is to produce wire with identical properties to those produced in salt/lead with no additional required bed length or operating cost. This arrangement also eliminates any environmental impact. Figure 5 shows a typical system arrangement.

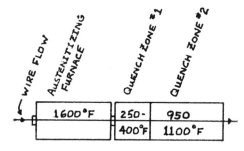

FIGURE 5
Typical Wire Patenting
System Arrangement

V. Austempering Marquenching, Direct Hardening

Among the most common heat treatments are austempering, marquenching and direct hardening of medium to low allow steels. We are defining these materials as having an Isothermal Transformation diagram with a "nose" of 10 seconds or less.

As we saw in Section II, a normal, operating fluidized bed had insufficient heat transfer characteristics to be useful with these grades of material.

But, recent developments have been made which dramatically enhance the quenching speed of the bed and, allow austempering, marquenching and direct hardening of a large variety of alloys.

We also saw in Section II that the two most significant factors affecting the quenching rate of a fluidized bed was the conductivity of the fluidizing gas and the gas flow rate. These two factors were combined with the fact that the fluid bed can operate over a wide temperature range without solidifying like salt.

These concepts opened the way to develop an economic and ecologically desirable replacement for salt.

Two changes were made to enhance the quenching speed of fluid beds. Nitrogen was replaced by helium which has a gas conductivity nearly 6 times nitrogen and the fluidization rate was doubled.

The primary drawback was cost. Helium is up to 20 times more expensive than nitrogen. Therefore using helium to fluidize the bed during the entire quench cycle was uneconomical.

The Isothermal Transformation diagram for a typical medium alloy steel such as 4340 (see figure 6) shows the critical portion of the cooling cycle is the first 10 seconds. Once the part is cooled past the nose of the curve the time allowed to finish cooling is greatly increased.

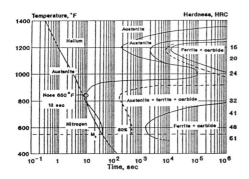

FIGURE 6
ISOTHERMAL TRANSFORMATION
DIAGRAM

This led to the use of a short pulse of helium (30-60 secs) at the initiation of the quench cycle to drive the load past the nose of the cooling curve. The quench bed is then switched back to nitrogen for the remainder of the cycle.

The first large scale use of this technology was in the austempering of 4340 steel tools. Salt processing involved austenitizing at 1625°F and quenching into salt at 540°F and holding 30 minutes. The successful switch to the helium quenching system required only a slight reduction in the austemper quench temperature. (From 540°F to 500°F).

The Fluid Bed using helium did not provide the same hardness results at the same temperature in all cases, but it was found that slight decreases in the fluid bed temperature resulted in a corresponding increase in hardness.

Figure 7 below is a summary of the types of material grades and processes that can be accommodated by fluidized bed quenching.*

Material	Typical Applications	Current Quench Method	Typical As Quenched Hardness	Dyna-Quench As Quenched Hardness
52100	Bearing Races	Oil	62.0 Rc	62.0 Rc
4340	Wood Router Bits	Austemper at 560°F	42.5 Rc	43.0 Rc
Modified S-2 F	Screw Driver Bits	Austemper at 540°F	55.5 Rc	55.0 Rc
O-1	General Tooling	Marquench Salt at 350°F	62.0 Rc	61.0 Rc
Ductile Iron	Crank Shaft	Austemper at 500°F	42.0 Rc	40.5 Rc
86B45	Forging	Oil	59.0 Rc	59.0 Rc
5150	Machined Parts	Oil	58.0 Rc	59.0 Rc

FIGURE 7
COMPARISON OF DYNA-QUENCH
TO CONVENTIONAL QUENCHING
METHODS

The combination of helium gas pulsing, higher fluidization rates and the ability to make bed temperature adjustments make for an extremely flexible quenching system.

REFERENCES

1.   M.A. Delano and J. Van Den Sype, Heat Treating (Reprint) (1988)

2.   American Society for Metals, "Heat Treater Guide", p. 331, 377, ASM Materials Park, Oh, (1982).

Proceedings of the First International Conference on Quenching & Control of Distortion, Chicago, Illinois, USA, 22-25 September 1992

HIP Quencher for Efficient and Uniform Quenching

C. Bergman
ABB Autoclave Systems
Västerås, Sweden

S. Segerberg
Swedish Institute of Production Engineering Research
Göteborg, Sweden

ABSTRACT

The HIP Quencher is a further development of the usual HIP units (Hot Isostatic Press). The improved cooling capacity and the high dense gas of the HIP Quencher give a high heat transfer coefficient. This, in combination with the active cooling mechanism, results in uniform quenching and consequently less variation in temperature between different parts of the components and naturally thus also less distortion. The HIP Quencher has been compared with the heat transfer coefficient calculated for a vacuum furnace, a salt bath and a fluidized bed. As the HIP Quencher uses only gas as the cooling media it is a clean and environmentally friendly technique.

THE HIP TECHNIQUE is used for densification of ceramics and metal powder and for improvement of castings. The HIP units of today can be provided with an powerful heat exchanger which allows faster cooling. This reduces the HIP process cycle time drastically, see Figure 1, and thus increases system productivity.

This new rapid cooling feature offers the possibility to combine densification and heat treatment in one operation thus reducing the cost [1]. It also gives the possibility to heat treat components of steel of higher hardenability such as high speed steel and tool steels.

HIP Quenching

The base of the HIP Quencher is the extremely high heat transfer coefficient in the gas under high pressure. This gives a uniform cooling of the component which results in less distortion. The control of the cooling rate will improve the material structure and thereby the quality. The typical pressure in a HIP Quencher is 1000 to 2000 bar.

Theoretically the heat transfer coefficient at 6 bar, as in vacuum furnaces, could be raised by blowing the gas by a fan over the surfaces of the components. However, the speed needed to achieve the same heat transfer is of the order of 110 m/s. As it is impossible to have a gas flow of that speed sweep every square mm of the charge, the heat transfer can differ up to 100 times between the different surfaces of the batch, resulting in very uneven cooling.

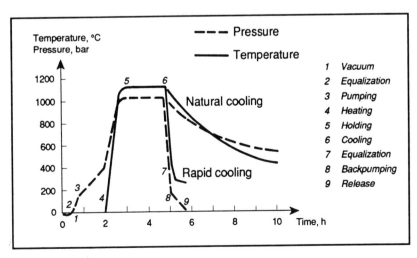

Fig. 1. Example of a HIP cycle with natural cooling versus rapid cooling. Note the reduced cycle time.

Table 1. The characteristics of the different methods of quenching.

Method of quenching	Temperature of quenchant, °C	Gas	Pressure, bar
Vacuum	60	Nitrogen	4
Salt bath	230	-	-
Fluidized bed	20	Nitrogen	1
HIP Quencher	Decreasing from 1000	Argon	800-1800

With HIP Quenching, the very high and uniform heat transfer coefficient assures that the surface temperature of all components of the batch will follow the temperature of the high pressure gas very closely.

Efficient quenching

The cooling capacity for the HIP Quencher was compared with vacuum furnace, salt bath and fluidized bed. The comparison was made with a ø 12.5 x 60 mm cylindrical probe made of Inconel 600 with the thermocouple in the geometrical centre. The test probe is endorsed by The International Federation for Heat Treatment and Surface Engineering, IFHT, [2]. The different parameters for the methods are summarized in Table 1.

The values used are the normal ones but they can naturally be varied within wide limits for each method. E.g. the salt bath contained 1 % water which strongly increases the cooling power and the fan in the vacuum furnace worked at its highest speed.

Before quenching the test probe was heated to 1000 °C and the heat transfer coefficient, h, is calculated from the cooling curves, see Figure 2.

The order of precedence between salt bath, fluidized bed and vacuum is what can be expected from the practical hardening results in each of them. The heat transfer coefficient for the HIP Quencher is of the same magnitude as for the fluidized bed and about three times higher than for the vacuum furnace. The high pressure in the HIP Quencher brings the gas atoms closer together and increases the number of atoms which remove the heat from the hot steel surface. The heat from the batch is dissipated through a powerful heat exchanger situated in the high pressure vessel but outside the hot zone.

Close coupling between gas and part's surface temperature

The gas temperature has to be lower than the surface temperature of the component. The difference in temperature is dependent on the cooling rate, thickness, density and specific heat of the metal and the heat transfer coefficient. Figure 3 is showing the temperature difference as function of the thickness of a Nickel-base plate at a cooling rate of 100 °C/minute.

Due to the high heat transfer coefficient for the HIP Quencher the temperature differs is about 100 °C between gas and plate surface while there is several hundreds of degrees difference at 6 bar pressure. Using a fan at 6 bar the temperature difference is less but is still higher than at 1000 or 2000 bar.

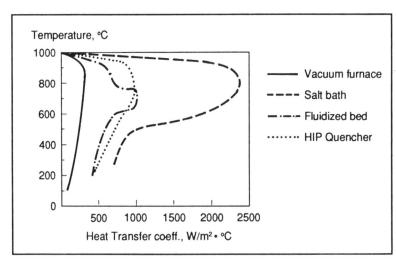

Fig. 2. Heat transfer coefficient for the different methods of cooling. Calculated by the program "ivf quenchotest".

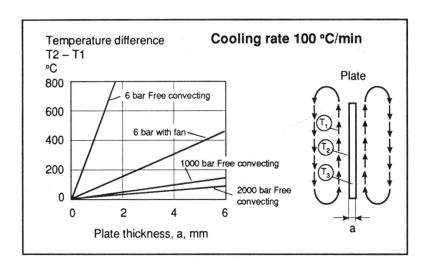

Fig. 3. Difference between gas temperature and plate surface temperature.

Distortion

Distortion is always a risk if the temperature differences in the component are too large. If it has different thicknesses it is specially important to have a high h-value to lower the distortion risk. A high h-value yields smaller differences in temperature between thick and thin sections of the component. The furnaces for quenching at 6 bar normally incorporate a fan to increase the h-value, but this will depend very much on the following factors:

- The shape of the components
- The positioning of the components
- The design of the charge basket
- The design of the batch

The complicated laws of gas flow dynamics makes it very difficult to predict the true h-value. Components of more complicated shape will cool very unevenly because of disturbances in the high velocity gas flow, which increases the distortion risk.

When quenching in a HIP Quencher, working at 2000 bar, it is not essential to place the components in the same position each time. Here self convection of the highly dense gas alone will assure very high h-values and thus minimize temperature differences and little distortion risk.

Quenching of a component according to Figure 4 has been computer-simulated to illustrate the influence of the heat transfer coefficient on the temperature differences in the component.

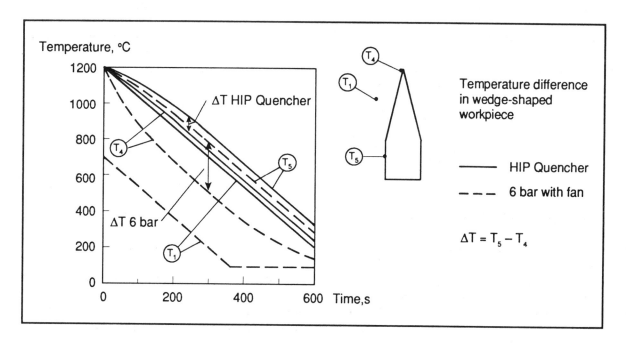

Fig. 4. Effect of variable plate thickness when quenching in gas quencher at 6 bar with a fan to raise the h-value and quenching in the HIP Quencher at 2000 bar.

The diagrams in Figure 4 illustrate the temperature in the thinnest and the thickest section and in the cooling gas as a function of the time. To get the desired and needed cooling rate the gas in the 6 bar quencher has to have a much lower temperature than the component. The coupling between the gas temperature and the temperature of the component is much better in the HIP Quencher, which obviously will give a lower distortion risk.

This has been verified by hardening a wedge-shaped workpiece, "blade", in the HIP Quencher and in a vacuum furnace [3]. The dimensions of the "blade" are shown in Figure 5, left side.

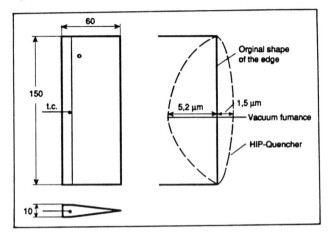

Fig. 5. Left: Dimension of the wedge-shape "blade" made of the steel AISI A2. Right: Distortion of the "blade" edge after quenching from 955 °C in a HIP Quencher and a vacuum furnace, respectively.

The "blades" were heat treated hanging in a hole, near the top, and the temperature was measured by means of a thermocouple, t.c., inserted in the thickest part of the "blade".

The HIP Quencher and the vacuum furnace were programmed so that the temperature curves were the same during both heating and quenching.

Three "blades" were heat treated simultaneously and the resulting distortion is shown in Figure 5, right side. The distortion is about three times larger in the vacuum furnace than in the HIP Quencher. This possibility can be very valuable for critical or expensive components.

Further improved quenching

When quenching in a vacuum furnace, salt bath or fluidized bed, the quenching takes place in media with rather low temperature. There is thus a large temperature difference between the surface of the component and the quenching medium, which results in faster cooling than in the HIP Quencher. The cooling in the HIP Quencher takes place through a heat exchanger which cools the hot gas.

In one model of the HIP Quencher, the cooling power is drastically increased so that the entering gas immediately has only 200 °C. The cooling curve for this HIP Quencher, see Figure 6, was calculated using the heat transfer coefficient shown in Figure 2 as input data.

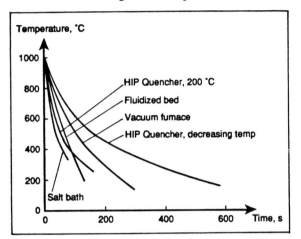

Fig. 6. Cooling curves for a 25 mm bar of high speed steel for different quenching methods.

The calculation was made for the centre of a bar of high speed steel with a diameter of 25 mm. From the figure it can be seen that cooling with a gas temperature of 200 °C gives the HIP Quencher almost the same cooling curve as the salt bath and somewhat faster than the fluidized bed.

Conclusions

HIP Quenching opens up new quenching process control possibilities due to the following:

- High heat transfer coefficient assures small temperature differences between gas and component surface
- Low distortion due to high heat transfer coefficient
- Gas temperature can be accurately computer-controlled as a function of time
- High cooling power is available
- Environmentally friendly as only gas is used as cooling medium

References

1. A. Traff, *Metal Powder Report*, 45, (1990), 4.

2. *Industrial Quenching Oils - Determination of Cooling Characteristics - Laboratory Test Method.* Draft international standard ISO/DIS 9950, International Organization for Standardization (submitted 1988).

3. S.Segerberg, *IVF-rapport 92021*, IVF, Göteborg, Sweden

Proceedings of the First International Conference on Quenching & Control of Distortion, Chicago, Illinois, USA, 22-25 September 1992

Distortion Control with Martempering Oils

R.J. Brennan
E.F. Houghton & Co. Inc.
Valley Forge, Pennsylvania

The use of interrupted quenching to reduce the danger of cracking while obtaining full hardness has been performed for many decades. This was usually accomplished by quenching in water followed immediately by oil quenching. In 1942, the term "Martempering" was first used to denote a method where steel was quenched from austenite in a bath heated to just above the Ms temperature.[1] The part was held in this bath just long enough for temperature equalization and then allowed to air cool, thus reducing the stresses usually encountered by uneven surface and core transformation. The term martemper is actually a misnomoner as the process is not a tempering operation and gives an untempered martensitic structure essentially the same as standard direct quenching.

The martempering process was usually carried out in molten salt or liquid metal (such as lead) due to the Ms temperatures for most steels being above the temperatures where most oils can be safely used. It was found that a modified martempering at a lower temperature just below the Ms could also eliminate many of the quenching stresses and obtain dimensional stability. Figure 1 (a, b, and c) depict the direct quenching, martempering and modified martempering cooling cycles.[2]

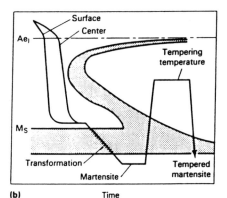

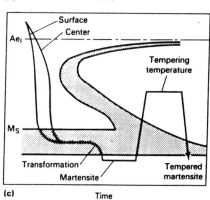

Fig. 1 - Time-temperature transformation diagrams with superimposed cooling curves showing quenching and tempering. (a) Conventional process. (b) Martempering. (c) Modified martempering.

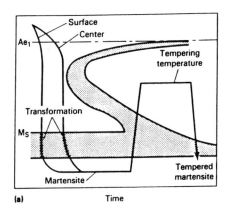

The use of martempering oils has increased over the last 30 or so years for several reasons:

•Industries push toward near-net shaped components leaving the hear treat department.

•The trend to look at total cost of a component rather than just the heat treating costs.

•The need to reduce scrap and rework due to dimensional or cracking problems related to quenching.

•The reduction or elimination of straightening or fixture tempering operations which are labor intensive.

Martempering oils are used in almost all types of heat treating equipment including: continuous furnaces, batch type, integral quench furnaces, open quench tanks, vacuum oil quench furnaces, plug and die type press fixtures, and immersion type quenching of induction heated components. The types of components normally martempered in oil include: bearing races, gears, shafts, flat tools such as hand saws, high alloy wear plates, and other distortion prone products.

Steel Selection

Martempering oils can usually give the same quenching characteristics as normal speed quenching oils and even approach faster oil's "as quenched" metallurgical properties with speed improvers and increased agitation. It is extremely difficult to use a martempering oil for quenching of low hardenability steels such as 1045 (or lower carbon steels in the 1000 series). Higher carbon 1000 series steels such as 1090 steel can be successfully martempered in oil due to the lower M_s temperature and the additional time before upper transformation products transform. (See Figures 2 and 3) [3]

Even with the 1095 steel, size limitations are reduced if maximum core hardnesses are required. Steels such as 4340 with the nose of the S curve moved to the right by alloying elements lend themselves readily to quenching in martempering oils. Figure 4 shows various alloy steels and their maximum diameters that will fully harden to the center in various quenchants;[2] and shows the maximum diameters where a core hardness of 10 HRC below the maximum hardness attainable is used as the criteria.[2] High alloy steel castings and white iron castings used in wear plate applications, are routinely quenched in martempering oils to eliminate rejections due to cracking and distortion.

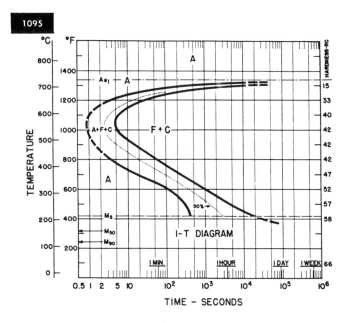

Fig. 3

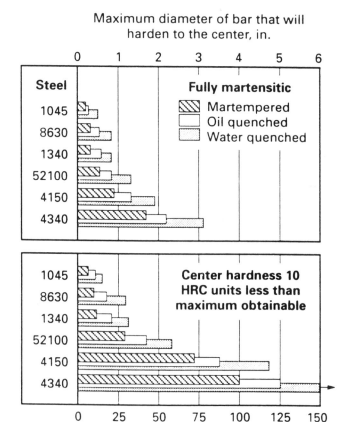

Maximum diameter of bar that will harden to the center, in.

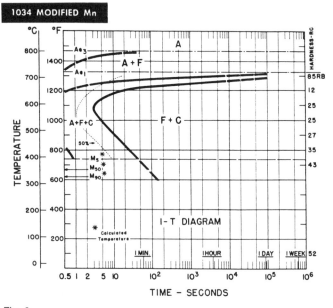

Fig. 2

Fig. 4

84

Oil Selection

Most martempering oils are formulated using highly refined paraffinic base stocks. This material is used due to its stability while operating at higher temperatures. Proprietary additives are incorporated to provide oxidation resistance and to obtain the desired quenching speed. The choice of oil depends upon the use parameters such as:

- Ms temperature of alloy
- Hardenability of alloy
- Oil temperature
- Geometry of components
- Cleanliness requirements
- Distortion requirements

The oil selection must consider the physical properties of the oil and one of the most important from a safety standpoint is the flash point. Usually the flash point increases with increases in viscosity, i.e., the thicker the base oil, the higher the flash point. Oils should not be used within 55.5°C (100°C) of their flash point. When the oil is covered by an protective atmosphere blanket, it can be used safely within 25°C (46.8°F) of its flash point.

Usually a straight base oil without additives will give the brightest finished part, but will oxidize rapidly and then discolor the work. Antioxidant additives will give a consistent finish while extending the oil's usage.

Leaner alloyed parts usually require a lower viscosity martempering oil with speed improvers added. Higher alloyed parts usually use higher viscosity oils with speed improvers which allow higher oil temperatures where distortion control is maximized. Thin distortion prone components are best martempered in the slowest speed martempering oil which will obtain the metallurgical properties required. Many times the higher viscosity oils with additives will give quenching speeds approaching that of medium-fast quenching oils while almost eliminating the vapor phase encountered with thinner quench oils (See Figure 5).

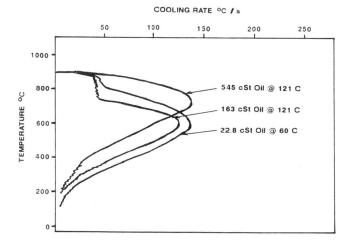

Fig. 5 - Cooling rate curves of various oils.

Usage

The modified martempering operation uses the fluids temperature to interrupt the heat extraction rate at an elevated temperature and thus equalize the parts temperature prior to slower continued cooling. The higher operating temperature, the contact with hot parts, and the forced heating of colder oil during heat up all contribute to the martempering oil's degradation. Usually the hotter the oil's temperature, the better the distortion control but this also accelerates degradation. This can be minimized with antioxidants and a protective atmosphere cover. Make-up of new oil to replace drag-out also reduces degradation.

When heating up the martempering oil from ambient temperatures, it seems that fast heating with high efficiency burners in radiant tubes, or high wattage electric resistant heaters would be the most economical means to quickly start production. Both methods accelerate the oil's degradation and greatly reduce the life of the martempering oil. Low velocity burners, or resistant heaters with a maximum rating of 0.016W/mm2 (10W/in.2) should be used to heat the oil.

Cooling systems should be sized to remove the heat extracted from the austenitized parts (and heated fixtures). Whether batch, single part, or continuous operation is employed, the closer a quenchant's temperature is controlled, the greater the potential for controlling distortion. Cooling systems (as well as other components of the quench tank) should be free of copper and other materials known to be catalyst for the oxidation of oil products. Distorted or cracked components after heat treating have historically been blamed on the heat treating process. Many times this is true, but the entire history of the component must be reviewed to ascertain the real reason for rejection.

Changing Quenchants

Changing from aqueous or fast, medium, or slow quenching speed oils to a martempering oil will definitely reduce the distortion of a component. This would mean straighter or more controllable dimensions. Care must be taken when contemplating such changes, if the component has been machined or formed to a give size due to historical data for parts being heat treated in the present quenchant. Bore sizes, TIR, outside diameter, growth, hardness, tensile strengths, and other dimensional characteristics may change due to a change in the quenchant's heat extraction capabilities. When changing from one quenchant to another in a quench tank, all of the previous product should be removed. This is especially true when converting to martempering oils as water or lower flash materials can be hazardous.

Applications

Steel Foundry. This manufacturer of high alloy steel and white iron wear plates had a high incidence of cracking prior to switching to a martempering oil. The high alloy steel is quenched from 1048°C (1920°F) and averaged 5672 kg (12,500 lbs.) gross loads including 1360 kg (3,000 lbs.) of fixturing. The white iron is quenched from 871°C (1600°F) and is similar in weight distribution. The average monthly output is 223.3 metric tons (246 short tons) of castings. A thinner type martempering oil was initially used and gave fair service life but as the proportion of high alloy steel increased, the thinner oil degraded quicker. A switch to the higher viscosity martempering enabled extended life while not sacrificing metallurgical properties. The use of a martempering oil at 121 to 176°C (250-300°F) has eliminated distortion and quench cracking.

Bearing Manufacturer. The control of distortion was critical to this division of a large bearing manufacturer. If raw material, grinding, and scrap costs could not be controlled the facility was in danger of being closed. A robot transfer mechanism, a new press quench apparatus and a low viscosity, fast martempering oil was installed. The capital expense and the martempering oil enabled drastic production and raw material savings. Forgings and tubing could be purchased at closer to finished sizes thus reducing the weight of steel purchased. The parts could be machined closer to final size while soft and thus reduce hard grinding times; and scrap was reduced.

Automotive Powertrain Components Manufacturer. This mid-western division of a major automobile manufacturer used a martempering oil for the control of distortion but the oil's physical and chemical characteristics changed quickly, forcing changes in agitation and increases in the oil's temperature to maintain the desired metallurgical properties and distortion control. Within eight to ten weeks, the oil had to be replaced. Switching to a more stable martempering base oil and additive package eliminated the need to constantly change the agitation and oil's temperature; and allowed consistent dimensional and metallurgical control. The use of the original martempering oil produced the required dimensional and metallurgical properties but was expensive due to its short life in the system. The second oil greatly extended the life of the oil without the changes to the oil's use parameters.

SUMMATION

Martempering oils, when used with an alloy of appropriate hardenability, can control distortion and eliminate the rejections due to cracking. They can reduce the overall manufacturing costs by the reduction or elimination of expensive hard grinding or straightening operations. When switching from another quenchant to a martempering oil, tests must be run to obtain the best dimensions in the soft state to produce the desired hardened dimensions. Martempering oils can usually replace other quenchants with minimal capital expense due to equipment modification. Safety precautions for the use of hot oils are similar to the use of other quenching oils.

REFERENCES

1. Shepherd, B.F., Iron Age (January, 1943)
2. ASM Handbook Volume 4, Martempering of Steel (1991)
3. U.S. Steel - Atlas of Isothermal Transformation Diagrams (1951)

Proceedings of the First International Conference on Quenching & Control of Distortion, Chicago, Illinois, USA, 22-25 September 1992

Salt Bath Quenching

R.W. Foreman
Park Metallurgical Corp.
Detroit, Michigan

Abstract

Martempering and Austempering using molten salt baths for the quenching of steel and cast iron have been and continue to be important commercial processes. The minimization of distortion, cracking, and residual stress is the major attraction of salt bath quenching. There has been a steadily expanding number of applications involving both salt and atmosphere austenitizing coupled with salt quenching. The use of energy efficient automated equipment has led to the increasing popularity of such processes. While the only suitable alternative quench medium, "hot oil" can be used in selected cases, salt quenching has many advantages and is often the only practical choice. Some typical applications, including relatively recent ones, are discussed.

There are many newer developments relating to salt quenching. Three important areas are selected for more detailed discussion. One of these is the recent development of quantitative data on the influence of salt quench variables, such as part mass, agitation, water addition, and salt composition. These can be correlated with steel or cast iron hardenability. A second area is that of salt reclaim and recycle, including chemical treatments, to assist in same. Disposal problems are of less concern than generally believed. A third and most promising area is that of Austempered Ductile Iron (ADI). There have been great strides made in the control of composition and proper heat treating of this material. ADI has properties that are comparable or superior to forged steel while offering substantial cost savings. The quenching to produce ADI is invariably done in salt.

Martempering and Austempering

Martempering (often referred to as Marquenching) involves the interrupted quench of austenitized steels or cast iron. This is done near the martensitic transformation temperature to equalize the temperature throughout the part before cooling through the martensite range. This minimizes the distortion, cracking, and residual stress compared with conventional water or oil quenching. Austempering is done similarly to martempering, except the part is held above the martensite and below the pearlite transformation temperatures sufficiently long to achieve either bainite or ausferrite (ADI). This produces parts having increased ductility, toughness, and strength with much reduced distortion. Hardnesses are generally below those of martempered parts. Fig. 1 shows the time-temperature relationship of these processes superimposed on an alloy steel TTT diagram. As noted martempering quench temperature can be somewhat above or below the M temperature depending mostly on the required cooling rate for a given part.

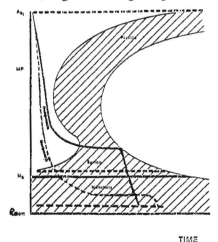

Fig. 1. TTT DIAGRAM

Molten salt is the preferred quench medium for either martempering or austempering. It offers the widest operating temperature range, has the highest thermal conductivity and heat capacity, and produces almost no vapor stage during quenching. While petroleum based quench oils (so-called hot oils) can be used in the lower temperature range for martempering, they have several disadvantages. Other quench media, such as fluid beds, have been investigated but are not commercially used to any significant extent. Table 1 compares basic composition and properties of quench salt and martemper oil.

Table I. Salt vs. Oil for Martempering

	SALT	OIL
Composition	KNO_3, $NaNO_3$, $NaNO_2$	Hydrocarbons
Operating Range	150-400°C	130-220°C
Density (200°C)	1.92	0.82
Viscosity, C.P.		
200°C	7.5	2.9
315°C	2.9	- -
Specific Heat	0.37	0.5
Thermal Cond. g-cal/sec/cm	1.8×10^{-3} °C/cm	0.39×10^{-3} °C/cm
Thermally Stable	to 540°C	to 230°C

The lowest melting quench salts are blended with potassium nitrate and sodium nitrite. The higher melting salts contain sodium nitrate which is the least expensive of the three raw materials. The lowest melting salts melt around 135°C (275°F) and are sufficiently fluid to be used above 177°C (300°F). The useful range of quench salts can be lowered to cover the entire martempering range if water addition is employed. Since salts are inherently thermally stable up to temperatures around 538°C (1000°F), they can be used throughout the martempering and austempering quench temperature range. Martempering oils are fluid at the very lowest end of the martempering range and can be used up to their thermal stability limit of around 220°C (430°F). Special precautions are needed to use these oils at the high temperature range to avoid oxidation and thermal breakdown. The fluid density of quench salts is more than double that of martemper oils, while viscosities of both are relatively low in their operating ranges (H O vis. is 1). The specific heat is somewhat higher for oil, but the thermal conductivity for salt is nearly five times greater on a weight basis and over ten times greater on a volume basis.

Salt Bath Equipment and Processes

The earliest martempering and austempering were performed in batch furnaces usually using an austenitizing salt bath (neutral chloride salts) and used by job shops for low volume or special work. In the past 25 years increasingly automated and comput high volume production systems have come into use. These

fall into two general types of systems. The first type utilizes atmosphere austenitizing (carburizing or neutral) followed by salt quench. The second type employs austenitizing salt baths with rapid transfer to the quench salt.

The atmosphere furnaces used with salt quenching come in a wide variety of designs which are either batch or continuous in nature. Nearly every variety of atmosphere furnace that can be used with conventional oil or water based quenching can be and is used in salt quenching. (Ref. 1) Some minor design changes are necessary when using continuous furnaces to be sure the atmosphere is properly interfaced with the salt. Failure to do this in some of the early atmosphere/salt quench furnaces led to violent reactions. The probable culprit was soot from the atmosphere which nitrate salts can oxidize rapidly. All modern designs are such as to avoid this problem.

Batch and semicontinuous atmosphere furnaces are very popular for use with salt quenching. These can accommodate a large variety and size of parts. An example of such a system is a rocket case martempering system recently installed which uses a large gantry atmosphere furnace for austenitizing the high cobalt/nickel forged steel rocket cases and the very large salt quench bath shown in top view in Fig. 2.

TOP VIEW

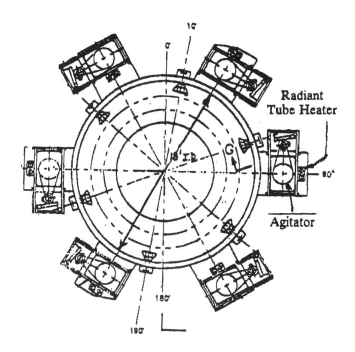

Fig. 2. ROCKET CASE SALT QUENCH

The bath is 5.5 m. (18 ft.) in diameter and 7.0 m. (23 ft.) deep with an operating salt level of about 5.2 m. (17 ft.). Extensive agitation and radiant tube heaters are an integral part of the design. There was about 295,000 kg. (650,000 lbs.) of salt required to fill this bath. Also of special interest are the semicontinuous atmosphere-to-salt quench systems used in producing ADI. These often utilize salt baths in which unusually high percentages of water are maintained. This will be discussed further under Newer Technology.

Austenitizing salt baths using neutral chloride salts also come in a wide variety of designs. (Ref. 2) Most modern baths are internally electrically heated which provides high energy efficiency. There is often a preheat bath of similar design used. The quench baths are either heated by radiant gas-fired tubes or by electrical resistance heaters. Provision for proper agitation, temperature control, and in many cases water addition are required. In the larger automated systems, the preheat, austenitizing bath, and quench bath operate independently but in sequence with sophisticated rapid transfer mechanisms which are computer controlled. Suitable air cool, temper, and wash stations are also provided.

Carburized steels are sometimes martempered or austempered in salt to provide high surface hardness coupled with more ductile core properties. Most of this processing is now done in atmosphere-to-salt systems. The use of cyanide-based carburizing salts was once an important procedure, but both the safety aspects of quenching into nitrate salts and the disposal problems of cyanide have greatly diminished their use.

Specific Applications

There have been and still are a very large number of specific applications of martempering and austempering using salt quenching. A sampling of these is listed in the ASM Handbook Vol. 4 in the chapters on martempering and austempering (Ref. 3,4). Table II lists a few of these to show the broad range of alloys, section size, weight of part, and temperatures employed.

Table II. Martemper and Austemper Applications

Part	Steel (AISI)	Max. Thick. (CM)	Salt Weight (KG)	Salt Temp. (°C)	Min. Time (Sec.)	Hard. (R)
Chain Links	1045	0.56	0.11	204 (+H O)	60	45 (as Q'd)
Gears	4350	0.89	0.36	246	120	54
Spindles	8620 (carb.)	1.02	6.36	204	180	60
Bearing Races	52100	1.27	13.3	220	180	64
Lawnmower Blades	1065	0.32	0.67	315	900	50
Bolts	10B20	0.64	0.01	420	300	38

A limited survey of current users and equipment manufacturers disclosed that over 10 new installations using salt quenching are in progress. These will be used to heat treat such parts as chain links for conveyor belts, aerospace parts, ice skate blades, chain saw blades, a wide variety of fasteners, and various gears. Add to this the newer applications involving ADI, and it must be concluded that salt bath quenching is growing in popularity.

Advantages/Disadvantages Of Salt Quenching

Some of the advantages of salt quenching were referred to earlier. The following summarizes the advantages for salt quenching when making comparisons between salt and all oil or water-based quenching:

- Reduced (often markedly) distortion and residual stress with little or no hardness sacrifice in martempering. Greatly reduced distortion with high ductility, toughness, and strength with some sacrifice of hardness in austempering.

- A wide choice of quench temperatures and agitations without concern for large viscosity changes, excessive vapor stage, thermal instability, oxidation (fire), or gas entrainment.

- A wide choice in the range of quench severities (quench rates) through adjusting quench variables without changing the basic quenchant or concen-tration (water based).

- Excellent thermal and chemical stability with relative insensitivity to contaminants.

- Totally water soluble coating on parts that can be easily washed off and either disposed of as a non-hazardous waste in many locations or readily recovered if needed or desired.

- Particularly suited for very large parts and high production systems that occupy minimum floor space.

The following disadvantages/limitations are noted:

- Since salt freezes around 135°C (275°F) it can not be used for lower temperature quenching where needed.

- Freezing can make intermittent use complicated and energy inefficient.

- Since nitrates are oxidizers, greater precautions to avoid introduction of oxidizable substances, such as soot, oil, or cyanides must be made. Drastic over-heating must be avoided, or salt can even oxidize steel or cast iron.

- Since salt is a costlier quenchant on a volume basis, more care to avoid excessive "drag-out" must be taken. Efficient washing and recycle can in turn negate this.

- Certain contaminants removal can be more complicated than for conventional quenchants though means for removal are well established. In some cases, salt has the advantage over conventional quenchants in regard to contaminants.

That salt quenching offers metallurgical and economic advantages can be illustrated by considering a specific case of a bearing manufacturer who switched from atmosphere, oil quenching to an automated salt-to-salt system. Table III summarizes the process steps and consequent reduction in energy consumption, process time, and other costs. A major overall cost reduction with superior quality parts was realized in this case.

Table III. Hardening Process Comparison for 52100 Bearings

	Batch Atmos/Oil Q	Autom. Salt Martemper
Manual Steps - 5:	(load/unload + after *)	2: (load/unload)
Process Steps - 6:	Austenitize	7: Preheat
	Oil Q *	High Heat
	Wash *	Martemper
	Air Draw	Air Cool
	Air Cool *	Salt Draw
	Grit Blast	Air Cool
		Wash
Process Energy	2.956 KCal/KG	1.025 KCal/KG
Process Time for 3200 KG	24-30 Hrs.	7 Hrs.

Major Savings for Martempering: 2.2 KW/KG of Parts
Labor/Fringe Benefits
Reduced Scrap
50% Less Grinding

Newer Technology

There have been many advancements in salt bath technology in recent years. Most are evolutionary and ongoing while a few could be considered revolutionary. Among these advancements are the following:

- Robot handling systems that are computer controlled and which move parts much faster and more smoothly from austenitizing to quench and in the total systems.

- More energy efficient austenitizing and salt quench furnaces which employ better insulation and in the austenitizing baths superior electrode designs.

- Agitation and heater systems in the quench baths that combine to provide more uniform flow and temperature control coupled with safe water addition.

- More reliable purity quench salts supplied in more convenient and safer to use pelleted form.

- Water addition monitoring systems using fast quench rate determination. Continuous monitoring is also on the horizon.

- Development of quantitative salt quench rate-quench variables data which can be correlated with steel hardenability.

- Development of quench salt reclaim and treating systems that enable economic recycling and reduce disposal problems.

- The establishment of a family of austempered ductile irons (ADI) that out perform steel castings and forgings and other cast irons for both structural and wear applications.

While all of these are important, the last three are less well known. They are still undergoing development, and promise to make salt bath quenching even more attractive in the near future. These are discussed in further detail as follows:

Quantitative Quench Rate Data

The qualitative influence of such salt quench variables as part mass, agitation, bath temperature water additions, and salt bath contaminants have been known for some time. Fig. 3 illustrates the influence of water addition as an example.

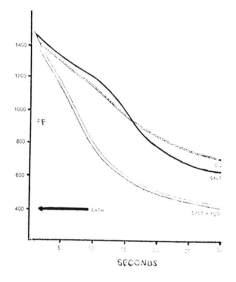

Fig. 3. QUENCH CURVES FOR 2.5 cm STEEL CYLINDER WITH EMBEDDED T.C.

However, there has been no standardization of the equipment or procedures for developing such plots. Because of this, the data from different sources can not be properly compared. Recent work (unpublished research by R. W. Foreman at Park Metallurgical Corporation; Detroit, Michigan) has made a start in developing quantitative data relating quench rates to controlled quench variables.

The procedure was patterned after that developed for measuring quench variables for conventional oil and water-based quenchants by Dr. Charles Bates. The key equipment was a steel U tube fitted with a propeller that produces measurable linear flow rates. This is shown schematically in Fig. 4.

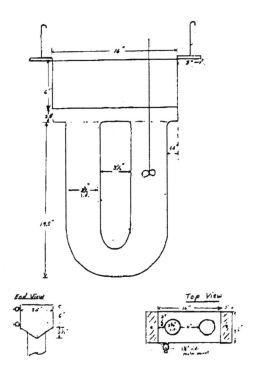

Fig. 4. SALT QUENCH VARIABLE APPARATUS

The U tube is immersed in a larger salt bath to ensure uniform and adequate heating. 304 SS probes of different diameters and centrally embedded thermocouples were heated in a tubular air furnace to a standardized austenitizing temperature of 871°C (1600°F). A high speed Bascom-Turner electronic storage recorder was used to record the time-temperature when the probes were transferred to the salt quench. The controlled variables were mass (section size of probes), temperature, salt flow rate, % water in the salt, and % contaminants, such as chlorides and carbonates. The quench rate was provided by using the recorder to plot the derivative of the time-temperature curve.

The experiments showed a high reproducibility even at high quench rates. A typical time-temperature plot is shown in Fig. 5 (the recorder provides temperature in °F only).

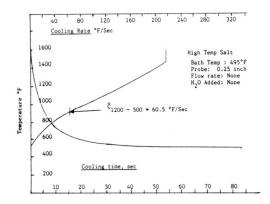

Fig. 5. TYPICAL SALT VARIABLE QUENCH CURVES
HIGH SPEED RECORDER PRINTOUT

An average cooling rate was determined by summing the cooling rates at close intervals between 549°C (1200°F) and 260°C (500°F) and dividing by the number of intervals. Since the curves are nearly linear in this range and it covers the pearlitic nose of most steels, these average cooling rates are useful for correlating with quench variables. In dealing with a specific steel or cast iron, a narrower cooling rate range could be used.

Major responses to mass (section size), flow rate, and water additions were found as expected. The salt bath temperature, nitrate/nitrite composition, and presence of chlorides or carbonates was found to have less influence than expected. Fig. 6 shows the mass effect under one set of conditions.

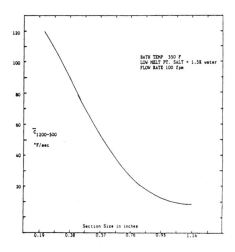

Fig. 6. SECTION SIZE (MASS) EFFECT
ON SALT QUENCH RATE

The influence is substantial and was found to be so regardless of bath temperature, water added, or flow rate.

Agitation (flow rate) was also an important factor as illustrated in Fig. 7.

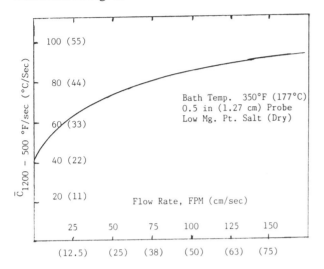

Fig. 7. EFFECT OF AGITATION
ON COOLING RATE

Even in dry salt, the average cooling rate is doubled by imposing a modest flow rate of 38 cm/sec. (75 fpm). The effect is even larger when water is added. An important consequence of this finding is that uniform flow around quenched parts is very important to minimize differential cooling and, thus, distortion.

Water addition also increases quench rate markedly. Fig. 8 shows the influence of low percentages of water added with relatively high agitation.

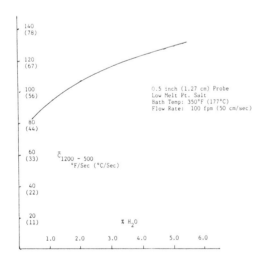

Fig. 8. INFLUENCE OF WATER
IN SALT QUENCH

Commonly used percentages of 2 to 3% water at the

indicated salt temperature increase the cooling rate by half again. When low agitation is present, the addition of water has an even greater effect.

The influence of bath temperature was found to be relatively small. Only about a 10% lower cooling rate was found when bath temperature was raised from 177°C (350°F) to 260°C (500°F). The mix of nitrates/nitrite other than changing the lowest practical operating temperature because of melt point had almost no influence on cooling rates. Likewise, the presence of chlorides (carry in from salt austenitizing baths) had very minimal effect on average quench rates.

There are a number of ways these data can be used:

- To select quenching conditions for specific alloys   and section size.

- To determine maximum section size that can be through hardened for an alloy of known hardenability.

- To assess the salt hardenability of an unknown hardenability alloy.

- To compare commercial quench salts that may have undesirable impurities or composition.
- Provide a means for monitoring the quench rate for a commercial quench bath.

A limited study was performed with three commonly salt quenched alloys wherein different section sizes were quenched in the U tube under the same conditions imposed for quench rate determination. The results are summarized in Fig. 9.

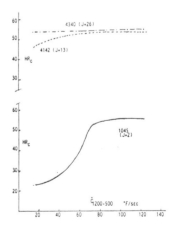

Fig. 9. HARDNESS COOLING RATE CORRELATION
FOR SALT QUENCH

As expected, the low alloy steel requires relatively high cooling rates to achieve maximum hardness.

Much more work is needed to extend the data

developed in this study. Coverage of more ferrous alloys and means for studying water addition/agitation interactions beyond the range covered so far are particularly needed. Also mathematical correlation to enable interaction could be developed.

Reclamation/Recycle

Nitrate/nitrite quench salts are highly water soluble and common practice for years has been to dilute wash water and discharge it to the sewer system. Municipal treating systems can actually benefit from the fact that the salt is a fertilizer and helps digest waste. However, where the discharge is to ground waters, severe restrictions are now in force in most localities. Both to solve the latter problem and as a means for reducing overall costs of salt quenching, increasing recycling of wash water is being used. To keep the energy cost of boiling off water minimal, recyclers operate with two or three stage wash systems. By washing in water containing up to 20% salt in the first stage, and then finishing the wash in more dilute (5% and then 1% e.g.) only the most concentrated wash needs to be evaporated for recovery. The more dilute washes are moved to the more concentrated washes. Those using water additions can even use some or all of the concentrated wash to make up water in the quench salt. One possible penalty is that minor impurities, such as alkaline carbonates, tend to build up. However, these can be counteracted by acidification of the wash water with dilute nitric acid.

One of the problems with salt-to-salt quenching is the build-up of chlorides in the quench bath. These do not interfere with quenching, as discussed earlier, but they can accumulate beyond their solubility limit. Deposition on heating surfaces and handling equipment then becomes a problem. Recently developed continuous filtering systems utilizing strategically located baskets equipped with properly sized screens are available and in use to greatly minimize this problem. Furthermore, a proprietary process has been developed and used successfully to recover over 90% of the nitrates present in the precipitate collected in these baskets. Since the precipitates (often referred to as sludge) are more than half nitrate salt and the final solid waste is classified as non-hazardous, unlike the basket precipitate, this processing is both economically and ecologically attractive.

In some salt quenching systems, usually those quenching heavy loads or more massive parts, build-up of carbonates, which are the end product of nitrate/nitrite thermal decomposition, can exceed solubility limits. This problem can be counteracted by use of a proprietary process using a nitric acid-based solution that can be added directly to the molten quench salt.

Austempered Ductile Iron

ADI is fast becoming a widely used and important engineering material. There are at least 100 current applications in the automotive, agricultural, heavy industrial, transportation, military, and construction industries. A great deal of information has been published on ADI in recent years. A few of these publications are referred to here and can provide leads to many more. ADI is an alloyed and heat treated (using salt quenching) ductile cast iron. In addition to the typical C, Si, and low Mn, S, and P of conventional ductile iron, alloying elements, such as Ni, Cu, and limited Mo are often added to enhance heat treatability. ASTM recently adopted standards for five grades of ADI. These are shown in Table IV.

TABLE IV

THE FIVE ASTM STANDARD ADI GRADES (ASTM 897-90)

GRADE	TENSILE* STRENGTH (KSI)	YIELD* STRENGTH (KSI)	ELONGATION* (%)	IMPACT ENERGY** (FT-LBS)	TYPICAL HARDNESS (BHN)
1	125	80	10	75	269-321
2	150	100	7	60	302-363
3	175	125	4	45	341-444
4	200	155	1	25	388-477
5	230	185	N/A	N/A	444-555

*MINIMUM VALUES
**UN-NOTCHED CHARPY BARS TESTED AT 72 +/- 7F

THE FIVE ASTM STANDARD ADI GRADES (ASTM 897M-90)

GRADE	TENSILE* STRENGTH (MPa)	YIELD* STRENGTH (MPa)	ELONGATION* (%)	IMPACT ENERGY** (Joules)	TYPICAL HARDNESS (BHN)
1	850	550	10	100	269-321
2	1050	700	7	80	302-363
3	1200	850	4	60	341-444
4	1400	1100	1	35	388-477
5	1600	1300	N/A	N/A	444-555

*MINIMUM VALUES
**UN-NOTCHED CHARPY BARS TESTED AT 22 +/- 4C

The physical properties of these grades are plotted in Fig. 10.

The heat treating regime required to produce ADI is somewhat demanding in control of temperatures and times both for the austenitizing and quenching. The austenitizing times must be long enough to saturate the austenite with carbon. The quench rate has to be high enough to get by the pearlitic nose, yet provide essentially isothermal conditions for the proper time and temperature in the needed 230-400°C (450-750°F) to produce the correct microstructure. The latter is an acicular ferrite dispersed in a carbon rich austenite which provides the remarkable properties of ADI.

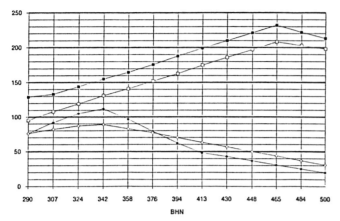

Fig. 10. TYPICAL ADI PROPERTIES AS A
FUNCTION OF THE BRINELL HARDNESS

It is not bainite and should properly be called
"ausferrite". The various grades of ADI are produced
primarily by varying the quench temperature. The lower
quench temperatures produce the higher strength/lower
ductility grades (3-5) and the higher quench
temperatures produce the lower strength but more
ductile grades (1 & 2). An important feature of ADI is
that mechanical working can lead to surface hardening,
thus excellent anti-wear properties.

There have been negative reports on ADI which can
be traced to poor casting composition and/or improper
heat treating. There are producers of ADI who can now
achieve over 99% of their production with proper
properties. One of the key factors in quenching
relatively unalloyed ADI is that high quench rates are
needed to miss the pearlitic nose. At least one producer
uses unusually high levels of water added to the quench
salt to achieve the needed quench rates. Both
atmosphere-to-salt and salt-to-salt systems are in use for
producing ADI.

Given the low cost, lighter weight, near net shape
design options, remarkable strength, ductility, and wear
resistance of ADI and the fact that its properties can be
modified considerably by selecting different heat treating
programs, it is most likely that its use will expand
considerably in the next several years and that salt
quenching will be enhanced by same.

Summary

Martempering and austempering using salt bath
quenching are important and growing heat treating
processes. Many improvements in equipment,
procedures, and safety coupled with new applications,
ecological problem solutions, and better process control
promise to keep salt bath quenching viable for the
foreseeable future.

References

1. "ASM Handbook, Vol. 4, Heat Treating" pp 465-
474, ASM International, The Materials Society,
Materials Park, OH (1991)

2. "ASM Handbook, Vol. 4, Heat Treating" pp 475-
483, ASM International, The Materials Society,
Materials Park, OH (1991)

3. "ASM Handbook, Vol. 4, Heat Treating" pp 148
ASM International, The Materials Society,
Materials Park, OH (1991)

4. "ASM Handbook, Vol. 4, Heat Treating" pp 157
ASM International, The Materials Society,
Materials Park, OH (1991)

Proceedings of the First International Conference on Quenching & Control of Distortion, Chicago, Illinois, USA, 22-25 September 1992

Quench Tank Agitation

K.S. Lally and M.W. Kearny
Lightnin Inc.
Rochester, New York

G.E. Totten
Union Carbide Chemicals and Plastics Co. Inc.
Tarrytown, New York

ABSTRACT

Rapid and uniform cooling is critically important for the optimum formation of mechanical properties of many metals. This is facilitated by various forms of agitation. Without agitation, natural convection and vaporization of the quenchant limit the heat transfer rate through the fluid film boundary layer at the surface of the part. Increasing agitation rates greatly reduce the resistance to heat flow during the quench. Quench system agitation may be provided by mechanically moving the parts during the quench or by inducing fluid flow. Fluid flow may be controlled by various methods including: recirculation pumps and impeller stirrers. Of these, perhaps the most cost effective means of optimizing fluid uniformity during the quench is with the use of impeller stirrers. This paper describes the most common impeller systems used for quench system design, principles of agitation and practical recommendations for mixer selection.

INTRODUCTION

Quenching is a critically important, but often neglected, part of heat treating. Proper agitation of the quenchant is often the single parameter that dictates the success of the quenching process. One of the most cost effective methods of agitation is with impeller stirrers. Other mixing technologies, e.g. recirculation pumps, often require as much as ten times greater power to provide the same agitation.

The classic reference that is still consulted for quench tank agitator design was originally published in 1954 by U.S. Steel (1). Mixing technology has made great advances since that time. Currently, more efficient impellers providing substantially greater flow with less

power resulting in substantially lower cost system designs have been developed.

The objective of this paper is to provide a brief overview of how quenchant agitation affects quench severity. This will be followed by a description of the most common impellers used in quenching applications today. The calculation methods used to properly size impeller agitators will then be described.

DISCUSSION

Effect of Agitation on Quenching Mechanisms

There are at least three different cooling mechanisms that occur when hot metal is immersed into a vaporizable liquid medium (2,3). These mechanisms are designated as: vapor blanket cooling, nucleate boiling and convective heat transfer. Each mechanism is associated with a uniquely different heat transfer process. These different cooling mechanisms are shown in Figure I for an aqueous poly(alkylene glycol) quenchant.

All quenchants exhibit increased heat transfer rates with increasing flow by the hot metal surface. The increase in heat transfer rates due to agitation is dependent on many factors such as the turbulence of flow, physical properties of the quenchant, presence of additives, etc.

The necessity for uniform heat transfer to reduce surface thermal gradients can be understood by examining the quench process shown in Figure 1. Cooling rates are controlled during the quenching by the formation of an insulating film around the metal after immersion into the quenchant. If the film formation is either discontinuous or has a variable thickness, substantial non-uniformity in localized heat transfer rates may be produced;

which may result in increased propensity for cracking and distortion (4). However, agitation will produce a substantial improvement in film formation by enhancing the uniformity of the heat transfer process.

Mixer Impeller Types

A mixer is used to provide fluid motion and shear. Flow-controlled mixing is strongly dependent on the fluid turnover produced by the impeller. Therefore common mixing processes which occur in quenching practice, fluid blending and heat transfer, are strongly related to impeller flow.

Mixing impellers used for quenching operations are either "open-impeller" or "draft-tube" systems (Figure 2). Open-impeller mixers do not have a flow-directing surface encasing the impeller. This system relies on the impeller itself, or a baffle, to direct flow into the quenching region of the tank.

A common impeller used for open systems is an "axial-flow" impeller such as the marine propeller shown in Figure 3 (4). An axial-flow impeller directs fluid flow parallel to the impeller shaft. Axial-flow impellers may be used in various configurations including: "top-entering", "side-entering" or "angled top-entering".

Currently, the impeller most often recommended for angled top-entering and vertical top-entering mixers is the high efficiency airfoil-type illustrated in Figure 3. A 40% greater flow efficiency is achieved with this type of impeller versus the traditional marine impeller operating at the same speed.

The impeller blades used for side-entry mixers have a different shape than those used for top-entry mixers (Figure 3). The different shape is required to balance the mechanical forces acting on the impeller in a side-entering mode. This is primarily due to the higher operating speeds that are typically required in the side-entering design. An advantage of this impeller over conventional marine impellers is that the bolted blades permit easier installation.

Basic Mixing Principles

The power (P) delivered by the mixer is also related to the flow rate or pumping capacity (Q) of the mixer and the velocity head (H) as shown in equation (I).

$$P = Q \times H \tag{I}$$

Performance of impellers in a draft-tube may be compared with a recirculation pump. For both systems, the volume of flow delivered is dependent on the resisting head. A typical head-flow capacity curve is shown in Figure 4. This curve was developed by plotting head as a function of flow. The region of instability when the curve flattens out will result in a "stall condition". System design should always result in selection of an operating point to the right of this position.

A series of system resistance curves are shown in Figure 4. System resistance (K_v) is defined as:

$$K_v = 2GH/V_D{}^2 \tag{II}$$

where:

H = system head,

V_D = velocity through the draft-tube,

K_v = system resistance constant which is a function of system geometry.

The head-flow and system resistance curves are then superimposed as shown in Figure 4. The objective is to select an impeller such that the system curve will intersect the head-flow capacity curve without creating a stall-condition. An appropriate value of K_v can be determined empirically for a given geometry. This, together with process flow and head requirements, will establish the appropriate mixer design.

An application of the use of these curves is illustrated in Figure 5 for a pitched-blade axial-flow impeller and an airfoil impeller. The slope of the curve represents the impeller's capacity for resisting changes in system head. This figure shows that the pumping capacity for the airfoil impeller is reduced 30% when the K_v is increased from 1 to 5. The axial turbine impeller, under similar conditions, exhibits a 35% decrease in pumping capacity. Therefore, at high system resistance, the airfoil impeller produces nearly 16% greater head. This data shows that the advantages of an airfoil impeller are:

- Higher head capacity

- Steeper head-flow curve

- More resistance to stall

- Greater operating efficiency

The total system head should be minimized in order to maximize flow developed by the impeller as shown in Figure 6. While it is not practical to change most of the characteristics of the quench tank system, the draft-tube can often be designed to minimize the head losses. For example, the lack of an entrance flare and insufficient liquid coverage can decrease the flow rate by 20%.

Quenching is a flow-controlled mixing operation, and different impeller types give different values of flow volume per unit power. Therefore, the "power per unit volume" mixer sizing approach is only applicable to one impeller type at one operating speed. Flow rate per unit power can be calculated from Equation V which is derived from the equations for flow (III) and power (IV) shown below.

$$Q = N_Q N D^3 \qquad \text{(III)}$$

$$P = N_p \rho N^3 D^5 \qquad \text{(IV)}$$

$$Q/P = (N_Q/N_p)(1/\rho N^2 D^2) \qquad \text{(V)}$$

where:

D = Impeller Diameter

N = Impeller Speed

N_p = Dimensionless power number

N_Q = Dimensionless flow number

ρ = Fluid Density

Q = Flow

P = Power

The dimensionless flow number, N_Q characterizes the flow producing capability of the impeller. The dimensionless power number, N_P, characterizes the power consumption characteristics of the impeller. Equation V shows that the relative flow per unit of power is dependent on the impeller type, the installation geometry, speed and diameter.

Equation V does not completely specify the mixer since there are other significant design elements such as torque that are imposed on the mixer. The relationship of torque (T) to the parameters shown in Equation V are given in Equation VI.

$$T = N_p \rho \frac{N^2 D^5}{2\pi} \qquad \text{(VI)}$$

Capital cost of the complete mixer is strongly dependent on torque and therefore must be examined as rigorously as flow and power. An illustrative comparison of flow per unit power (Q/P) and torque as a function of impeller diameter is shown in Figure 7.

Comparison of the speed and diameter options for more than one type of impeller is a more complex problem. For flow-controlled processes, such as quenching, it is probably best to compare the different mixing parameters at constant flow and mixer speed. This case is represented by Equation VII.

$$Q/P = \frac{(N_Q^{5/3})}{N_P} (I/N^{4/3} Q^{2/3} \rho) \qquad \text{(VII)}$$

At a constant flow and impeller speed the flow volume per unit power can be expressed as:

$$Q/P \; \alpha \; \frac{(N_Q^{5/3})}{N_P} \qquad \text{(VIII)}$$

Application of Equation VIII requires that the value of N_Q and N_P be known. These values can be either experimentally determined or obtained from the mixer supplier. The power number is obtained by measuring the mixer torque and speed on a mixer dynamometer. The power number is calculated by dividing the dynamometer power by speed according to Equation VI. The power number is then calculated from Equation IV.

Measurement of the impeller flow number (N_Q) is more difficult. Flow measurement methods such as streak photography, pitot-static tubes and hot-wire anemometers have been described by Oldshue (6). In addition to being time consuming and difficult to use, their presence affects the flow pattern. Currently, the laser doppler velocimeter (LDV) is the preferred method for flow measurement. In view of the relative difficulty in measuring the flow number of an impeller system, it is recommended that these values be obtained from the mixer supplier.

Table 1 provides the flow numbers and power numbers for a representative marine and airfoil impeller. It is important to note that the change in flow number, and therefore flow rates, will affect the flow per unit power when the impellers are used in a draft-tube. Impellers operating in a draft-tube develop less flow for the same power consumption, but have the advantage of directional flow control.

A common mistake in the selection of draft-tube mixers is to use flow numbers reported for open-impellers. Draft-tube mixers will be undersized if the mixer manufacturer has not adequately defined the draft-tube flow number. The draft-tube flow number should be used for calculation and will result in the selection of a significantly larger mixer. The values shown in Table 1 are given for a "typical" draft-tube. The draft-tube flow number varies considerably with the length and geometry of the system. Therefore, it is essential that the draft-tube flow number (and therefore K_v) be properly determined. Mixer manufacturers should be able to demonstrate that the draft-tube flow number has been properly determined upon request.

Multiple Mixers

Many quench systems in use today were not properly designed to provide optimal uniformity of flow. This can be easily shown by the use of computational fluid dynamics (CFD) (7).

To illustrate the potential use of CFD in quench tank agitator selection and design, the fluid velocity vectors for a rectangular tank containing a single draft-tube was calculated. The results, shown in Figure 8, show that there are relatively quiet zones near the corners of the tank, except where the draft-tube is located. Clearly, one draft-tube agitator used in this manner will not provide adequate quenching uniformity throughout the total quench zone.

In order to optimize uniform heat transfer in the quench zone, it is desirable to design a quench system that will have minimal fluid flow velocity variation throughout the total quench zone. This often requires the use of multiple mixers.

Although, there are no rigorous quantitative solutions to predict if multiple mixers are required or their placement in a quench tank, there are a number of general rules that can be followed. For example, in cylindrical or rectangular tanks where the length to width ratio is less than 2:1, a single properly designed mixer is usually sufficient. However, if the length to width ratio is greater than 2:1, then multiple mixers are recommended. Figure 9 shows possible multiple draft-tube and side-entry mixer arrangements for rectangular tanks when the length to width ratio exceeds 2:1. (Top-entering mixers may be arranged differently.)

To size a mixer in a multiple arrangement, it is first necessary to determine the total power requirement based on the tank volume using Table 2. The power per mixer is determined from Equation IX.

$$\frac{\text{Power}}{\text{Mixer}} = \frac{\text{Total Power}}{\text{Number of Mixers}} \qquad \text{(IX)}$$

The impeller sizing for each mixer is determined from Table 3.

Tank Baffles

Tank baffles eliminate vortexing and convert swirl motion to more productive top to bottom fluid motion. In rectangular tanks with properly placed multiple mixers, the combined effects of tank corners and interference between mixer flow patterns generally eliminate the need for baffles.

Vertical cylindrical tanks with top-entering mixers use a standard baffle configuration as shown in Figure 10. Side-entering mixers generally do not require baffles due to the asymmetric flow pattern. Draft-tubes require internal baffles or flow straightening vanes as shown in Figure 11.

CONCLUSIONS

It has been shown that agitation exhibits a substantial impact on the cooling rates of all of the major liquid quenchant media, oil, water, and aqueous polymers. In quenching applications, it is critically important that uniform and directional flow in the quench zone be provided. This is usually done with the use of draft-tube impeller agitation. This paper has provided an update on the use of use of impeller mixers for quench tank agitation. A basic primer on the use of quantitative methods to assist in the design of optimal agitation was also provided. Finally, the potential use of computational flow dynamics (CFD) to facilitate future quench tank agitation design was introduced.

REFERENCES

I.   United States Steel Bulletin, "Improved Quenching of Steel by Propeller Agitation", Pittsburgh, PA, 1954.

2.   A. Rose, *Archiv Eisenhuttenwessen*, Vol.13, 1940, p 345-354.

3.   C.E. Bates, G.E. Totten and R.L. Brennan, *Quenching of Steel*, ASM Handbook Vol.4: Heat Treating, ASM International, 1991, p 76-80.

4.   R.K. Zhelokhovtseva, *Steel in the USSR*, Vol.15, 1985, p 238-239.

5.   The term "impeller" is generic. Propeller refers only to a marine-type impeller, although propeller is used outside of the mixing industry to refer to all types of mixing impellers.

6.   J.Y. Oldshue, *Fluid Mixing Technology*, McGraw-Hill, New York, 1983.

7.   B.J. Hutchings, R.J. Weetman and B.R. Patel, *Computation of Flow Fields in Mixing Tanks with Experimental Verification*, ASME Winter Annual Meeting, 1989.

Table 1[1]
Typical Draft-Tube Impeller Flow Numbers and Power

Impeller	Open Impeller			Typical Draft-Tube ($K_v = 4$)		
	N_Q	N_P	Relative Q/P	N_QDT	N_PDT	Relative Q/P
Marine Propeller (3 blade, 1.0 pitch ratio)	0.46	0.35	1.0	0.40	0.40	0.69
Airfoil Type (Lightnin A312)	0.56	0.33	1.47	0.45	0.36	0.94

Table 2
Recommended Quench Tank Mixer Sizes[3]

Mixer Type[1]	Std. Quench Oil	Water or Brine
Open Impeller Mixers- top entry or side entry at 280 RPM[2]	0.004	0.003
Draft Tube Mixer at 280 RPM[2]	0.006	0.0045

1. Table based on use of high pumping efficiency impellers. Increase power levels by 50% for other impeller types.

2. Adjust power level for other mixer output speeds by: $P \propto N^{4/3}$.

3. Recommended mixer sizes provide higher agitation levels than originally recommended in the 1954 U.S. Steel article (See Reference 1).

Table 3
Sizing of Impeller Mixers

Motor[1,2]		Impeller Size[3,4]	
Hp	kw	inch	cm
0.25	0.19	13	33.0
0.33	0.25	14	35.6
0.50	0.37	15	38.1
0.75	0.56	16	40.6
1.0	0.75	17	43.2
2.0	1.49	20	50.8
3.0	2.34	22	55.9
5.0	3.73	24	61.0
7.5	5.59	26	66.0
10.0	7.46	28	71.1
15.0	11.19	30	76.2
20.0	14.92	32	81.3
25.0	18.65	33	83.8

1. The power requirements were calculated from Equation V assuming 280 rpm, the specific gravity to be 1.000 and an airfoil impeller with N_P = 0.33. (Airfoil and marine propeller power numbers are nearly identical.)

2. The shaft horsepower (sHp) is equal to 80% of the motor horsepower (mHp), (0.8 x mHp = sHp)

3. These are the power requirements for an open impeller operating at 280 rpm.

4. When used in a draft-tube, the impeller size should be reduced by 3%.

Figure 1

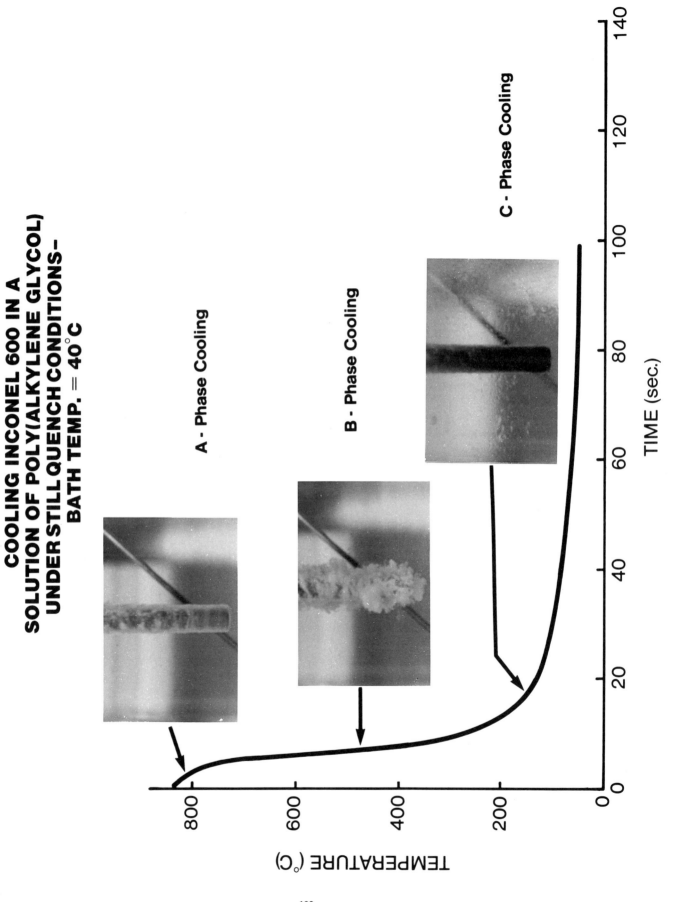

COOLING INCONEL 600 IN A
SOLUTION OF POLY(ALKYLENE GLYCOL)
UNDER STILL QUENCH CONDITIONS–
BATH TEMP. = 40°C

A - Phase Cooling

B - Phase Cooling

C - Phase Cooling

TEMPERATURE (°C)

800

600

400

200

0   20   40   60   80   100   120   140

TIME (sec.)

Figure 3
Typical Impeller Used for
Quench Tank Applications

High efficiency airfoil-type impeller
(Lightnin A310 fluidfoil impeller)

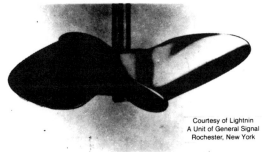

Marine propeller
(1.0 pitch ratio)

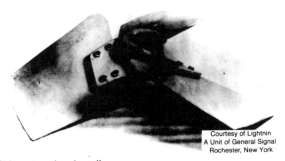

Side-entry mixer impeller
(Lightnin A312 fluidfoil impeller

Figure 2
Open impeller and draft-tube
impeller mixer

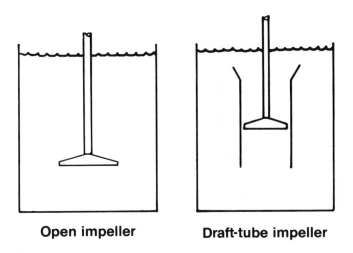

Open impeller   Draft-tube impeller

Figure 4
Draft-tube impeller mixer
operating points

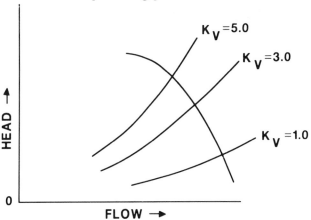

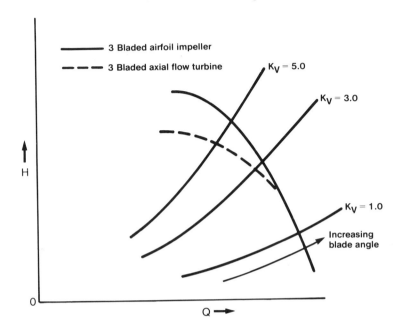

Figure 5
**Total head-flow capacity curves
for draft-tube impeller**

—— 3 Bladed airfoil impeller

– – – 3 Bladed axial flow turbine

$K_V = 5.0$

$K_V = 3.0$

$K_V = 1.0$

Increasing
blade angle

H

0

Q →

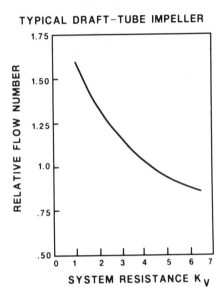

Figure 6
**Effect of system resistance
on flow**

TYPICAL DRAFT–TUBE IMPELLER

RELATIVE FLOW NUMBER

SYSTEM RESISTANCE K_V

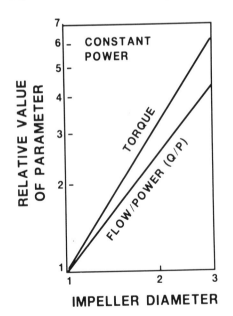

Figure 7
**Illustration of incremental power
requirement due to torque in system**

RELATIVE VALUE OF PARAMETER

CONSTANT POWER

TORQUE

FLOW/POWER (Q/P)

IMPELLER DIAMETER

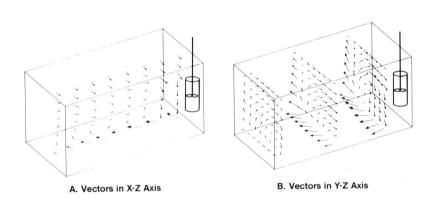

Figure 8
**Computational flow dynamics prediction of
3-dimensional flow field in quench tank**

A. Vectors in X-Z Axis

B. Vectors in Y-Z Axis

Figure 9
Typical draft-tube mixer placement
for tanks where L:W exceeds 2:1

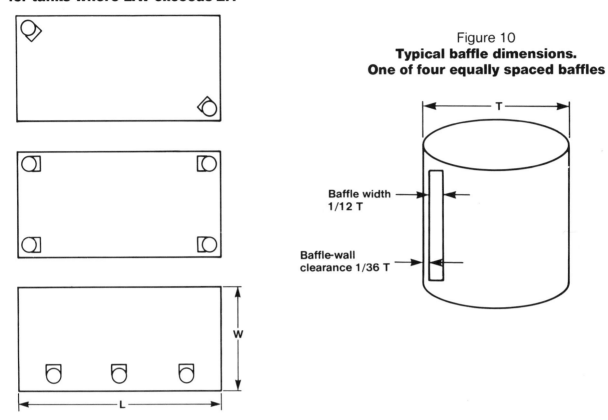

Figure 10
Typical baffle dimensions.
One of four equally spaced baffles

Baffle width
1/12 T

Baffle-wall
clearance 1/36 T

Figure 11
Typical draft-tube impeller
system characteristics

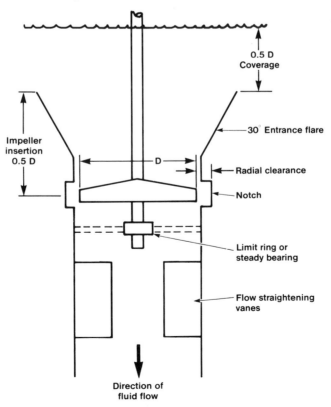

0.5 D
Coverage

30° Entrance flare

Impeller
insertion
0.5 D

Radial clearance

Notch

Limit ring or
steady bearing

Flow straightening
vanes

Direction of
fluid flow

Proceedings of the First International Conference on Quenching & Control of Distortion, Chicago, Illinois, USA, 22-25 September 1992

Poly (Alkylene Glycol) Concentration Control and Reclamation Using Membrane Technology as Applied to Steel Heat Treatment

R.D. Howard
Despatch Industries Inc.
Minneapolis, Minnesota

L.F. Comb
Osmonics, Inc.
Minnetonka, Minnesota

ABSTRACT

Reverse osmosis or ultrafiltration principles may be used to separate water from a polyalkylene glycol (PAG) and water solution by means of membrane technology. The feed solution consisting of PAG, water and other ionic, organic and suspended impurities is separated into two streams by collecting fluids from both sides of a pressurized membrane. A semipermeable reverse osmosis (RO) or ultrafiltration (UF) membrane, under sufficient, effective pressure, allows passage of purified water while retaining the larger PAG molecules. The three streams are referred to as the feed, permeate (water), and concentrate (retained PAG). The membrane is most effectively packaged in a spiral wound configuration with a turbulent flow design. The semipermeable membrane is wrapped around a perforated central tube which collects the permeate (water) which has passed through the pores in the membrane.

Choice of membrane material and permeation size are functions of the PAG in solution. If the characteristics of the quenchant to be separated are not known, a representative sample should be tested to determine appropriate membrane selection. Type I and Type II PAGs may be separated by the membrane method provided the permeation size is determined by the smallest PAG molecule. Typically, membrane material is cellulose acetate or polyamide, and membrane penetrations are sized to retain PAG molecules with characteristics similar to those of a molecular weight of approximately 300 - 1,000.

PRODUCTIVITY ADVANTAGES TO THE HEAT TREATER who has the ability to change polyalkylene glycol (PAG) concentration upon demand include the following:

1. The capability to adjust the heat transfer coefficient accurately in order to influence the metallurgical properties of the product, and to reduce distortion.
2. The means to vary the mediation rate of cooling of the product by the quenchant, particularly in the 400-200°C range.
3. The production versatility to change the PAG concentration within a quench tank infinitely over a wide concentration range.
4. The flexibility to run low PAG concentrations for such processes as through hardening of 1045 alloy, then to increase concentrations for processes requiring slow cooling.
5. Reclamation of both the PAG (concentrate) and the water (permeate) used in the process.
6. Proactive management of the quenchant and the metallurgical processes it supports.

The spiral-wound membranes and permeate tubes are contained in stainless steel cylindrical housings. Each housing may contain multiple membranes, each with the permeate tubes coupled together and sealed with O-Rings. Housings may be piped in series or parallel, depending upon flow. Feed solution is pumped at high pressure into the membrane assembly into side inlets in order to increase turbulence at entry. Retained concentrate flows lengthwise through the space between the membranes and exits at the opposite end of the tube from the feed inlet. Permeate is passed through the membrane permeations into the permeate tube and is discharged from a port located on the centerline of the membrane assembly. See Figure 1 for an illustration of a membrane assembly.

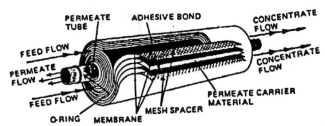

Figure 1 illustrates a membrane assembly. Figure courtesy of Osmonics, Inc.

Permeate Rate (Separate Water Rate) [Qp] equals flow rate of purified water which is passed through the membrane expressed in gallons/minute (gpm) or liters/minute (lpm). Specified permeate rates are normally at 77°F (25°C).

Concentrate Rate (Qc) equals flow rate of retained PAG solution in gpm or lpm.

Feed Rate (Qf) equals flow rate of incoming quenchant in gpm or lpm. Feed rate equals permeate rate plus concentrate rate.

Recovery equals permeate rate divided by feed rate and is expressed as a percent. For example, 60% recovery means that out of a given feed rate, 60% is separated out as pure water.

$$Recovery = \frac{Qp}{Qf \ x \ 100} \qquad (1)$$

Concentration equals the total PAG in solution with water and is expressed as a percent.

C_f = Feed Concentration
C_p = Permeate Concentration
C_c = Concentrate Concentration
C_{avg} = Average Concentration

$$Passage - \frac{C_c}{C_{avg}} \qquad (2)$$

$$Retention = \frac{C_c}{C_{avg}} \qquad (3)$$

$$C_{avg} = \frac{C_f + c_c}{2} \qquad (4)$$

Salt (Ionic) Rejection equals the percent dissolved salt rejected by the membrane calculated on an average concentration over the membrane.

Salt (Ionic) Passage - equals (100 - % rejection) or the percent dissolved salts passed through the membrane.

Salts (calcium, magnesium, sodium, etc.) are introduced into the quench through process salts (salt baths for example) or from salts in the make-up water. By increasing salt concentration, the chance for calcium carbonate precipitation and permeate flow loss is increased. As water is separated from the process volume, feed concentration (C_f) increases, concentrate (C_c) concentration increases, permeate (C_p) concentration remains essentially zero, permeate flow rate decreases asymptotically, feed flow rate decreases proportionate to the flow/pressure curve of the pump supplying the membrane array, salt passage as a percentage increases, and the effective pressure required to separate the water from the PAG increases. See Figure 2 for representative curves.

Flow Description - The feed solution is presented to the high pressure separation pump by a constant volume process pump sized to satisfy the feed flow rate at an inlet pressure of 20-60 PSI.

The feed solution passes through a replaceable 5 micron cartridge filter which removes large suspended solids. Filtered solution then flows to an inlet control valve. (This solenoid or solenoid controlled diaphragm valve opens when the system is turned on, allowing quenchant to flow to the pump inlet.) When the system is turned off, the inlet valve closes, preventing non-turbulent flow through the membranes when the pump is not operating.

The high pressure separation pump feeds PAG solution to the membrane housings at 450-650 PSI in series or in parallel configuration, depending upon design flow rates. The water is separated from the PAG feed stream and leaves the membrane array as the permeate stream. PAG in the concentrate stream is retained.

Permeate from each membrane housing is collected at a permeate manifold where a check valve presents back flow into the membranes. A pressure relief valve is installed to prevent excessive back pressure. The permeate then flows through a flow meter and a conductivity monitor to the permeate drain outlet on the machine.

The concentrate leaves the last membrane housing into a concentrate manifold and through a flow meter before passing through a valve back to the process container.

Maximum effective PAG reconcentration is in the 55% - 70% range. At 55% concentration, C_f and C_c are essentially equal, and an effective separation pressure of approximately 650 PSI should be expected. Cellulose

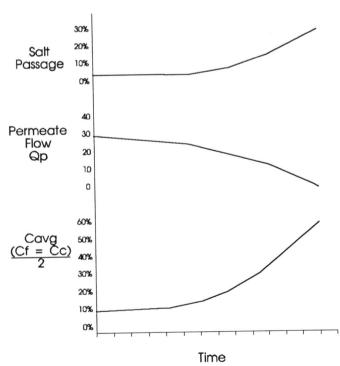

Figure 2 illustrates representative curves for salt passage, permeate flow and C_{avg}.

108

acetate or polyamide membranes have a thermal limitation of 105°F. Significant heat is generated, particularly at higher concentrations due to the viscosity of feed solution; therefore, a heat exchanger in the feed loop is typically required. At the conclusion of a separation cycle, the membranes should be forward flushed with pure water in order to flush the gelatinous layer of concentrated PAG from the membranes. Periodically the membranes should be flushed with a diluted concentration of phosphoric or citric acid to dissolve and remove calcified precipitates which have been trapped in the membranes. When these procedures are followed, extended membrane life can be expected.

Crossflow membrane filtration is best defined as very fine filtration where pressurized feed stream flows across the surface of a semi-permeable membrane. A portion of the feed passes through the membrane and the balance sweeps tangentially along the membrane and exits the system.

The stream which passes through the membrane is called the permeate and the stream which exits the system without passing through the membrane is called the concentrate.

Because the feed and concentrate flow parallel instead of perpendicular to the membrane, the process is called crossflow filtration.

The membrane's pore size is the determining factor in defining which crossflow membrane technology is being used. The differences between reverse osmosis (RO), nanofiltration (NF), ultrafiltration (UF) and microfiltration (MF) are the pore size and operating pressure.

Reverse Osmosis (RO)

RO was the first crossflow membrane separation process to be widely commercialized. RO removes virtually all the organic compounds and 90 to 99% of all ions. Pressure, on the order of 200-1000 PSIG, is the driving force of the RO purification process. A number of RO membranes are available to meet varying rejection requirements.

Nanofiltration (NF)

NF membranes remove organic compounds in the 300-1500 molecular weight range, reject selected salts (typically divalent), and produces higher permeate flow rates at lower pressures, 150-250 PSIG, than RO systems.

NF is relatively new to crossflow filtration, but offers unique softening characteristics on water supplies and organic desalting capabilities on process applications.

Ultrafiltration (UF)

UF is similar to RO and NF but is crossflow filtration which does not reject ions but is more like sieve filtration. UF rejects contaminants based on molecular weight (MW), typically 5000 to 200,000 MW. Because of the large membrane pore size, UF requires much lower operating pressure, 10 to 100 PSIG.

Microfiltration (MF)

MF membranes are absolute filters typically rated in the 0.1 to 2 micron range and operate at pressures of 1-25 PSIG.

Crossflow MF substantially reduces the frequency of membrane replacement, but are typically of higher capital cost than disposable MF cartridge filtration.

Polyalkylene Glycol Concentration

Reverse osmosis technology is being used in the recovery and concentration of polyalkylene glycol (PAG). RO was chosen based on the PAG type, PAG molecular weight and the high percentage of PAG rejected by the RO membrane. The RO membrane passes purified water while rejecting and concentrating all the PAG.

The RO machine design is based on the size and turnover specifications of the heat treatment/quenchant system. Knowing the targeted PAG concentration and volume to be processed, the necessary permeate, concentrate and feed flow rates are calculated for the design of the RO.

The flow of fluid through the machine begins with a feed stream at 20 to 60 PSIG, passing through 5 micron cartridge prefilters for the removal of large suspended particles. From the filters, the feed solution enters a multi-staged centrifugal pump where the pressure is boosted to 400-600 PSIG. The pressurized feed enters the stainless steel housings containing the spiral wound membranes.

As the feed flows through the machine, the permeate is continuously drawn off of each housing. The permeate is manifolded and the flow rate and conductivity (purity) are measured as it exits the machine. Since the RO membrane rejects ions as well as the PAG, the permeate will have a lower total dissolved solids (TDS) concentration than city water and, therefore, can be used for rinse make-up.

The concentrate containing the PAG flows to the next set of housings. After the concentrate leaves the final housing, it is manifolded and passes through an orifice which sets the concentrate flow rate and reduces the discharge pressure.

As the PAG is recirculated and concentrated, the viscosity increases and the temperature rises. To maintain the temperature at less than 95°F, the concentrate passes through a cooling heat exchanger before returning to the process tank. The RO does incorporate a high temperature alarm and shut down to prevent damage to the membrane.

The system is operated continuously until the PAG concentration reaches the set point, typically 50-55%. Once the set point is reached, the machine is diverted to the flush mode. The flush consists of city water entering the feed of the machine to flush the viscous PAG solution from the membrane surface and is intended to increase membrane line.

Periodically the machine is cleaned with a mild phosphoric or citric acid solution which removes any remaining PAG as well as calcium carbonate or ferrous precipitate which may have formed during the concentration process.

The machine cleaning, which takes 1-2 hours, is determined based on the increase in pressure drop across the membranes and a decrease in the permeate flow rate. Typically, a cleaning is necessary after every second or third concentration cycle.

PAG quenchant concentration control using membrane separation technology is fundamentally a process of removing a calculated volume of a known concentration of PAG and replacing that volume with a known higher concentration for concentration increase, or with the same volume of water for concentration decrease. The original volume of PAG quenchant removed to allow for the concentration change is then reconcentrated to a higher concentration and stored for subsequent quenchant concentration increases. The separation of water from the removed PAG quenchant and subsequent storage of the reconcentrated higher concentration of PAG is typically performed off-line from the quench process in order to allow for rapid quench tank concentration changes. Formulas for volumes of quenchant to be removed and processed are as follows:

PAG Concentration Increase

$$\text{Gallons of quenchat to be removed} = \frac{Q_{wlc} - Q_{whc}}{\text{PAG \% to be added} - \text{PAG \% removed}} \quad (5)$$

Q_{wlc} = Gallons of water at lower concentration
Q_{whc} = Gallons of water at higher concentration

Example: Increase PAG concentration in a 10,000 gallon quench tank from 10% PAG to 20% PAG by replacing a calculated volume of 10% PAG with 55% PAG.

Table I Calcualted Volumes

%	PAG	Water	Total
10%	1,000 gal.	9,000 gal.	10,000 gal.
20%	2,000 gal.	8,000 gal.	10,000 gal.

$$\text{Gallons of 10\% PAG to be removed and replaced with 55\% PAG} = \frac{9,000 \ gal. - 8,000 \ gal.}{0.55 - 0.10} = 2,222 \ gal. \quad (6)$$

PAG Concentration Decrease

$$\text{Gallons of quenchant to be removed} = \frac{Q_{chc} - Q_{clc}}{\text{PAG \% removed}} \quad (7)$$

Q_{chc} = Gallons PAG at higher concentration
Q_{clc} = Gallons PAG at lower concentration

Example:
Decrease PAG concentration in a 10,000 gallon quench tank from 20% PAG to 10% PAG by replacing a calculated volume of 20% PAG with water.

$$\text{Gallons of 20\% PAG to be removed and replaced with water} = \frac{2,000 \ gal. - 1,000 \ gal.}{0.20} = 5,000 \ gal. \quad (8)$$

Time required for off-line concentration of the PAG of the quenchant removed is a function of the membrane area which determines permeate flow rate (Q_p) over the range of reconcentration. Solving the equations for volumes of quenchant to be removed can be done manually, but is ideally performed by a PLC or computer based controller. Such a controller also receives inputs as to actual PAG concentration, initiates actions of the valves and pumps which perform the transfers, and alarms upon non-permissible conditions. Measurement as to actual quenchant concentration may be made by hand held refractometer; however, automated PAG concentration control systems utilize either density, refractive index, or viscosity based monitors which provide analog outputs to the system controller and display. Typical density based concentration monitors provide an accuracy of measurement of 0.0001 specific gravity units. Density based monitors do not distinguish PAG in solution from dissolved salts; therefore, the concentration signal will be distorted by the percentage of salts by weight in solution.

Reconcentration of the removed quenchant to storable concentration is performed off-line in a process tank. Determination as to when sufficient water has been removed from the quenchant in the process tank is achieved either by measuring totalized permeate flow (Qp total) or by monitoring process tank concentration using a density or viscosity based concentration monitor. Capacities of process and storage tanks are functions of volumes of the quench tank and PAG concentration range in the quench tank. Typical PAG concentration ranges accurately maintained and controlled in the quench tank are 10-40%. See Figure 3 for PAG concentration control schematic.

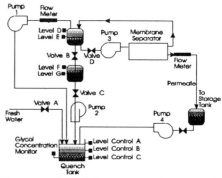

Figure 3 illustrates the PAG concentration control schematic.

Multiple quench tanks may be controlled by a single centralized PAG Management System. The PC or PLC monitors the conditions of all tanks within the control network and automatically maintains the PAG concentration at set point or changes concentration upon demand. For instance, multiple induction hardeners or integral quench furnaces can be piped and wired to a single reverse osmosis unit and controller. Each tank may include its own concentration monitor for maximum accuracy or multiple tanks may be piped to a single monitor but accuracy may be less. Each tank may be controlled at a different concentration. Reverse osmosis permeate rate and process/storage tank capacities are determined based upon maximum system requirements; however, the controller may be programmed to stagger or cascade changes so as to minimize component sizes.

PAG concentration is automatically monitored by one of three methods:
1) Density,
2) Viscosity,
3) Refractive Index.
All three methods are accurate and provide milliamp or millivolt output for display and process calculations. The density method is susceptible to bias from contamination and air bubbles. The viscosity curve is not linear with respect to concentration, thus requiring signal characterization. The refractive index method is affected by oil and metallic salts. As the monitor outputs are used for display as well as for process calculations, periodic calibration should be performed. All three methods include span and zero adjustments and temperature compensation. Other necessary monitors include pH, conductivity, pressure and temperature.

PAGs by themselves are not normally considered to be hazardous substances except as components of total toxicity within an enviro-system. Most heat treaters have determined the most responsible attitude toward PAG management is to avoid dumping PAG and to reclaim whenever possible. Overflow quenchant and rinse water containing PAG may be processed through the reverse osmosis unit for reclamation of both PAG and water. Reclaimed PAG and water may be conditioned and stored for later use.

In ferrous heat treating applications, sludge, oil and oxides are often present in the quenchant and must be separated and removed from the solution before presentation to the semi-permeable membranes. Sludge is filtered by conventional sieve strainers utilizing 50-100 mesh cartridges. Such strainers require fairly frequent cleaning. Oil is separated from the PAG and water solution using coalescing technology in which micron size oil droplets are combined into larger particles big enough to separate by density from the main fluid. The oil is then separated into a stream when passed through coalescing elements and collected for disposal. Iron oxides are removed from the solution by pumping the feed stream through housings which contain catalysts for retaining the iron, both ferrous and ferric.

The principle driving force for coalescing action in a liquid stream is the interfacial tension of the droplets. Interfacial tension is the excess free energy due to the existence of an interface at the surface of a droplet, arising from unbalanced molecular forces. A relatively small interfacial tension value is typically required to obtain a coalescence rate low enough for practical application.

In a carrier stream of dispersed liquid droplets, the total interfering effect of surface active agents, particulate masking, or electrical charge is not great enough to render the dispersion permanent. The interfacial tension value between the two liquids is neither drastically reduced nor destroyed. Therefore, the dispersed droplets can be physically induced to agglomerate, and the natural process of fluid coalescing can be mechanically accelerated to economically separate the liquids making up the emulsion. This provides the basis for liquid/liquid coalescing technology.

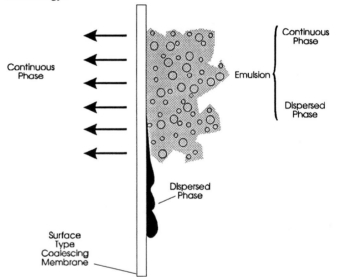

Figure 4 illustrates the sequence of operation of liquid/liquid surface coalesing. Figure courtesy of Osmonics, Inc.

Industry uses a variety of mechanical means to effect fluid coalescing. A settling tank will reduce the velocity of a liquid emulsion and provides a quiescent zone. At low velocity, the dispersed droplets agglomerate and form a second continuous phase due to differences in specific gravity.

Additional techniques are used to improve the coalescing rate in settling tanks, including directional flow inducers and baffles. System modifications may include recycling the excess dispersed phase and flowing the emulsion through beds of coarse, porous media, such as wire mesh or fiberglass.

Similar methods are used to effect liquid coalescing. Surge tanks are used to reduce the velocity of the stream, encouraging the agglomeration of liquid droplets. After the droplets settle, they are removed from the system. In many instances, vessels use devices to induce centrifugal flow and create abrupt changes in the direction of flow. This causes liquid droplets to agglomerate, form a separate phase and leave the system.

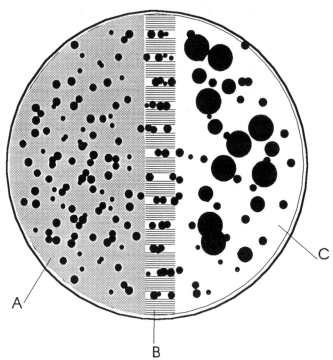

Figure 5 illustrates depth coalescing through porous section of rigid coalescing element. Figure courtesy of Osmonics, Inc.

Section A illustrates the structure of a typical liquid system during flow conditions. The dark spheres (or dispersed phase) represent liquid droplets of various sizes. These spheres are encapsulated with a particulate film which prohibits coalesing of dispersed droplets when collisions occur before conditioning.

Section B shows the liquid droplets passing through the exagereated openings of the porous media. During this travel, two actions occur.

1. The ecapsulated film is removed from the liquid droplets by the filtering action of the porous media's surface.
2. Coalescing of the conditioned liquid droplets accelerates because of forced droplet contact.

Section C illustrates the coalesced liquid system with liquid droplets continuing to coalesce. Note the continuous formation of larger droplets by additional collisions as they travel through and away from the porous element.

Conclusion

Reconcentration of PAG concentration in water upon demand using membrane technology is a proven, viable process. Cleanliness of the quenchant is a prerequisite for long term functionality. Contaminants must be removed, pH must be controlled, conductivity must be monitored, and all forms of iron must be removed from the process. Membranes must be flushed and chemically cleaned to eliminate salts and metallic precipitates which reduce membrane life.

The same disciplines which prolong membrane life also promote effectiveness of the quenchant. The elimination of dissolved salts, iron and sludge from the quenchant improves uniformity of the bath and allows the PAG to effectively mediate cooling rates of the product.

The capability to reconcentrate polyalkylene glycol in water to any concentration upon demand gives the heat treater the ability to adjust the heat transfer coefficient of the quench to match the metallurgical characteristics of the product. This capability enhances the heat treater(s) product range and production flexibility as well as providing the monitors and control necessary to effectively manage the quenchant.

Proceedings of the First International Conference on Quenching & Control of Distortion, Chicago, Illinois, USA, 22-25 September 1992

Vacuum Quenching Improvements Through Controlled Atmospheres

R. Holoboff
Liquid Air Corp.
Countryside, Illinois

B. Lhôte
L'Air Liquide
France

R. Speri
Vide et Traitement
France

O. Delcourt
Peugeot S.A.
France

ABSTRACT

Alnat V is a newly developed atmosphere control process for vacuum furnaces which allows heat treaters to enhance their quenching capabilities by optimizing quenchant composition and furnace parameters with helium.

Small cylindrical parts and larger turbine blades made from nickel based superalloys were selected to determine the gas composition needed to maximize the cooling rate and produce the desired mechanical properties. By optimizing the quenchant gas composition with a helium/argon blend, an aerospace manufacturer was able to heat treat parts previously impossible with argon alone.

A similar study was conducted on transmission main shafts for an automotive manufacturer. A strong correlation between improving furnace operating parameters and quenchant composition is demonstrated.

In the introduction, furnace parameters and relevant thermodynamic principles will be examined.

THE ELIMINATION OF FLAMMABLE and environmentally unsafe quenching baths has made quenching an attractive alternative for heat treaters. However, the slower cooling rate when compared to quenching baths has limited gas quenching to metals having a moderate critical rate of quenching. As furnace technology improves, the use of gases will be extended to metals having higher critical cooling rates. In an effort to expedite this process, L'Air Liquide has explored the use of helium as a quenchant.

The techniques employed in optimizing cooling rates with helium will be discussed along with production test results from manufacturers Vide et Traitement and Peugeot. We will begin by examining the relevant thermodynamic principles and furnace parameters. A practical guide on optimizing these parameters is also provided.

Using Helium

The use of helium for quenching provides many advantages specific to it as well as all the advantages inherent to gas quenching generally. Gas quenched parts are clean and bright, do not require an additional cleaning step, and present no disposal problems. In addition, the cooling rate can be adapted to each individual treatment thereby providing flexibility to the process.[1,2] For its part, helium has the second highest thermal conductivity of any gas (second only to hydrogen which is rarely considered as a quenchant in vacuum furnaces because of its explosion hazard). This high thermal conductivity translates into faster cooling and a more homogeneous furnace environment. Because helium picks up the heat and releases it to the heat exchanger faster than nitrogen or argon, it recirculates at a cooler temperature promoting faster conduction of heat through the part. Ultimately, this means larger loads and bigger parts may be treated. Because helium is a light gas, the power required to recirculate it during the quench is far less than either nitrogen or argon. In addition, helium is both non flammable and inert making it an ideal gas for heat treatment.

An increase in productivity and/or technical performance may be realized with helium. The question then becomes one of how to optimize these properties for production applications.

The first lesson to be learned when using helium is that its use should be combined with other quenchants and minimized when possible.

There are a couple of reasons for this:

1) Helium is expensive.
2) When considering cooling rates, there is a competition between gas density and thermal conductivity.

The first reason will not be discussed here. The second will require a brief thermodynamic analysis.

The rate at which the surface of the part is cooled, Q, is defined by the following relationship:

$$Q = h \cdot A \cdot \Delta T, \qquad (1)$$

where

h = the heat transfer coefficient,
A = the surface area of the part, and
ΔT = the difference in mean temperature of the part and quenchant.

Expanding h, we find that for forced convection (over cylinders):

$$h = C \cdot (\Lambda/d) \cdot (rho \cdot S \cdot d/\mu)^\alpha \cdot (Pr)^\beta, \qquad (2)$$

where

Λ = thermal conductivity of the gas,
rho = density,
S = velocity,
μ = kinematic viscosity,
Pr = Prandtl number,
d = characteristic length (in our model d = cylinder diameter), and
C, α, and β are constants dependent on furnace design and load configuration (perpendicular or parallel to cylinder axis).

Because Pr and μ vary little with gas species, we will suppress them in the constant. The heat transfer coefficient simplifies to

$$h = K \cdot d^{\alpha-1} \cdot \Lambda \cdot (rho \cdot S)^\alpha \qquad (3)$$

Since Λ decreases with increasing molecular weight and rho increases, they compete as functions of molecular weight. The result of this competition is that there are mixtures of helium and the original quenchant that will have higher heat transfer coefficients than pure helium. In Figures 1 and 2, we provide a theoretical example of this phenomenon. Using Colburn's formula for the heat transfer coefficient, assuming the necessary parameters, and assuming parallel flow over cylinders for the sake of example, we find that mixtures of 55-85% of helium in argon have higher h values than pure helium. Our tests at Vide et Traitement bear witness to the need for optimizing gas composition and support theory.

So, in summary, reason 1 and 2 suggest that we need to optimize our use of helium so that only the minimum amount is used to achieve the desired mechanical properties in our part.

Furnace Evaluation and Optimization

In order to take full advantage of helium's high thermal conductivity, the furnace must be evaluated for its effectiveness in removing heat. This includes a thorough look at the heat exchanger, quenching pressure, use of the recirculating fan, as well as the introduction of gas into the furnace.

Heat Exchanger

The need for an effective heat exchanger is made clear in equation (1). The higher the temperature difference between the part and gas, the faster the cooling rate.

The efficiency of the heat exchanger is affected by several factors: surface area of the heat exchanger, the nature of the heat exchanger fluid, and the temperature of this fluid.

The most effective heat exchanger will have a surface area larger than the load plus the inner surface of the furnace. Therefore, the heat exchanger should be finned.

A water-cooled heat exchanger will have better cooling rates than an air-cooled one.

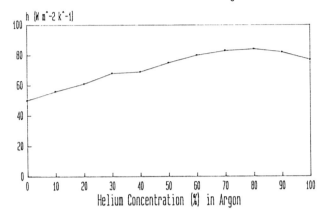

Heat Transfer Coefficient (h) vs. Helium Concentration in Argon

Figure 1: Theoretical Calculation of heat transfer coefficient as a function of helium concentration in argon. Assumes Colburn's formula and parallel flow. T = 400 C.

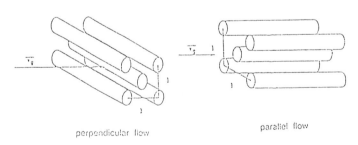

TYPES OF GAS FLOW

perpendicular flow   parallel flow

Figure 2: Graphical model of two standard types of load configuration for theoretical calculation of heat transfer coefficient.

PART AND PLACEMENT	HELIUM 100%	HELIUM 68% ARGON 32%	HELIUM 43% ARGON 57%	HELIUM 16% ARGON 84%
CYLINDERS				
Left	1708	1562	1359	929
Middle	2291	2039	2039	1310
Right	2295	2448	2369	1597
TURBINE BLADES				
Left	742	680	628	472
Middle	1094	1147	1048	742
Right	979	979	941	742
TEMP$_{max}$ WATER(°F)	149	133	129	124
FAN CURRENT (amps)	79	93	97	113
PRESSURE (bars rel)	3.2	2.9	2.9	2.9

Table 1   Cooling rate from 2370° to 1290° in °F/min by helium concentration and placement of the part in the basket for small cylinders and larger turbine blades at Vide et Traitement

If water is used, the temperature and pressure need to be monitored. The temperature should never exceed the boiling point of water. Steam is not as efficient as a heat exchange fluid and the increased pressure presents a potentially hazardous condition.

If argon or nitrogen are currently used, check to make sure the water temperature during the quench is well below the boiling point. Because the temperature will increase with helium. (More heat is being removed in less time.)

Furnace Pressure

The effect of furnace pressure on cooling rates can be seen in equation (4). From the ideal gas law we know pressure is proportional to density and by suppressing the other variables into the constant, we find

$$h = K (PS)^{\alpha}, \qquad (4)$$

where P = gas pressure.

So, higher pressures translate into faster cooling rates. The trend to 5 bar furnaces from 1 and 2 bar furnaces is clear.

In tests conducted by L'Air Liquide it has been shown that helium may optimize a furnace's pressure as well and, in some instances, provide an alternative to the capitalization of a new furnace. For example, helium at 1.5 bars relative pressure has been shown to provide the same metallurgical results as quenching with argon at 3.5 bars relative pressure.[2] However, at least 3 bars relative pressure, even with helium, is required to reproduce the cooling rate of an oil or salt bath for small parts. For larger parts, a higher pressure and/or optimization of the gas flow patterns may be required.

Of course, furnace specifications will limit the pressure that can be used. Sometimes, however, the power consumption of the fan is the limiting factor. Since it takes more power to recirculate the gas as the pressure is increased, in some instances, the fan may cut out, especially at start-up, where the power transient may be 8-10 times that of normal running. If this is the case, switching from argon or nitrogen to helium will reduce the power requirement and allow the furnace pressure to be increased.

At Peugeot, for example, the fan's power consumption under nitrogen at 2.5 bars was 57 kW. When the nitrogen was replaced with helium, the pressure was raised to the maximum allowed (4 bars relative) and the power consumption was reduced to 11 kW.

Recirculating Fan

Helium will optimize a furnace's recirculating fan, but in a not so obvious fashion. As the rotational speed of the fan increases, the gas velocity increases as well. From equation (4), we see that this increase in velocity translates into higher cooling rates. However, computer studies at L'Air Liquide have shown that at any particular fan speed, the velocity field is fairly independent of gas species.[3] So it would appear there is no apparent advantage to helium when considering fan speed.

The advantage, as we have seen earlier, is in lower power consumption. Heavier gases require more power to achieve the same rotational speed.

There is a complication, however, that needs further clarification. In some instances, merely changing the gas species will have an effect (test 1, Peugeot) even though the fan's rotational speed and gas pressure are kept constant. This effect is a function of pressure.

At high pressures, gas (or external) parameters dominate load (or internal) parameters with regard to cooling rates. In other words, the quenching gas will remove heat from the surface faster than it is being conducted from the interior of the part. In this situation, different gas species are on an equal footing, i.e. their cooling rates are similar. Conversely, at low pressures, one has lower external (gas) parameters and therefore

GAS	PRESSURE (bars rel)	COOLING RATE UPPER STAGE (°F/min)	% OF ORIGINAL COOLING RATE UPPER STAGE	COOLING RATE LOWER STAGE (°F/min)	% OF ORIGINAL COOLING RATE LOWER STAGE	FAN POWER CONSUMPTION (kW)
N₂	2.5	178	100	121	100	57
He	2.5	236	132	176	145	7
He	4.0	349	196	275	228	11

Table 2   Cooling rate between 1470°F and 750°F and fan power consumption as a function of gas and quenching pressure for Trial #1 at Peugeot.

internal parameters (load size, dimension, and configuration) dominate the cooling rate. In this case, changing the gas species may improve or increase the cooling rate even though pressure and rotational speed are held constant.

This phenomenon is evident when comparing tests 1 and 2 at Peugeot.

The furnace manufacturer should be consulted to determine the speed limitations of the fan. In general, a better fan can be made if it does not need to be located in the hot zone.

Gas Introduction

When attempting to increase the cooling rate, it is important to backfill as quickly as possible. Helium has an advantage in this category too. Because of its lower molecular weight, helium will fill three times faster than argon (Graham's law).

To minimize the filling time, the surge tank must be of adequate size and pressure rating and be in close proximity to the furnace. The piping connecting the surge tank to the furnace should be sized so that the furnace will fill in 10 seconds or less. This is dependent on the distance between the surge tank and furnace, but a 2 inch diameter pipe is normally adequate.

Case Studies

Case Study # 1 Vide et Traitement. The tests at this site were designed to optimize quenchant mixtures.

Testing was done in a furnace (BMI 63T) having an 88 cf internal volume at a pressure of 3 bars relative. The cooling rate was measured from 1460°C (2660°F) to 700°C (1290°F). Thermocouples were placed in parts located in the left, center, and right of the basket to determine whether the part's location influences its cooling rate. Two types of parts were used: a small cylinder 0.63 cm (1/4") in diameter and a larger turbine blade. The parts were lined up in the basket in three rows. Each row contained both cylinders and turbine blades with 8 parts to a row.

Four different gas mixtures were used containing between 16% and 100% helium in argon.

Since there are different types of turbulent flow around the load in the furnace, the optimized gas mixture can only be determined through actual testing. Theoretical calculations were performed, however, on two standard types of turbulent flow: parallel and perpendicular flow over cylinders (see Figure 2). Taking various mixtures of helium in argon or nitrogen into account, theoretical calculations show that 100% helium provides the best heat transfer for perpendicular flow.

Concerning parallel flow, however, the calculated optimum is near 70% helium in nitrogen or argon.

The cooling rate for each type of part in each of the three baskets is given in Table 1 for the four gas mixtures used. The cooling rate varied somewhat according to part placement and significantly, as expected, according to part size. The mixture containing 16% helium provided a much lower cooling rate than did the higher percentages. When this data (Table 1) was plotted, it became obvious that the optimum percentage of helium for maximum cooling was between 60% and 100%, with 50% being almost as good and more cost effective. This was true for both the cylinders and the turbine blades.

By optimizing the helium/argon mixture, L'Air Liquide gave Vide et Traitement the capability of heat treating parts they were previously unable to do with argon alone, but at a fraction of the cost of pure helium.

Case Study # 2 Peugeot. L'Air Liquide used a combination of computer modelling and production line testing to determine the effects of pressure (test 1) and velocity (test 2) on the cooling rate for both nitrogen and helium.

Test 1. The furnace (ECM PF500) used had an upper and lower level to hold parts and an interior volume of 88 cf. The test pressures were 2.5 bars and 4.0 bars relative. Under nitrogen, 2.5 bars was the maximum allowed. Higher pressure would have overloaded the fan. The test load was approximately 200 kg (440 lbs).

Theoretical calculations predicted that an increase from 2.5 to 4 bars relative in helium would double the cooling rate.

The data for this test is summarized in Table 2. It contains the average cooling rate between 800°C (1470°F) and 400°C (750°F) for parts in both the upper and lower levels of the furnace and the percent increase in cooling rate over nitrogen at 2.5 bars relative pressure. The power consumption of the fan is also given.

When operating under the same pressure (2.5 bars relative), there was an increase in cooling rate when helium was merely substituted for the nitrogen. A 32% increase was seen in the upper stage and a 45% increase in the lower stage. The difference in cooling rates between levels is easily explained. Since the gas flows from top to bottom, it warms up before reaching the bottom tray. In equation (1), we see the qualitative effect of a smaller temperature difference, i.e. lower cooling rates. However, even this temperature difference was minimized with helium. Under nitrogen, the parts in the lower tray had a cooling rate of 67% of those in the upper tray. With helium, the furnace environment is more homogeneous. The lower tray was increased to 75% of the upper tray.

The fan's power consumption was reduced from 57 kW with nitrogen to 7 kW helium at 2.5 bars. Even though the furnace was rated for 4 bars relative pressure, the power limitations of the fan (nominal power rating of 45 kW)

QUENCHANT GAS	FAN SPEED (rpm)	COOLING RATE (°F/min)	% OF ORIGINAL COOLING RATE	FAN POWER CONSUMPTION (kW)
N₂	3000	554	100	16
He	3000	544	98	4
N₂	3300	709	128	20
He	4500	941	170	7

Table 3   Cooling rate between 1470°F and 750°F and fan power consumption as a function of gas and fan rotation speed for Trial #2 at Peugeot

prevented operation above 2.5 bars with nitrogen. Changing to helium, made it possible to increase the pressure to 4 bars and still use only 11 kW of power. When this was done, the cooling rate doubled from that of nitrogen at 2.5 bars, confirming the results predicted by the model.

Test 2. In test 2, the effect of increased gas velocity on cooling rate and hardness was investigated.

By using a different furnace (ECM PF300), the furnace pressure was increased to 4 bars relative for both nitrogen and helium. The furnace volume was 81 cf and test load was 40 Kg (88 lbs). Rotation speed was increased from 3000 rpm to 4500 rpm.

Theoretical predictions made by computer indicate at least a 50% increase in the cooling rate of helium over nitrogen when the fan speed was increased.

Table 3 contains the data from this test including the fan's rotational speed and power consumption, the cooling rate, and the percentage increase in cooling rate over that of nitrogen at 3000 rpm (normal operating speed).

As expected, there was no increase in cooling rate experienced by a simple exchange of helium for argon (high operating pressure).

The trend in power reduction is clear.

The extra power made available with helium was used to increase the rotational speed of the fan. With helium, the only limitation placed on the fan speed was the upper limit allowed by the fan. At this limit, 4500 rpm with helium, the cooling rate was increased by 70%, close to the predicted rate. The power consumption was also decreased substantially.

The 28% increase in cooling rate (to 709°F/min) provided by nitrogen at 3300 rpm and 4 bars relative pressure developed a part Brinell hardness of 363 HB. With helium at 4500 rpm, the cooling rate increased to 941°F/min providing a hardness of 415 HB. This higher hardness was within Peugeot's specifications and gave this leading French car manufacturer the capability of gas quenching parts that previously could only be quenched in oil.

Summary

There can be many advantages realized when switching from nitrogen or argon to helium for gas quenching. Depending on the pressures in the furnace, the cooling rate can be increased simply by changing gases (a 30% increase in test 1 at Peugeot). Because of the fan's lower power requirements with helium, the furnace can be optimized to its maximum pressure and fan rotation speed. The cooling rate increased by 70% in test 2 at Peugeot by increasing the fan rotation speed from 3000 to 4500 rpm. It was doubled in test 1 when nitrogen was replaced by helium and the pressure was increased from 2.5 to 4.0 bars relative.

The type of gas flow around the parts will determine the best helium mixture to use. Since the flow patterns may not be obvious, this is best determined through actual testing to optimize the cooling rate and costs. Simply changing gases at Vide et Traitement, on a standard 2 bar furnace, allowed them to heat treat parts they couldn't heat treat before.

Superior results are obtained when using helium to optimize parameters such as pressure or gas velocity. A more refined computer model of gas flow and heat transfer as a function of pressure and velocity is now in development at L'Air Liquide. The different gases and their mixtures will be modelled to provide the heat treater a better understanding of the dynamics of gas quenching in vacuum furnaces.

Acknowledgment

A special thank you is extended to Ms. Marcia Phillips of Liquid Air for the background information she provided, to Mr. Ben Jurcik of American Air Liquide for his insight into convective heat transfer, and to Ms. Shelly Hahn also of Liquid Air for her help in the preparation of the manuscript.

References

1.   E.J. Radcliffe, "Gas Quenching in Vacuum Furnaces - a Review of Fundamentals," Industrial Heating, November 1987.

2.   B. Lhote, O. Delcourt, "Gas Quenching with Helium in Vacuum Furnaces," 1st ASM/Europe Heat Treatment Conference, Amsterdam, May 22-24, 1991

3.   J. Friberg, J. M. Mergoux, "Ventilateurs, Soufflantes, Compresseurs," Techniques de l'Ingenieur B 4210a.

Continuously Variable Agitation in Quench System Design

S.W. Han and S.G. Yun
Sam Won Industrial Corp.
Ahson, Korea

G.E. Totten
Union Carbide Chemicals and Plastics Co. Inc.
Tarrytown, New York

ABSTRACT

The quench severity of all quenchants including air, water, oil, and aqueous polymers is dependent on agitation. In general, quench severity increases with increasing agitation. Optimal hardness is typically achieved by increasing agitation during the initial stages of the quenching process. Conversely, distortion and cracking is typically reduced by minimizing the agitation rates as the martensitic transformation temperature is approached. This technology can now be readily applied in either batch or continuous systems with the recently developed "immersion time quenching system". The application of this system in the production of AISI 1043 and 4140H crankshafts using a polymer quenchant will be discussed.

INTRODUCTION

The performance of any quench medium is critically dependent on its heat transfer (removal) characteristics at the metal surface throughout the quenching process. A number of factors can impact on heat transfer during quenching including quenchant temperature and agitation (both linear flow and turbulence), additives, etc. Of these, only agitation can be varied within the time frame of the quenchant process.

The remainder of this paper will discuss:

- The effect of agitation on quench severity.

- Describe the concept of the **Immersion Time Quench System** (ITQS).

- Describe the use of cooling curve analysis to quantify quench severity variation.

- Illustrate the application of an ITQS process in the production of polymer quenched crankshafts.

Although the focus of this paper will be on the use of polymer quenchants, the same fundamental properties described here also apply equally for other vaporizable liquid quenchant media, e.g., oil, water, brine, etc.

DISCUSSION

Effect of Agitation on Quench Severity - The classic method of quantifying quench severity is with the Grossman H-factor. The relative impact of polymer quenchant concentration, bath temperature and agitation on a specific poly(alkylene glycol) - PAG quenchant is shown in Figure 1 (1). Depending on the particular selection of quenching parameters, it is possible to produce a wide range of quench severities ranging from an H-factor of approximately 0.2 (slow oil) to >1.0 (still water). However, although all of these parameters may produce significant variations in quench severity, only agitation rates can be effectively varied in a controllable manner throughout the quenching process.

In order to understand the impact of agitation on the severity of a particular quenchant, it is first necessary to understand the mechanism of the quenchant process itself. To facilitate this understanding, the mechanism of the quenching process for a poly(alkylene glycol) quenchant will be examined (Figure 2). Figure 2 shows that there are at least three different cooling regimes that occur when the hot metal is immersed into the quenchant solution (2). Each of these regimes is associated with a uniquely different heat transfer process.

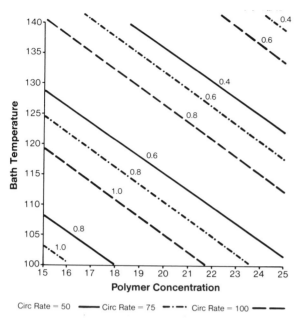

Fig. 1 - Grossman Hardenability Factors for Poly(Alkylene Glycol) Quenchants

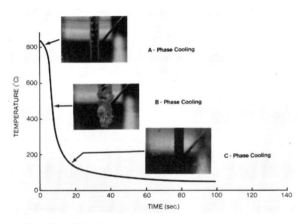

Fig. 2 - Film Rupture During Quenching of a Poly(Alkylene Glycol) Quenchant

A-phase (vapor blanket) cooling results when the hot steel is initially immersed into the quenchant. The stability of the vapor blanket surrounding the metal part may be reduced by increasing the quenchant agitation which will reduce the cooling time in this region. (The thickness of the polymer film surrounding the hot surface is also controlled by the quenchant concentration. Cooling rates will decrease as the film thickness increases.) The duration of this cooling phase will dictate whether it is possible to harden a particular alloy. The ability of a particular quenchant medium to harden steel is inversely proportional to the duration of

A-phase cooling.

The second phase of cooling is designated as "B-phase" cooling and is characterized by nucleate boiling. Further increases in heat transfer rates can be achieved in this region by increasing the agitation rate.

C-phase cooling results when the quenchant temperature at the cooling metal interface is less than the boiling point of the liquid. Cooling occurs by convective and conductive processes. Increases in agitation will also cause a corresponding increase in cooling rates in this region.

It is generally desirable to maximize agitation during the initial stages of the quenching process in order to avoid pearlite formation and thus optimize the as-quenched hardness as shown in Figure 3. However, the martensite formation is accompanied with a volumetric expansion. In order to reduce the accompanying stresses, it is generally recognized that the cooling rates should be minimized in the region of the M_s temperature.

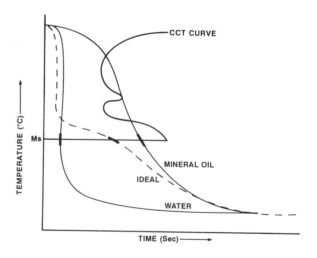

Fig. 3 - "Ideal" Cooling Curve

From the discussion thus far, it is now obvious that quench severity of a polymer quenchant is controlled by the polymer concentration (film thickness) and agitation (film durability).

Another critically important role of quenchant agitation is to provide a **uniform** film around the metal during the quenching process (3).

If insufficient agitation is used during the quench, it is possible that a polymer film with variable thickness or even discontinuous film formation around the total surface of the metal may be encountered. This may produce substantial variation of localized heat transfer rates which may result in thermal gradients across the surface and from the surface to the core. If the thermal gradients are sufficiently high, increased cracking or distortion may result. However, "uniform" agitation will produce substantial reduction in the formation of these thermal gradients thus producing a corresponding reduction in cracking and distortion.

Immersion Time Quench System (ITQS) - A batch and a continuous quench system has been developed which permits the variation of agitation, and thus, quench severity sequentially during the quenching process. The agitation for the batch quench system shown in Figure 4 is provided by two continuously variable impeller stirrers with directional draft-tubes or by pump agitation. The linear flow rates in the quench zone were quantified at various motor frequencies using a Mead Velocimeter. The calibration data is summarized in Table 1.

Table 1
Flow Calibration for Batch
ITQS System

Motor Frequency (Hz)	Flow Rate (m/s)
10	0.03
20	0.25
30	0.4
40	0.5
50	0.6
60	0.7

Two variations of the batch system that have been used thus far are:

• Timed variable speed propeller agitation.

• Timed variable speed pump agitation.

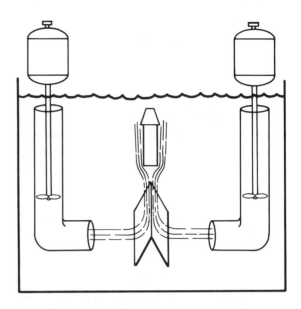

Fig. 4 - Batch Immersion Time Quench System

A continuous quench system has also been developed and is illustrated in Figures 5a and 5b. The features of this system are:

• The upper conveyer is used during A- and B-phase cooling.

• Cooling rates are controlled on the upper conveyor by both agitation and quench time (conveyor speed).

• The lower conveyor is used for C-phase cooling

Accurate determination of the cooling time for the initial cooling stage (AR I) is critical to assure the success of the system. Currently, this is done by one of two methods. One method is to calculate the required H - factor to obtain the desired amount of through - hardening and then quantify the AR I cooling time from a graph correlating H - Factor, workpiece diameter and cooling rate. The data obtained by this method is summarized in Table 2. Once calculated, Table 2 is then consulted to determine the required amount of time for AR I. For example, the cross-section size is 1 inch (25.4 mm), and the desired H - factor is 0.8, then the AR I cooling time is 14 seconds. Experience has shown that the best flow rate in this region is approximately 1.5 m/s.

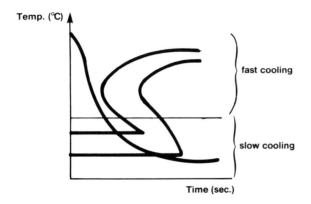

Fig. 5a - Correlation of Cooling Curve and CCT Curves for Alloy of Interest

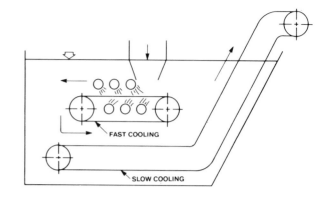

Fig. 5b - Continuous Immersion Time Quench System

Table 2
Grossman H-Factors as a Function of Stage 1 Cooling Times and Cross-Section Size

Stage 1 Cooling Time (s)

Dia.	H - factor (H)			
mm	0.8	1.0	1.5	2.0
15	5 ± 1	4 ± 1	--	--
20	9 ± 2	7 ± 1	--	--
25	14 ± 3	10 ± 2	8 ± 2	--
30	17 ± 3	13 ± 2	10 ± 2	--
35	--	16 ± 3	12 ± 3	10 ± 3
40	--	18 ± 3	15 ± 3	12 ± 3
45	--	22 ± 4	19 ± 4	17 ± 4
50	--	26 ± 5	23 ± 5	20 ± 5

The second method is to determine the critical cooling rate to avoid undesired transformations and temperature where this occurs from a compilation of CCT diagrams. Thus, it will be necessary to quench the part at a rate equal to, or greater than, the critical value if the maximum amount of martensite is to be formed. If a flow rate sufficient to minimize A - phase cooling is used, the cooling time is then calculated as follows:

$$AR\ I = \frac{(T_A - T_{CR})}{CR}$$

where: T_A = austenitizing temperature, T_{CR} = temperature where the critical cooling rate of the quenching medium occurs and CR is the critical cooling rate which must be equal to, or

greater than, the critical cooling rate of the process. The cooling rate for the quenchant at AR I can be determined by cooling curve analysis for the cross-section size of interest.

For the continuous ITQS system, the AR I cooling time is controlled by varying the speed of the upper conveyor using a timer.

Cooling Curve Analysis - The impact of varying the agitation rate and the exposure time during the initial cooling phase and the cooling rate during the second phase of cooling on cooling rates was measured by cooling curve analysis. A 25.4 x 100 mm cylindrical stainless steel probe was used to measure the cooling rates in the quench chamber. An IVF Quenchotest instrument was used for data acquisition. A variety of quench conditions were selected for this work. These are summarized in Table 3.

Table 3
Selection of ITQS Quenching Conditions

Exp. No.	AR I (m/s)	AR I time (sec)	AR II (m/s)
1	0	--	0
2	1.5	2	0
3	1.5	4	0
4	1.5	6	0
5	1.5	8	0
6	1.5	--	1.5

Some illustrative cooling curve data is summarized in Table 4.

Table 4
Summary of Cooling Curve Properties
for ITQS Experiments

Exp. (No)	Max. Rate (°C/s)	Temp. @ Max. Rate (°C)	Rate @ (°C/s)
1	40	596	11.0
2	54	699	10.0
3	53	715	11.5
4	53	699	12.0
5	53	709	12.5
6	53	714	16.5

The effect of flow rate on A-phase cooling times was studied. The data in Table 5 shows that the duration of A-phase cooling times are inversely proportional to the flow rate.

Table 5
Effect of Flow Rate on
A-Phase Cooling Times[a,b]

Flow Rate (m/s)	A-Phase Cooling Time (s)
0	16
0.2	14
0.4	10
0.6	5
0.8	2
1.0	0

a. These times were determined in the quench zone of the batch system shown in Figure 4.

b. The data were obtained with a 25.4 mm, Type 304 stainless steel probe using UCON® Quenchant E at 15% and 40°C.

Referring to Figure 3, optimal quenching is often obtained when the cooling rate during the initial stages of the quench is 0.5 - 2.0 m/s and when the cooling rate during the second stage

of the quench (M_s - M_f) is 0 - 0.3 m/s. The duration of the initial phase of cooling is dependent on the cross-section size of the parts as shown in Table 2.

This data shows that a wide variety of cooling profiles may be obtained by simply varying the agitation rates in stages 1 and 2 and by controlling the quenching time in stage 1. Therefore, the ITQS procedure can be used to optimize the cooling process required by numerous steel alloys using a single polymer quenchant.

Crankshaft Production by the ITQS Procedure

A test quenching program was conducted with 17.8 kg and 18.4 kg crankshafts constructed from AISI 1043 steel (Figure 6).

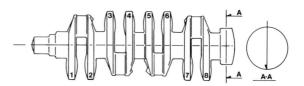

Fig. 6 - Crankshaft Used for Quenching Trials

The steel was austenitized at 850°C and then quenched and tempered. The tempering temperatures were 580°C and 610°C for the 17.8 kg and 18.4 kg crankshafts respectively. The required Brinell hardness is 93.8 - 105.5.

The hardness and distortion results are summarized in Table 6 and Figure 7 respectively. This data shows that the use of a polymer quenchant in conjunction with the ITQS procedure will produce hardness and distortion results superior to those achievable with a conventional quench oil.

A similar test quench program was conducted with 23.4 kg crankshafts constructed from AISI 4140H steel. The crankshafts were austenitized at 850°C, quenched and then tempered at 630°C. The required Brinell hardness is 105.5 - 109.

The hardness and distortion data obtained for these crankshafts quenched under various conditions according to the ITQS procedure are summarized in Figure 8 and Table 7 respectively. This data shows that quench conditions can be selected which will produce results significantly superior to a conventional quench oil.

CONCLUSIONS

It has been shown that:

- The performance of a polymer quenchant (like all quenchants) is agitation rate dependent.

- Quench severity variation in a quench tank can be modeled by cooling curve analysis.

- The use of a continuously variable procedure such as ITQS can substantially expand the flexibility of a polymer quenchant.

- Test quench results obtained with AISI 1043 and 4140H crankshafts showed that significantly more uniform quenching results are attainable with a polymer quenchant using the ITQS procedure than was obtainable previously with a conventional quench oil.

REFERENCES

1. Totten, G.E., Dakins, M.E., and Jarvis, L.M., *Heat Treating*, Dec. 1989, p 28 - 29.

2. Bates, C.E., Totten, G.E., and Brennan, R.L., *ASM Handbook - Vol. 4: Heat Treating*, ASM International 1991, p 67 - 120.

3. R.K. Zhelokhovtseva, *Steel in the USSR*, Vol. 15, 1985, p 238 - 239.

Figure 7
RESULTS: AISI 1043[1,2]

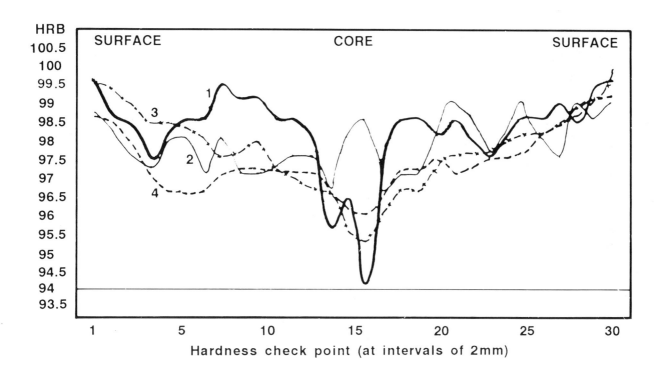

Hardness check point (at intervals of 2mm)

Table 6

Test Number	Conc. (%)	Temp. (C)	Agitation	Bending (mm)
——1	Super Quench	70	0.4 m/sec-Full-0.4 m/sec	0.6
——2	UCON QUENCHANT E 10	43	0.73 m/sec-110 sec-0.12 m/sec	0.95
—·—3	UCON QUENCHANT E 15.75	30	0.73 m/sec-20 sec-0.12 m/sec	0.5
–x—x–4	UCON QUENCHANT E 15	44	0.55 m/sec-20 sec-0.12 m/sec	0.5

1. Total Quenching Time is 4 Minutes
2. Total Bending Distortion Limit is 1.2 mm

Figure 8
RESULTS: 4140H[1,2]

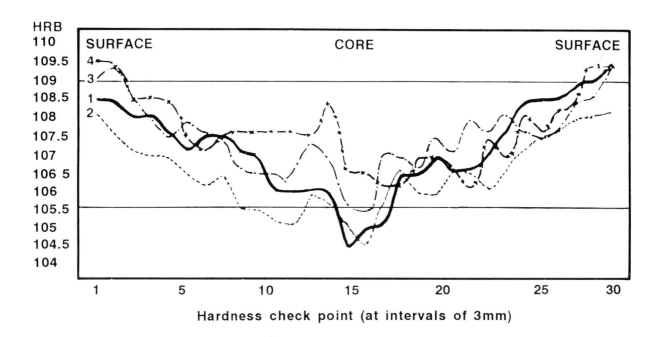

Hardness check point (at intervals of 3mm)

Table 7

Test Number		Conc. (%)	Temp. (C)	Agitation	Bending (mm)
———	1	Super Quench 70	70	0.4 m/sec-Full-0.4m/sec	0.96
············	2	UCON QUENCHANT E 20	45	0.55 m/sec-20 sec-0m/sec	0.4
—·—·—	3	UCON QUENCHANT E 10	38	0.73 m/sec-60 sec-0.12 m/sec	0.6
—x—x—x—x	4	UCON QUENCHANT E 7	39	0.73 m/sec-90 sec-0.12 m/sec	1.0

1. Total Quenching Time is 4 Minutes
2. Total Bending Distortion Limit is 1.2 mm

Proceedings of the First International Conference on Quenching & Control of Distortion, Chicago, Illinois, USA, 22-25 September 1992

HF Pulse Hardening in the Millisecond Range

G. Plöger
Hagenuk/Impulsphysik GmbH
Hamburg, Germany

HF Pulse Hardening in the
MilI isecond Range

BURGEONING DEVELOPMENT OF rapid methods of heat treatment is making the technologies for partial hardening of surfaces increasingly important.

Rapid hardening includes heat treatments that transform a partially austenitised volume into martensite by exploiting the self-quenching effect of the surrounding unheated volume. To achieve this the workpiece must be heated to the austenitising temperature sufficiently faster than it conducts heat in order to produce a steep enough temperature gradient. The heating rates required are achieved with high-energy heat sources, eg with electron or laser beams or inductive electric heating. The rapidly heated volumes are necessarily small compared with the total volume of the workpiece. The peak temperatures lie higher in the τ range than those involved with traditional methods of heat treatment.

The possibilities of rapid hardening are therefore determined by the characteristics of the heat sources mentioned, by the austenitising sequence with considerable superheating and very short dwell time in the temperature range above A_{c3}, and by the martensitic transformation under self-quenching conditions.

Basic principles of inductive heating

All materials that conduct an electric current can be heated inductively. HF induction heating works on the principle of a high-frequency alternating current flowing through an operating coil (inductor) with one or more turns. This inductor sets up an alternating electromagnetic field in its neighbourhood. If an electrically conductive workpiece is then brought into this alternating magnetic field, a voltage is induced in it that produces an alternating current. This eddy current (Fig. 1) causes the heating of the workpiece in the coil according to Joule's Law. The amount of heat produced $Q = I^2 \times R \times T$.

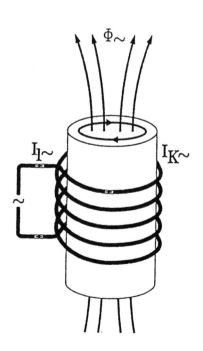

Fig. 1: Eddy current induced in a coil

According to Lenz's Rule the eddy current is in the opposite direction to the coil current producing the field. Like the coil current, the eddy current in turn sets up an alternating magnetic field, which is always opposed to the coil field. The two fields partially cancel each other inside the workpiece, so that a significant current only flows at its surface (skin effect). The depth at which the current density has dropped to 1/e is called the depth of penetration δ:

$$\delta = \frac{1}{\sqrt{\pi}} \sqrt{\frac{\rho}{\mu f}}$$

ρ = specific electrical resistance
μ = magnetic permeability
f = frequency

About 86% of the total energy induced is converted into heat in the skin layer characterised by the depth of penetration δ (Figure 2).

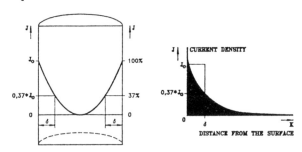

Fig. 2: Distribution of current density at the surface of workpiece (skin effect)

This depth decreases with increasing frequency and permeability and is proportional to the specific electrical resistance. Since this resistance and the permeability change markedly with increasing temperature, the depth of penetration is correspondingly temperature-dependent.

According to Joule's Law a temperature distribution proportional to the current distribution is set up initially in the inductively heated workpiece. However this temperature distribution changes with time in such a way that
- the heat flows away into lower temperature zones
- the heat is radiated from the surface into the surroundings
- the heat at the surface is carried away by gases or liquids flowing past.

The temperature pattern in the workpiece can therefore only be regarded as a snapshot, and is affected by the heating level and time as well as the frequency.

HF induction hardening

As a result of the heat generated in the material, the inductive skin heating characteristic is not as steep for the same energy input per unit time as that for indirect heating by means of radiation. This allows larger amounts of energy to be input without the material melting from the outside. HF pulse hardening employs a power density of up to about 300 W/mm², which leads to rapid heat treatment in the millisecond range. The high-energy pulses are produced by high-frequency induction at 27.12 MHz.
Another important criterion aside from the energy density and austenitising time is the energy doubling. This is particularly favourable in the case of HF induction hardening, despite the high frequency essential. The magnetic saturation produced by it causes what is termed a primary depth of thermal influence, which is independent of the thermal conductivity of the steel. The depth of penetration into the steel parts to be hardened can therefore not only lie in the range of a few μm, as given by the conventional formulae, but up to several millimeters.

The purpose of the inductor is to transmit the electric energy provided by the energy source to the workpiece in such a way that a heat treatment result as specified on drawings can be achieved. The design of the inductor depends on the shape of the heating zone required, and is therefore decisively influenced by the form of the workpiece. This gap between the inductor and the workpiece is only 0.3 to 0.5 mm, and the ionised atmosphere arising is blown away with a gentle flow of air to prevent sparkover.

The electric impedance of an inductor for pulse hardening is very low - of the order of several ohms - as a result of its relatively small dimensions. With the aid of a variable vacuum trimming capacitor the inductor (output circuit) is brought into agreement with the generator to achieve an optimum match with the workpiece (Figure 3).

The generator itself has a very high-Q resonant circuit and is excited by coupling a very-high-emission low-impedance oscillator tube. The oscillator tube receives rectangular pulses of power on the anode side. The pulses have to be this shape to allow the cooling to start spontaneously after the heating cycle has ended. This ensures a very short cooling time of 10^4 to 10^6 K/s, and the structure transforms to martensite.

As a result of the very high operating frequency the pulse hardening method is only to be used for small surface segments per pulse. The HF pulse hardening method is therefore sometimes only appropriate for small workpieces.

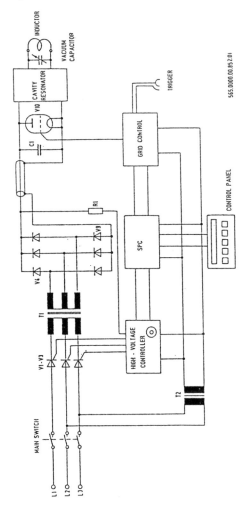

Fig. 3: Circuit diagram of an HF pulse hardening machine

Rapid austenitising

As a result of the non-diffusive transformation the most important characteristics of the martensite are determined by the state of the austenite. These include the distribution of the alloying elements, the content of the carbon determining the properties and the grain size and distribution of the austenite. The state of the austenite and hence the heat treatment conditions are accordingly significant factors influencing the subsequent martensite.

By contrast with the diffusion-controlled processes of austenitising and tempering, the non-diffusive martensitic transformation is faster than the cooling rates achievable with inductive skin hardening, which employs liquid quenching agents.

Thus the martensitic transformation for a Ck 50 takes about 70 ms, corresponding to a cooling rate of > 5×10^3 K/s, if there is a difference of 200°C between the beginning M_s and end M_f of the martensite formation. With the self-quenching of HF pulse hardening cooling rates > 5×10^4 K/s are achieved, although the cooling rates required for the martensitic transformation are of the order of about 4×10^3 K/s.

Partial supercritical quenching, which is also possible, of a component that is austenitic over the entire cross section basically depends on the hardenability of the material and the cooling conditions, and does not have any affects on the behaviour of the material peculiar to skin hardening.

With slow (conventional) heating and slight superheating, ie to austenitising temperatures about 20 to 30 K above A_{c3}, the transportation of carbon by τ-solid solution determines the rate. The process is controlled by diffusion and involves nucleation and nuclear growth at the interface between the carbide and the α-solid solution (Figure 4).

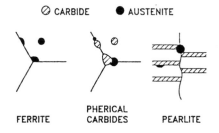

Fig. 4: Austenite nucleation

With increasing heating rates the transformation temperatures are pushed higher. The diffusion-controlled process is retained. The initial structure has an extraordinary affect on the position of the critical points that is dependent of the heating rate. Particularly with a martensitic initial structure; the formation of austenite follows a different mechanism to that with a ferritic carbide structure, since heating cannot be taken beyond the beginning of the transformation in this case. The end of the transformation also lies at lower temperatures than with a normalised structure.

At higher austenitising temperatures, ie with increasing superheating, the transformation takes place considerably faster. Account must also be taken of which of the steps of the phase transformation - nucleation, carbon transport, dissolving of carbide - affect the rate. According to findings to date austenite nuclei arise inside at the interfaces between solid solution and carbide. The incubation time depends on the superheating and with electric heating is under 1 ms, provided at least 100°C of superheating is used.

This means that with superheating to a sufficiently high temperature the nucleation phase is short compared with the transformation time, especially since the incubation time is very short and all carbides undergo nucleation and therefore release carbon. The nucleation phase does not determine the rate involved with electric heating.

Current findings indicate that the transportation of carbon does determine the rate, ie austenitising temperature, concentration gradient and initial structure, in the case of plain carbon steels. Heterogeneous initial structures require longer heat treatment than homogeneous ones. The dissolving of carbide however has no effect on this.

The dissolving of carbide dominates with alloyed steels. Overall it influences the rate of the α-τ transformation much more markedly than the carbon transport.

Martensitic transformation

The martensitic transformations have a series of characteristic features that clearly distinguish them from the transformations controlled by diffusion.
- They take place athermically at low temperatures, ie they start suddenly during cooling when a certain temperature is reached,

and cascade in fractions of a second as cooling continues.
- If cooling is stopped, further transformation also ceases, even though equilibrium has not yet been reached.

The martensitic properties are determined by the structure. Rather than changing the concentration of the alloying atoms, the martensitic transformation alters the lattice structure by coordinated movements of atoms. The distinguishing feature of the transformation to martensite is the tetragonal distortion of the lattice, which is reflected in the hardness (Figure 5).

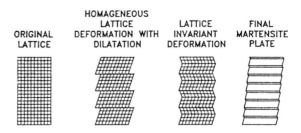

Fig. 5: Schematic of the formation of plate martensite by sliding or twinning

The hardness of the martensite increases with the amount of carbon dissolved. The hardenability depends solely on the carbon content, whose useful range should lie between 0.4 and 0.7%. The potential hardening depth, ie the zone of the material that still transforms martensitically under given quenching conditions, is small in the case of plain carbon steels. Alloying elements improve the potential hardening depth to allow larger cross sections to be transformed martensitically. Heat-treated base structures are to be preferred, as the carbon is homogeneously distributed compared with normalised or annealed structures.

The fact that pulse-hardened martensite is generally observed to be harder than martensite produced by conventional heat treatment is easily explained. The pronounced self-quenching effect suppresses the single-phase segregation of martensite during cooling (self-tempering effect), so that in the martensitically transformed skin marked internal compressive stresses are set up, which change into tensile internal stresses in the underlying tempered zone with a hardened initial structure, or in the untransformed soft structure. The hard structure produced by pulse hardening is extraordinarily finely acicular. The hardened skin is called "white layers" due to its appearance under an optical microscope (Fig.6)

Fig. 6:Hardened teeth of band saw
Although etching is sometimes made more difficult, a structure can always be developed even here with normal etching agents. In the region of the transition to the base material the structure is incompletely dissolved, as austenitising is only partial there.

In hypereutectoid steels, and particularly with a high chromium content, the carbides are sometimes only incompletely dissolved right up to the surface. The hard layer of these steels may also typically have high residual austenite contents as well as martensite. Surprisingly, this residual austenite content has no effect on the hardness profile. Many papers have shown that residual austenite in the skin layer of rapidly hardened steels may improve their mechanical and technological properties. However this requires unstable residual austenite that converts into martensite under the mechanical stresses.

The grain boundaries are generally where corrosion attacks. The very fine structure of the pulse-hardened material significantly increases the corrosion resistance as well as the hardness. Moreover with the HF pulse method the hardened zones of the structure and constituents are so evenly distributed that heterogeneities in the surface of the material very largely do not arise. Selective corrosion in the heat-treated zone is suppressed extremely effectively.

Application

The use of HF pulse hardening should be evaluated according to technical, technological and above all commercial criteria. The operating conditions can also be decisive. In many cases the method can more than rival laser and electron beam hardening and - particularly with "problem parts" - other conventional hardening techniques as well, such as flame hardening for example.

Pulse hardening is very suitable for automation and has advantages for high-volume production. As with other plant and machinery the most economic solutions are achieved with customised technology. Its maximum commercial benefit is often obtained through cutting finishing time, eliminating technological processes and reducing energy costs compared with conventional technology.

The power consumption with HF pulse hardening is extraordinarily low and an average output of 2 to 6 kW is sufficient for homogeneous hardening.

The method offers the following particular technological advantages:
- precise positional and chronological control of the energy input with high reproduceability
- profile of the hardness zone and hardening depth can be accurately defined
- low thermal stresses in the workpiece avoid distortion of the component
- longer service lives as a result of high hardness and abrasion resistance
- self-quenching, ie no additional coolants required
- high resistance of the hardened zones to corrosion caused by aggressive vapours and chemicals
- unaffected by mains fluctuations, stabilised pulse energies
- high energy efficiency

Pulse-hardened products find a wide range of applications where corrosive media arise. The very finely acicular hardened structure turns out to be very resistant to corrosion and is only attacked very slowly by dilute acids.

The method is used along with others for hardening saws, circular blades and other cutting and punching tools, precision engineering components in the textile and electrical industries for instance, and sections of surfaces for teeth (Figures 7a-d). HF pulse hardening is particularly suitable for full automation, and, once the workpiece guide has been positioned, can be used for continuous belts or rows of items.

Fig. 7a: Samples of automobile industry

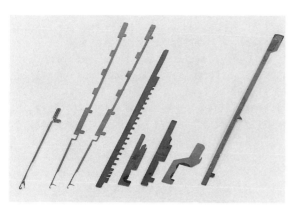

Fig. 7b: Samples of textile industry

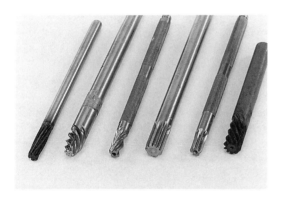

Fig. 7c: Samples of electrical industry

Fig. 7d: Samples of paper processing industry

132

Benchmark Testing of Computer Programs for Determination of Hardening Performance

J. Bodin and S. Segerberg
Swedish Institute of Production Engineering Research
Göteborg, Sweden

ABSTRACT

Computer models for simulation of heat treatment processes have been developed at several places around the world. With them, the result of heat treatment can be predicted, within certain limits, for various combinations of process parameters. The objective of this study has been to develop a procedure for evaluation of commercial computer models for heat treatment of steel components. The basis is a number of test heat treatments carried out on test pieces of different sizes, made of a number of steels with defined chemical compositions. Hardness has been determined after quenching in different quenching media.

Three computer programs intended for personal computers have been tested against each other and against the results from the test heat treatments. It is shown that the programs produce results which sometimes differ quite considerably from each other and from the experimental results. Obviously, the method of modelling the quenching process has been of great importance for the calculated results. In addition, the method of modelling phase transformation seems to have influenced the results.

GROWING MARKET REQUIREMENTS in respect of lead times, capital tied up in inventories, quality and costs put pressure on heat treatment companies to be able to deal with smaller batch sizes without impairing quality or increasing costs. To meet these requirements, it is essential that the company can predict the result of heat treatment as accurately as possible without having to make more or less extensive test runs. Today, there is an increasing number of computer models available on the market for prediction of the results of various heat treatment processes. Depending on the size and sophistication of the models, various properties such as hardness, micro-structural composition, distortion and residual stresses can be predicted.

Since computers are used to an increasing extent, even in small companies, it is essential for the presumptive buyer of programs to be able to judge beforehand which model gives the best value for money for his needs. The predicted results will depend partly on the quality and sophistication of the computer model of the process, partly on the quality of the input data presented to the program. The input data comprises data relating to the material, boundary conditions relating to the process and shape and size of the component being treated. The objective of this work has been to develop a procedure for systematic evaluation of computer models for heat treatment of steel components.

Benchmark testing procedure

The procedure of evaluating the computer programs is illustrated in Figure 1. It must be stressed that one has to be very careful in in determining all initial data and process data used in the experimental testing, since the test results will form the basis for evaluation and judgement of the calculated results. Some of these data are also used as input data for the calculations and, thus, there must be no question whatsoever about the accuracy of these data.

To obtain a thorough measure of the computer programs, the evaluation should be carried out under conditions that, in respecr of the hardenability of the steels used, test piece dimensions and quenching intensity, produce critical hardening conditions, i.e. the variation in hardness from the surface to the core should be as large as possible. This implies that the requirements for accuracy in determining the material and process data will be extremely high.

Hardness distribution and microstructural phase composition can be calculated by the programs that we have tested here. More advanced programs can also predict distortion and residual stresses.

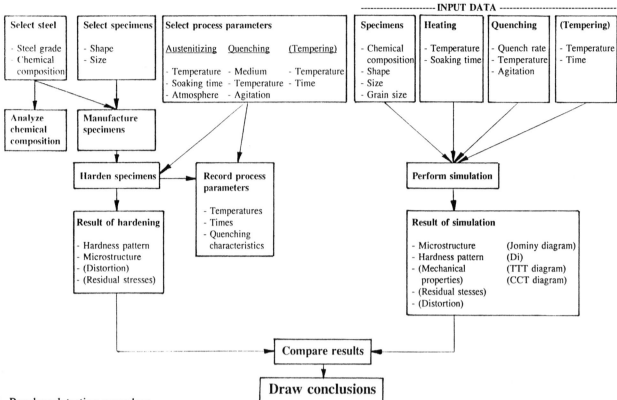

Fig. 1. Benchmark testing procedure.

Experimental procedure

Test heat treatments. The basis for the evaluation was a number of test heat treatments carried out on cylindrical specimens of different sizes, made of three steels, each with two chemical compositions representing low (L) and high (H) hardenability, see Table 1.

The diameters of the specimens were 10, 16 and 25 mm for the SS 1672 steels, 16, 25 and 44 mm for the SS 2225 steels and 25 and 50 mm for the SS 2258 steels. The length of the specimens has been three times the diameter.

The specimens were heated in a protective atmosphere to 860°C and held for 10 minutes.

Quenching was performed in two hardening oils (A and B) and in water at two temperatures. Quenchants were unagitated except for oil A, where agitation (0.3 m/s) was also used. Cooling curves were determined with the method defined in ISO/DIS Draft 9950 using a 12.5 dia.x50 mm cylindrical probe made of Inconel 600 [1], see Figures 2 and 3. The heat transfer coefficient versus time characteristic was calculated from the cooling curve using the finite difference method.

The hardness patterns of the quenched specimens were determined in a mid-section using the Vickers method with 1 kg load.

Table 1. Chemical composition (wt-%) of the steels used in the test.

Steel	C	Si	Mn	P	S	Cr	Ni	Mo	Cu	V	Al	W
SS 1672-L	0.43	0.24	0.64	0.013	0.029	0.172	0.203	0.026	0.291	0.002	0.006	0
SS 1672-H	0.49	0.23	0.72	0.017	0.030	0.100	0.076	0.012	0.128	0.004	0.004	0
SS 2225-L	0.22	0.34	0.68	0.008	0.012	1.07	0.207	0.162	0.173	0.005	0.027	0.01
SS 2225-H	0.36	0.27	0.80	0.027	0.023	1.15	0.162	0.196	0.219	0.009	0.034	0.01
SS 2258-L	1.02	0.24	0.36	0.008	0.015	1.49	0.233	0.035	0.209	0.003	0.031	0.01
SS 2258-H	0.97	0.25	0.26	0.009	0.011	1.80	0.12	0.20	0.139	0.008	0.032	0.01

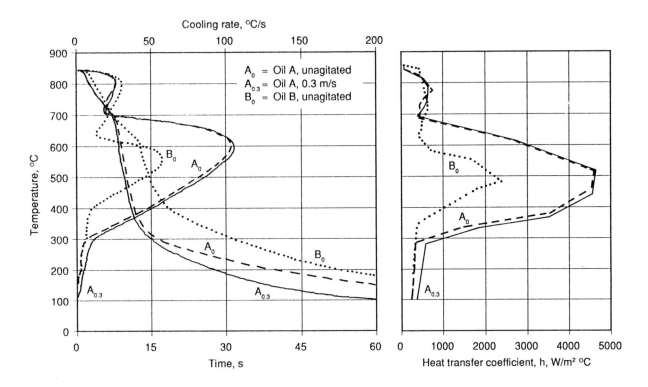

Fig. 2. Cooling curves and heat transfer coefficients for the oils used.

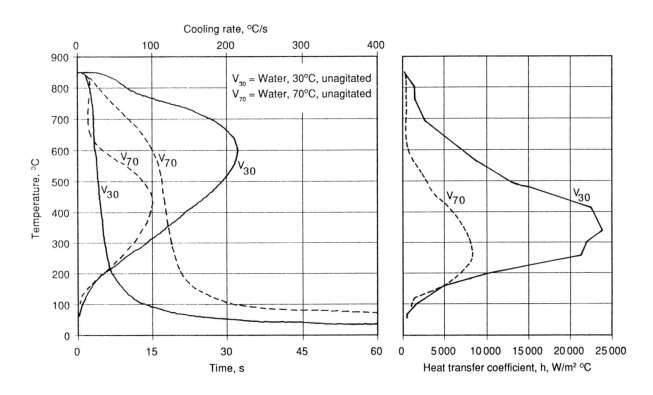

Fig. 3. Cooling curves and heat transfer coefficients for water at 30° and 70°C.

Computer models evaluated. Three commercial programs for personal computers, which are all in use in heat treating companies in Northern Europe, were selected. The relative market price range is approximately 1:5:15, from the cheapest (SteCal) to the most expensive (AC3) program. The fundamental construction of the programs is as shown in Figure 4. A few characteristics of the programs, some of them critical for the calculations, are shown in Table 2.

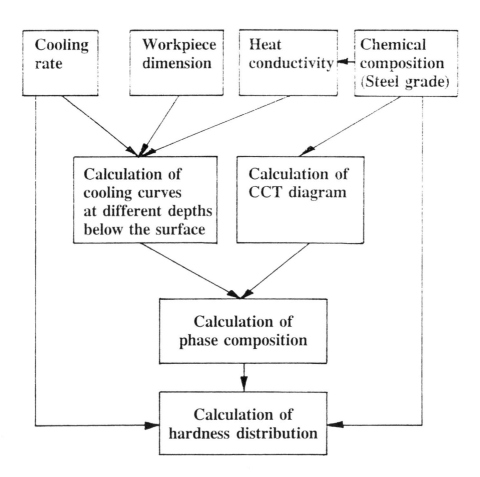

Fig. 4. Construction of the programs tested (schematic).

Table 2. Some characteristics of the computer programs used for hardening simulation.

Name of program	Austenitization		Quenching process			Calculated results		
	Temp.	Time	Medium	Temp.	Agitation	Hardness	Microstructure	Retained austenite
SteCal[1]	No	No	H value[4]	No	Indirectly	Yes	Yes	Yes
Database Hardenability[2] (DB H)	Yes	Yes	Water/Oil/Air[5]	No	No	Yes	Yes	No
AC3[3]	Yes	Yes	Heat tr. coeff.	Yes	Yes	Yes	Yes	Yes

[1] SteCal is a registered trade mark of Pascual Tarín-Remohí, licensed to ASM International. Version 1.1 (1989) was used.

[2] Database Hardenability is a trade mark of Dr. Sommer Werkstofftechnik GmbH. Version 2.1 (1992) was used.

[3] AC3 is a registered trade mark of Marathon Monitors, Inc. Version 1.2 (1989) was used.

[4] Grossmann's quench severity factor

[5] Only the type of quenchant can be defined, no further characteristics.

It should be added here that AC3 is the only one of these programs that is also capable of simulating hardening of carburised test-pieces. This function has not been tested here.

AC3 also offers two methods of calculating the cooling curves. The first (standard accuracy) is a classical solution assuming fixed quench media temperature, fixed heat transfer coefficient, fixed heat capacity and fixed thermal conductivity. The advantage is speed of calculation, while the disadvantage is some loss of accuracy. The second method (extended accuracy) is a finite element analysis allowing all appropriate variables to change during the quenching process. The advantage is higher accuracy, while the disadvantage is longer calculation times.

It was also the intention to test the second method. However, we were unable to find any combination of material, test piece dimensions and quenchant with which the program would operate without divergency when using the finite element method of calculation. Nor was it possible to change any of the input data, such as the time steps used in the finite element method, in order to be able to perform calculation with this accuracy.

Results and considerations

Four examples of calculated and measured hardness patterns in specimens are shown in the following figures. A full report of the evaluations is published in [2].

Influence of small and large difference in hardenability. Figure 5 shows that all programs separate the two heats of both steels to the same extent as in the measured specimens. However, for the SS 1672 steels, the absolute level was considerably overestimated by DB H and considerably underestimated by SteCal. For the SS 2225 steel, the estimates were better, but DB H underestimated the hardness level and SteCal overestimated it for the SS 2225-H steel. In both cases, AC3 makes fairly good estimates, except for the centres of the SS 1672 steels and the SS 2225-H steel.

Influence of small and large difference in quenching intensity. Figure 6 shows the result when the quenching intensity was varied. For the SS 2225 steel, the estimates are quite good. For the SS 1672 steel, the deviation is larger. It should be noted that DB H does not allow the intensity to be varied for an "oil" and that SteCal could not make any estimate for Oil B since the H value was below its lower limit 0.3 (here 0.18).

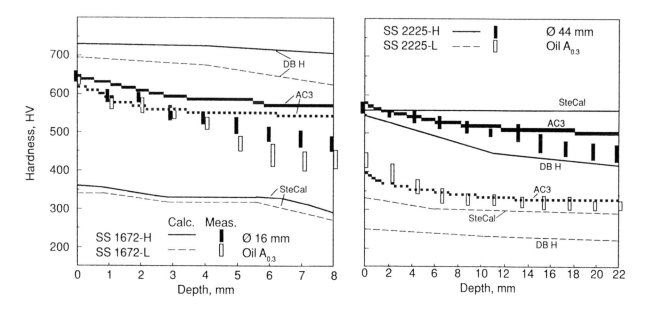

Fig. 5. Calculated and measured hardness patterns in specimens of SS 1672 and SS 2225 steels representing small (left) and large (right) difference in chemical composition.

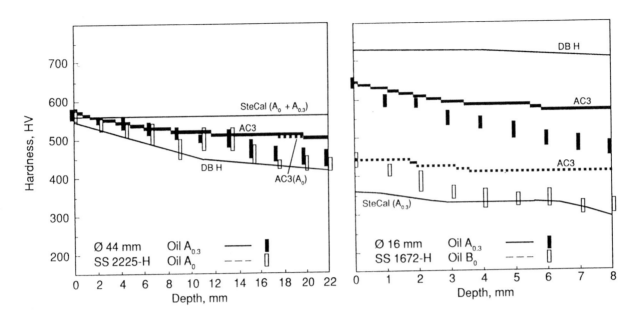

Fig. 6. Calculated and measured hardness patterns in specimens quenched in oils under conditions representing small (left) and large (right) variation in quenching intensity.

Influence of the water temperature in water quenching. An increase in water temperature from 30° to 70°C has a drastic influence on the quenching intensity as could be seen in Figure 3. Testing with SS 1672-H steel test pieces verified this large influence.

The difference in quenching intensity can be considered with both AC3 and SteCal. The H value was appr. 2.2 at 30°C and 0.7-0.8 at 70°C. DB H cannot separate these variants since no quenchant temperature can be specified. Figure 7 shows the calculated and measured hardness patterns.

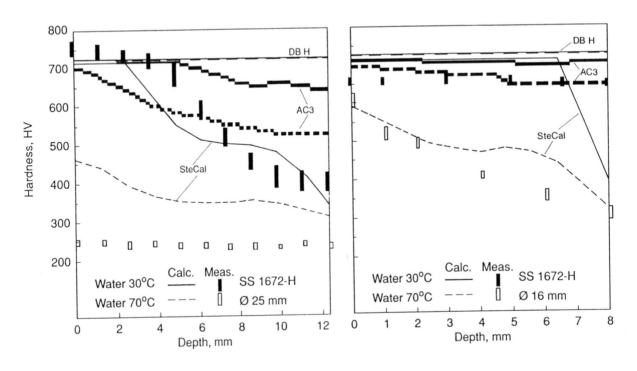

Fig. 7. Calculated and measured hardness patterns in specimens quenched in water at 30° and 70°C respectively.

Steels with high carbon content. SteCal and DB H accepts carbon contents in steel up to 0.65 and 0.60 % respectively. AC3, on the other hand, accepts carbon levels up to 2.00%. In order to evaluate this possibility, testing was done in the slow-hardening oil B with SS 2258 bearing steel with carbon contents about 1.0 %, see Table 1. The calculations showed that 25 and 50 mm dia. test pieces were through-hardened with a hardness level of about 860-880 HV, while experimental testing showed considerably lower hardness, from 675 to 540 HV, on 25 mm dia. test pieces of SS 2258-L steel and from 540 to 390 HV on 50 mm dia. test pieces of SS 2258-H steel, see Figure 8. Obviously, these kinds of steel present problems for AC3.

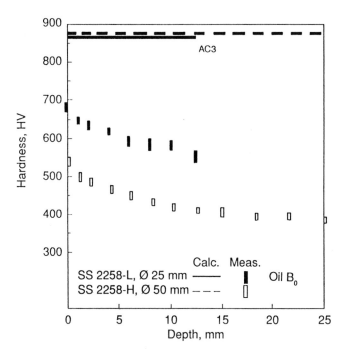

Figure 8. Calculated and measured hardness patterns in SS 2258 steel specimen quenched in a slow-cooling oil.

NB. It must be observed that the programs selected, and other similar programs, are usually under continuous development, and improved versions may be released. Therefore, it is <u>not</u> the purpose of this paper to stress those discrepancies in the evaluated models which have been found, but to point out the need and usefulness of common procedures and data from practice which facilitate objective judgement of computer models marketed to the heat treating industry. Also, it must be noted that only a limited part of the capacity of the programs has been evaluated here.

When considering the possibilities of predicting hardening performance one must also allow for difficulties in determining the quenching conditions in industrial hardening. It has been shown, for example by the authors, that the heat transfer may vary considerably over the surface of components being quenched in a batch or individually [3] and that

the surface condition (for example, oxidation) may also influence heat transfer considerably [4].

Conclusions

Computer models for prediction of hardening performance may be an important tool for the industry in developing understanding of the influence of various process parameters. However, it is essential that programs of interest for the potential purchaser be evaluated with regard to his experience of actual applications. One of the most important features to check, obviously, is the way quenching is defined in the programs considered. This feauture does not explain all the differences in the predicted results, however, so it is obvious that the model for phase transformation prediction has also influenced the results.

References

1. *Industrial Quenching Oils - Determination of Cooling Characteristics - Laboratory Test Method.* Draft international standard ISO/DIS 9950, International Organization for Standardization (submitted 1988)
2. J. Bodin, S. Segerberg, E. Troell, *IVF-skrift 92807*, IVF - The Swedish Institute of Production Engineering Research, Gothenburg, 1992
3. S. Segerberg and J. Bodin, in *Proc. 3rd Int. Seminar: Quenching and Carburising, Melbourne, 2-5 Sept. 1991*, Institute of Metals & Materials Australasia Ltd. Parkville Vic 3052, Australia
4. S. Segerberg, *IVF-resultat 88605*, Sveriges Mekanförbund, Stockholm, 1988

The Use of Glutaraldehyde for Microbiological Control in Polymer Quenchants

J.F. Kramer
Union Carbide Chemicals and Plastics Co. Inc.
Bound Brook, New Jersey

Abstract

Polymer quenchants from heat treating operations were found to contain Gram-positive and Gram-negative bacteria, including anaerobic sulfate-reducing bacteria. No fungi were found. The use of glutaraldehyde for the preservation of polymer quenchants is outlined. Laboratory studies indicated that glutaraldehyde effectively controlled bacteria in a polymer quenchant at low doses (50 ppm active ingredient) for greater than six weeks. Glutaraldehyde had no effect on the properties of a poly(alkylene glycol)-based quenchant and was compatible with nitrite-based additives. The efficacy of glutaraldehyde was confirmed in the field where a polymer quenchant system was treated with glutaraldehyde for an extended period of time. Microbial populations in this system were reduced to undetectable levels over the course of the trial.

UNCONTROLLED GROWTH OF MICROORGANISMS in a polymer quenchant system can lead to a variety of process and quality problems, resulting in downtime and loss of production. Microorganisms can enter a bath from a number sources such as the air, make-up water, floor sweepings, and human wastes. When unchecked, microbiological contamination can result in foul odors, staining of parts, and the formation of slime deposits and biofouling.

In addition to these outward signs, there are unseen consequences of significant importance. Biofouling of reverse osmosis membranes can reduce the performance of these systems and may result in damage to the membranes. Microbiological activity can potentially result in the degradation of the polymers used for quenching. When this happens, the performance of the quenchant can be affected. If microbiological activity begins to present a problem it can usually be controlled by the addition of an appropriate biocide.

It is important that care be taken to ensure that the organic moieties, e.g. polymer and additive systems, are chemically compatible with the biocide that is selected. For example, many effective biocides contain active chlorine compounds which will spontaneously react with poly(alkylene glycol) polymers producing vinyl and chloroethyl ether by-products. In addition, several biocides commonly used to control microorganisms in quenchant baths are amine-based products, which can react with nitrite corrosion inhibitors. The use of such biocides in quenchants containing nitrite corrosion inhibitors is not recommended, due to the possible formation of nitrosamines. Some nitrosamines have been shown to cause cancer in laboratory animals (1,2). Although there is no direct evidence firmly linking human cancer to nitrosamines, it seems unlikely that humans should be uniquely resistant to their action (3).

Chemical incompatibility, in addition to possibly degrading the quenchant or resulting in the formation of hazardous compounds, will also affect the performance of the biocide. A biocide which has reacted with a quenchant cannot exert any control over microorganisms. Because of these issues, an extensive program was undertaken to examine the chemical compatibility and antimicrobial efficacy of glutaraldehyde in both the laboratory and in the field.

The properties of glutaraldehyde and its efficacy against a wide variety of microorganisms have been well documented for many years (5). Glutaraldehyde has been used for over twenty years as a commercial biocide. Its antimicrobial activity was first documented in 1957, in a patent concerning control of sulfate reducing bacteria in water (4). Its efficacy against microorganisms of medical interest led to its use as a sterilant for surgical instruments which could not tolerate the more traditional approach of steam treatment. In more recent years, glutaraldehyde has been used for the control of microorganisms in such diverse environments as oil fields, cooling towers, farm animal housing, and metalworking fluids.

Materials and Methods

Analysis of Glutaraldehyde. Glutaraldehyde concentrations were determined on a standard gas chromatograph equipped with a flame ionization detector and integrating recorder. Separations were carried out isothermally at 200°C on a 4 ft x 4 mm i.d. glass column packed with silanized poly(divinylbenzene)(6).

Cooling Curves. Cooling curves were measured using a laboratory quenching rig. An inconel 600 probe (0.5 x 4 inches) equipped with a thermocouple was heated to 844°C and then plunged into the quench bath. Temperature of the probe was monitored by a computer. The quench bath contained approximately 2.5 liters of quenchant. The quenchant was maintained at 40°C. Flow rate in the bath was 20 feet per minute.

Viscosity. Viscosity was measured with a Cannon-Fenske viscometer. Measurements were performed in triplicate at 100°F and the average time was used to calculate viscosity.

Identification of Bacteria. Bacteria were identified by isolating the colonies on brain heart infusion (BHI) agar. After an incubation period of 48 hours at 37°C the bacteria were stained by the Gram Stain Technique. The Gram-negative rods were then streaked onto MacConkey's agar. After a 24 hour incubation period at 37°C the bacteria were identified using the Minitek[tm] System (7). Gram positive bacteria were identified by their morphological characteristics.

Sulfate-reducing bacteria were identified by inoculating API RP38 (with nail) media. After an incubation period of up to 28 days at 37°C the presence of sulfate-reducing bacteria was indicated by a blackening of the media.

Identification of Fungi. Fungi were identified by isolating the colonies on potato dextrose (PD) agar. After an incubation period of 7 days at 25°C the fungi were identified by their morphological characteristics.

Laboratory Efficacy Studies. *Inocula.* The microorganisms used in this experiment originated in contaminated industrial systems. They were grown for several weeks in the same quenchant used in the following experiment. The culture was swirled on a gyrorotary shaker maintained at 40°C. During culture maintenance, 20 percent of the liquid was drained once per week and replaced with a fresh mixture of quenchant. Evaporative losses were made up by occasional additions of deionized water.

Microbiocidal Assays. The efficacy of glutaraldehyde in a commercial poly(alkylene glycol)-based quenchant was examined. The quenchant was diluted with ten parts of tap water. This mixture was transferred to a culture flask. Fifteen percent (w/w) aqueous glutaraldehyde was added to the flask in an amount calculated to give the desired glutaraldehyde concentration. The flask was capped with a stopper which allowed air to enter and was swirled on a gyrorotary shaker overnight at 40°C. To the mixture was then added 1.0 mL of inoculum and swirling was continued at 40°C. Deionized water was added as necessary during the course of the experiment to compensate for evaporation and maintain fluid volume at 100 mL. Once each week, the mixture was reinoculated with 1.0 mL of inoculum. Bacteria were assayed several times on the day after the initial inoculation and then at one week intervals. Each experiment was carried on for 6 weeks.

Microbiological Assays. Bacteria were counted by standard pour plating of serially diluted samples on BHI agar. Plates were incubated for 48 hours at 37°C.

Fungi were counted by standard pour plating of serially diluted samples on PD agar. Plates were incubated for 48 hours at 25°C.

Results and Discussion

Our initial studies concerned the chemical compatibility of glutaraldehyde with typical polymer quenchant formulations. In these studies, the biocide was added to solutions of commercial quenchants, and the lifetime of glutaraldehyde in these media was determined gas chromatographically.

A variety of commercially available polymer quenchants were randomly selected and tested by this method. The resulting data are compiled in Table 1, and are presented in terms of glutaraldehyde half-life in the quenchant. This half-life may be defined as the amount of time necessary for the glutaraldehyde concentration to fall from its initial value of approximately 1000 ppm to half of that value.

An examination of Table 1 reveals several general trends. First, most poly(alkylene glycol) quenchants show excellent chemical compatibility (i.e. long glutaraldehyde half-lives in Table 1), although one was poor in this respect. The other quenchant types all had poor compatibility with glutaraldehyde. Thus, while general trends are evident based on quenchant type, the compatibility of glutaraldehyde with any single quenchant cannot be predicted, but rather must be determined on an individual basis.

The nature of the ingredients which are responsible for causing reactivity with glutaraldehyde has been investigated in the past (8). The results indicated that the only ingredients which caused incompatibility were primary and to a lesser extent secondary amines. Incompatibility with these components was not entirely unexpected based on the biochemical mechanism of glutaraldehyde's cidal action. Since the biocide kills microorganisms by chemically reacting with the primary amines present on the cell surface, it is not surprising that it would also react with a similar chemical present in a mixture. Nonetheless, many commercial quenchants are formulated without primary amines and are quite compatible with glutaraldehyde.

Good chemical compatibility of glutaraldehyde with a

Table 1. Compatibility of commercial quenchants with glutaraldehyde.

Quenchant Type	No. of Fluids Tested	No. of Quenchants Yielding Glutaraldehyde Half-Life of:		
		0-24 Hours	1 Day-2 Weeks	> 2 Weeks
Poly(alkylene glycol)	7	1	0	6
Poly(ethyloxazoline)	1	1	0	0
Poly(ethyloxazoline)/poly (vinylpyrrolidone)	1	1	0	0
Polyacrylate	1	1	0	0

quenchant indicates that glutaraldehyde at some level can be expected to function in controlling microorganisms, but leaves unanswered the question of dosage. The following laboratory studies were designed to examine the actual efficacy of the biocide in these media. Because there have been numerous examples, both in the scientific literature and in practical experience, that the susceptibility of laboratory microorganisms to biocides can vary greatly from that of microorganisms found contaminating industrial systems it was though wise to use microorganisms which originated from fouled industrial systems in this study.

Several poly(alkylene glycol) quenchants from industrial systems experiencing microbiological problems were examined and the major microorganisms were identified. The results of the microbiological testing are given in Table 2.

Table 2. Microorganisms contaminating industrial quenchant systems.

Industrial System	Bacteria	Fungi

Ferrous	Gram Negative
	    Pseudomonas pikettii
	    Pseudomonas vesicularis
	    Sulfate-reducing bacteria
	
	Gram Positive
	    Staphylococcus sp.
		None

Nonferrous	Gram Negative
	    Pseudomonas sp.
	
	Gram Positive
	    Corynebacterium sp.
		None

In systems in which ferrous parts were treated, Gram-negative bacteria of the genus *Pseudomonas* predominated. These organisms are very versatile in their nutritional capabilities and as such are commonly found contaminating industrial products and systems. They are also common biofilm formers and contribute to slime formation. Pseudomonads were also found in nonferrous systems; however, Gram-positive bacteria of the genus *Corynebacterium* predominated in these systems. Sulfate-reducing bacteria were found only in ferrous systems. Since these organisms have an absolute nutritional requirement for relatively high concentrations of iron, they may be more common in ferrous systems. Sulfate-reducing bacteria produce hydrogen sulfide which is corrosive and causes foul odors. Ferrous systems contained low numbers of Staphylococci. Of concern is the fact that some species of

Staphylococcus and *Corynebacterium* are human pathogens. Fungi were not found in either type of system. The pH and/or the temperature of the typical quenchant system may inhibit the growth of these organisms.

The efficacy of glutaraldehyde was evaluated in a poly(alkylene glycol) quenchant under laboratory conditions. Glutaraldehyde was added in the desired concentration to each quenchant sample. One day later the samples were "spoiled" by the addition of an inoculum which contained approximately 1.6×10^6 microorganisms/mL. Each sample was reinoculated once each week for the duration of the experiment. All experiments were conducted at 40°C. Adequate dissolved oxygen supplies were maintained by continuous swirling of the samples. The results of the microbial growth assays are given in Table 3.

Examination of these data reveal several interesting points. First, glutaraldehyde was rapidly cidal to the microorganisms found in quenchant systems. Viable microorganisms were reduced to undetectable levels within 1 hour with as little as 50 ppm glutaraldehyde. Furthermore, 50 ppm glutaraldehyde successfully inhibited microbial growth for greater than 6 weeks under these laboratory conditions. The results suggest that low levels of glutaraldehyde may provide rapid and prolonged control of microbial growth in quenchant systems.

Since we had shown that glutaraldehyde was effective in protecting a poly(alkylene glycol) quenchant from the unwanted growth of microorganisms, we thought it wise to investigate what effect the biocide would have on the quenchant. Cooling curves were run on a fresh poly(alkylene glycol) quenchant treated with different levels of glutaraldehyde. Time/temperature data was collected for three quenching cycles at each test concentration and the average plotted. The results are shown in Figure 1. The cooling curves for the samples treated with glutaraldehyde are very similar to the untreated control.

A plot of cooling rate versus temperature is shown in Figure 2. There was no significant change in the cooling rate of the glutaraldehyde treated quenchant relative to the untreated control. The results of the quench tests indicate that glutaraldehyde has no significant effect on the cooling properties of the quenchant even at concentrations far higher than would be used in actual systems.

Viscosity and pH of the glutaraldehyde treated quenchant was also checked after each series of three quenching cycles. The results are given in Table 4. As was seen with the cooling rate, there was no significant change in the viscosity of the quenchant after treatment with glutaraldehyde. The pH of the quenchant was not affected by the addition of the biocide.

Table 3. Microbial control by glutaraldehyde in a poly(alkylene glycol) quenchant

Glutaraldehyde (ppm active)	Microbial Populations (CFU/mL)				
	1 hour	1 day	1 week	3 weeks	6 weeks
0 (control)	2.9×10^3	3.1×10^4	3.5×10^5	1.3×10^5	2.2×10^4
50	<10	<10	<10	<10	<10
100	<10	<10	<10	<10	<10
300	<10	<10	<10	<10	<10

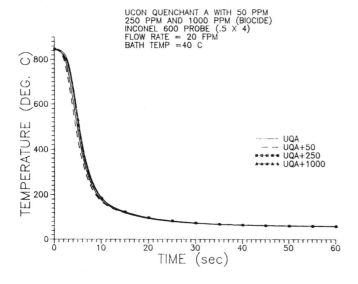

Figure 1. Cooling curves for glutaraldehyde treated quenchant.

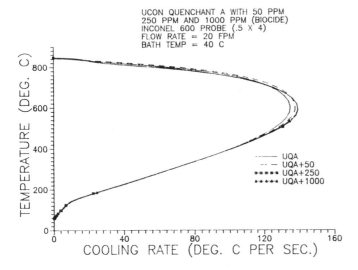

Figure 2. Cooling rate of glutaraldehyde treated quenchant.

Table 4. Viscosity and pH of a poly(alkylene glycol) quenchant treated with glutaraldehyde.

Glutaraldehyde (ppm active)	Viscosity (centistokes @ 100°F)	pH
0 (control)	1.87	9.9
50	1.90	9.9
250	2.00	9.9
1000	1.90	9.9

In a separate experiment, the chemical compatibility of glutaraldehyde with the sodium nitrite-based corrosion inhibitor in a commercial poly(alkylene glycol) quenchant was studied. In this test, the quenchant was diluted 1:10 with water, 1000 ppm of glutaraldehyde was added, and the sodium nitrite concentration was monitored over time using a Hach NitriVer® 2 test kit (9). The quenchant was maintained at 40°C

during the course of the experiment. The results are shown in Figure 3. The results indicate that glutaraldehyde is compatible with sodium nitrite-based corrosion inhibitors.

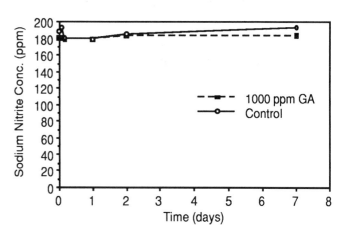

Figure 3. Sodium nitrite concentration in glutaraldehyde (GA) treated quenchant.

These various tests indicate that glutaraldehyde does not affect the properties or functioning of poly(alkylene glycol) quenchants. Extended treatment of a poly(alkylene glycol) quenchant with glutaraldehyde did not have an adverse effect on the quenchant's properties (data not shown). It would appear that glutaraldehyde is well suited to the preservation of poly(alkylene glycol) quenchants.

While these results were most encouraging, the most rigorous and instructive test of a biocide must be carried out in an actual industrial system. It is only in such a test that the effect of chemical contaminants, built up levels of microorganisms, contamination patterns, and local water conditions can be evaluated. A field trial was, therefore, carried out in an industrial quenchant system.

The system chosen for the trail was using a poly(alkylene glycol) quenchant at 18-22% concentration. The quench tank had a capacity of 5,800 gallons. Pre-cast and machined aluminum parts were processed in this system. The system experienced periodic episodes of microbiological activity resulting in foul odors, depletion of the sodium nitrite corrosion inhibitor, and staining of parts. When this occurred, the heat treater would thermally separate the quenchant and recover a portion of the polymer. The polymer would then be diluted with fresh water. Since the recovery process was not 100% efficient in recovering the polymer and did not recover any of the corrosion inhibitor, additional quenchant and sodium nitrite had to be added to the system. As a final step, a biocide was added to the system. In addition to the cost of the replacement chemicals, there was considerable expense due to lost production time, since the whole process normally required 2.5 days to complete. Immediately before the start of the trial with glutaraldehyde, and only three weeks after the system had undergone a thermal separation, it was again experiencing problems with microbiological activity.

Our first step in the investigation was to determine the chemical compatibility of the quenchant used at the plant with glutaraldehyde. As already noted, the biocide may be deactivated by certain quenchant components, and should not be used to treat systems containing such substances. Fortunately, samples of both the fresh and used quenchant were found to be quite compatible with glutaraldehyde, and we were, therefore, confident that the biocide would perform well in the system.

Since fouled systems typically show an initial demand for biocide due to accumulated chemical and microbiological contaminants, we expected to see a need for gradually decreasing glutaraldehyde doses. The system was initially treated with a 250 ppm dose of glutaraldehyde. The concentration of the biocide in the system was monitored via a simple colorimetric test (10). The results of the testing are presented in Figure 4. Glutaraldehyde showed excellent stability in the quenchant system. The system showed an initial biocide demand where glutaraldehyde concentration dropped rapidly; however, once this initial demand was satisfied, the rate of glutaraldehyde loss decreased and the concentration dropped relatively slowly. Because of glutaraldehyde's excellent stability in the system, doses were reduced to 80 ppm once per week after the first week.

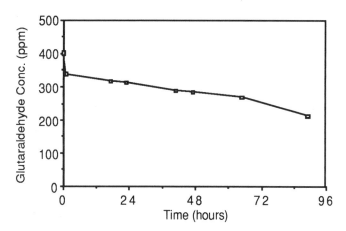

Figure 4. Glutaraldehyde stability in the quenchant system.

During the time of these treatments, the system was closely monitored for microbiological growth. The observed values are presented in Table 5.

Table 5. Microbiological Response in quenchant system.

Time After Initial Glutaraldehyde Treatment	Bacterial Counts (CFU/mL)
0	1×10^7
1 hour	1×10^3
2 hours	N.D.*
18 hours	N.D.
3 days	N.D.
8 days	N.D.
15 days	N.D.
29 days	N.D.
64 days	N.D.
96 days	N.D.

* N.D. = none detected ($<10^3$ CFU/mL)

Several interesting points are apparent from these data. First of all, glutaraldehyde caused extremely rapid kill of bacteria when it was added to the system. Bacterial populations were reduced four orders of magnitude within one hour of the addition of glutaraldehyde. Within 2 hours, bacterial counts revealed no colonies (detection level = 10^3 CFU/mL), indicating the system to be under excellent microbial control. The treatment regime maintained the system free of microorganisms, while maintaining a residual concentration of biocide in the system (compare Figure 4).

Operations were carefully monitored in the plant for several months after the trial began. No problems due to microbiological activity have been encountered since the treatment program with glutaraldehyde has been initiated. The biocide has had no effect on the properties of the quenchant, and no detrimental effects on metal quality or finish have been observed.

References

1. Issenberg, P., Fed. Proceed., 35, 1322-26 (1976).
2. Magee, P.N. and J.M. Barns, Advan. Can. Res., 10, 163-240 (1967).
3. Fribush, H.M., American Chem. Soc. Nat. Meet., Las Vegas, U.S.A., 1 April 1982.
4. Yonder, D.M. and D.C. Torgeson, U.S. Patent 2,801,216.
5. Gray, K.G., Austral. J. Hosp. Pharm., 10, 139-41 (1980).
6. Porapak PS 80/100 mesh.
7. BBL Microbiological Systems, Cockeysville, Maryland.
8. Leder, J., Lubrication Engineering, 43, 666-671 (1987).
9. Hach Company, Loveland, Colorado.
10. UCONEX Field Test Kit, Union Carbide Chemicals and Plastics Company Inc., Danbury, Connecticut.

Proceedings of the First International Conference on Quenching & Control of Distortion, Chicago, Illinois, USA, 22-25 September 1992

Analysis of Quenching and Heat Treating Processes Using Inverse Heat Transfer Method

J.V. Beck and A.M. Osman
Michigan State University
East Lansing, Michigan

ABSTRACT

The present paper describes the use of an inverse heat conduction method for the estimation of the varying film coefficient (h-factor) and heat flux in quenching and heat treating of materials. The method uses the function specification inverse heat transfer methodology to find the h-factor as a function of the surface or film temperatures. The inverse method uses measured cooling curves at one or more points inside the test sample during quenching experiments. A new and important feature of the method is the ability to have a variable number of future time steps for the analysis of a single experiment.

A computer program QUENCH1D based on the above method was developed and verified. The program can treat nonlinear, composite materials which may have different shapes such as plates, cylinders, and spheres. The method is applied to several problems including simulated Jominy end-quench test data for which the surface heat flux and film coefficient are estimated. The obtained results show that the program provides an efficient and accurate procedure for analysis of quenching and heat treating data.

THE INVERSE HEAT CONDUCTION PROBLEM

(IHCP) for extracting information from experimental data has received increasing attention in the last two decades. It has the potential of analyzing a wide range of practical problems in industrial processing, quenching, and heat treating of materials. In the IHCP, the unknown surface heat flux and/or the heat transfer coefficient are estimated by utilizing transient temperature measurements inside the solid.

Numerous numerical methods (1-4) have been proposed for the solution of the IHCP, many of which are restricted to the linear case. The basic method used herein is the one most widely used in the United States for the nonlinear case. This method is sequential and involves the use of future temperatures for each calculated component of the surface heat flux. The procedure permits much smaller time steps than making the calculated interior temperatures equal the measured values. Small time steps permit extraction of much more information regarding the time variation of the surface heat flux than with large time steps.

Some method to filter out the high frequency components and to stabilize the IHCP is needed. There are many ways of accomplishing this. One method is to specify a functional form for the surface heat flux (called the function specification method (1-4)). Another method is called regularization (1, 5-7). A combined function specification and regularization method was used for the solution of some two-dimensional inverse heat conduction problems, Osman and Beck (8-10).

The sequential function specification method is used in the present paper. A computer program QUENCH1D based on the function specification method was developed and applied to several test cases. Also, simulated data from the familiar Jominy end-quench test are analyzed for the estimation of the surface heat flux history. The surface temperature and heat transfer coefficient are also estimated. Program QUENCH1D is quite flexible in treating one-dimensional flat plate, cylindrical or spherical composite bodies. Multiple interior thermocouples can be employed.

A new and important feature of the present function specification method is the ability to have variable number of future time steps for the analysis of a single experiment. The total time interval of the experiment can be divided into several subintervals and different value of future time steps, r, can be used for each subinterval. This feature allows for smooth curves and reduces the biases of the estimated functions. It is particularly important for quenching, since there are both rapidly varying heat fluxes and slowly varying fluxes in a typical quench.

For brevity, the basic theory of the method is omitted but is contained in ref. 1. The main emphasis is upon some examples.

SAMPLE CASES

Four sample cases are given that demonstrate the use of the function specification method for the analysis of various inverse problems, including quenching and heat treating experiments.

The first case is for a homogeneous planar body having constant thermal properties. The body is exposed to a triangular heat flux history. The results are given in dimensionless form. The second case is for a composite planar body with temperature dependent thermal properties. The body is exposed to a heat flux which is linearly increasing with time. The third case is for a solid cylinder with temperature-dependent thermal properties. The data is from an actual quenching experiment. The cylinder is quenched (suddenly immersed) in a constant temperature bath. The film coefficient from the outer surface of the cylinder is calculated and displayed.

The fourth case relates to the simulation of end-quench hardenability test of Jominy bars in which there are side heat losses from the outer surface of the Jominy specimen.

Test Case of Triangular Heat Flux

In this case the body is a finite plate of thickness L. It is insulated at $x = L$ where the temperature sensor is located. See the insert of Fig. 1. The heat flux is triangular in shape as shown in Fig. 1. The heat flux is plotted versus the dimensionless time t^+,

$$t^+ = \alpha t/L^2 \qquad (1)$$

where α is the thermal diffusivity. The linearly increasing portion of the heat flux, i.e., $0 < t^+ < 0.6$, is described by

$$q^+ = t^+ \qquad (2)$$

where $q^+ = q/q_N$ and q_N is a nominal value of heat flux which is the value associated with t^+ set equal to unity in eq. (1). The dimensionless x coordinate is $x^+ = x/L$ and the dimensionless temperature is

$$T^+ = (T - T_0)/q_N L/k \qquad (3)$$

The simulated temperatures for $x^+ = 0$ and 1 are shown in Fig. 2. The temperature at the insulated surface

is delayed in response to changes on the surface heat flux. The heated surface starts to rise in temperature at the instant q is nonzero. The delay and damping of $T^+ (1,t^+)$ is such that simple inspection does not indicate changes in the surface heat flux. The dimensionless delay in the $x^+ = 1$ temperature corresponding to changes in the surface q is about $\Delta t = 0.1$. The use of future temperatures by setting r to some value greater than unity allows QUENCH1D to calculate values of q using interior temperature sensors.

The inverse calculations for the surface heat flux and temperature was performed using temperature "measurements" at $x^+ = 1$ with a time spacing of $\Delta t^+ = 0.06$. In order to investigate the effects of the finite difference approximations, data correct to 6 decimal places using exact solution were used (see ref. (1)). Twenty equally-spaced Δx's were sufficient to give good accuracy. Two future temperatures ($r = 2$) for the entire experimental times were used.

Some results of using QUENCH1D are displayed in Figs. 3 and 4. In Fig. 3 the solid circles (dots) are for the calculated heat flux for r values of 2, 3, 4 and 5. The $r = 1$ value is for exact matching of the experimental temperatures by the calculated values. The calculation is unstable and thus is not shown in Fig. 3. A comparison of results for the various r values is given in Table 1.

From Fig. 3, the estimated q values near $t^+ = 0.6$ have the greatest error and thus this is the region covered in Table 1. Even in this difficult region, the values are relatively accurate, with the largest error being about 10% for $r = 5$. Even more accurate than the heat flux values are the calculated surface temperatures which are shown in Fig. 4 for $r = 2$ and 5. The $r = 2$ curve is very close to the exact surface temperature except for the times just above $t^+ = 0$.

The differences between the sensor temperatures and the corresponding calculated temperatures by program QUENCH1D are called the residuals. The residuals, e_i, are calculated by,

$$e_i = Y_i - \hat{T}_i \qquad (4)$$

where Y_i is the measured temperature and $\hat{T}_i$ is the corresponding calculated temperature. Program QUENCH2D provides an output of the e_i's values as well as a running root-mean-squares of the residuals. Study of residuals is an integral part of the estimation problem and can provide the user with criterion for the selection and adjustment of the different input parameters of the function specification method.

It is also important to examine the effect of random errors in the measured temperature. The above analysis with errorless data implies that the best choice of r is 2, the

148

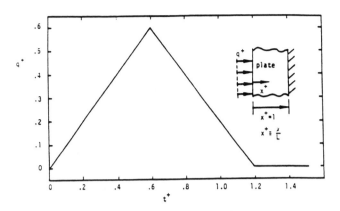

Fig. 1 Triangular heat flux test case for the finite insulated plate.

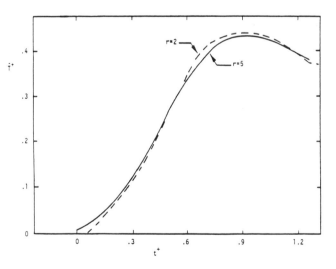

Fig. 3 Estimated heat flux values for the triangular heat flux test case with different r values and $\Delta t^+ = 0.06$.

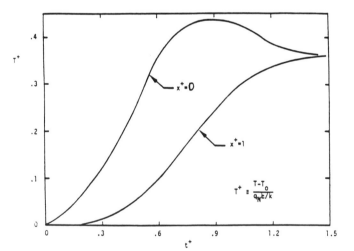

Fig. 2 Heated and insulated surface temperatures for the triangular heat flux test case.

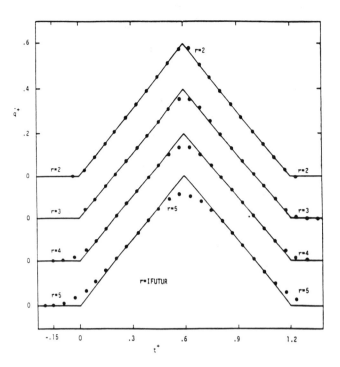

Fig. 4 Calculated surface temperature for the triangular heat flux test case.

minimum stable value. The introduction of measurement errors leads to larger r values that depend, among other things, upon the magnitude of the errors. An analysis has been performed for a case with errors that are additive, have zero mean and constant variance, are uncorrelated and are normal. (These are the first five standard statistical assumptions discussed in (1).) The dimensionless standard deviation of the temperature errors is denoted, σ_Y^+ and was selected to be 0.0017 C which is about 0.5% of the maximum temperature at x = 1 (see Fig. 5).

Results are shown for r = 3 and 4 in Fig. 5. The r = 3 case is shown to be considerably more sensitive to measurement errors than the r = 4 case. To be more specific, the standard deviations of the estimated heat flux are 0.34, 0.042 , 0.016, 0.009 and 0.0056 for r = 2, 3, 4, 5 and 6, respectively. These values can be compared with the maximum true q^+ values of 0.6. Notice that these standard deviation values go down with increasing r unlike the bias errors implicit in Table 1. The sum of the bias in Table 1 plus the $\hat{q}^+$ standard deviation for a given r is a minimum for r = 4 for the present example; hence r = 4 is best for this example with $\sigma_Y^+ = 0.0017$.

Test Case of Temperature Variable Thermal Properties Composite Body

The second test case involves a composite planar body with temperature-variable thermal properties. The body is composed of three materials and has two temperature sensors. The x = 0 surface is heated by a flux that is zero for t < 0 and that is 900,000 W/m² at t = 30 s when it suddenly drops to zero. (See Fig. 6.) The x = L surface is insulated. The first material is 0.02 m thick and is broken into three regions. It is broken into 3 because the first sensor is located at 0.005 m from the heated surface and the second one is at 0.01 m from the heated surface. The second material is broken into two regions, 0.005 m thick, and the third material has two regions of 0.02 and 1 m thick. The location of the interfaces (sensors) are at x = 0.005 and 0.01 m, respectively. The experimental time step is 1 s and the experimental to calculation time step ratio is 4. The input temperature "measurements" are the result of an accurate finite difference solution of the direct problem. The thermal conductivity of material 1 varies from 1 W/m-K at 300K to 10 W/m-K at 1300 K, a factor of 10. The computed temperatures cover nearly this range as shown Fig. 7 where a range of 300 to about 1200 K is shown for x = 0. The output results for this case are plotted in Figs. 6 and 7.

Figure 6 displays the surface heat flux for two different r calculations. The vertical axes are displaced so that each case is shown separately. The upper curve is for r = 3 and is the same case as for Fig. 7. For most of the curve the agreement with the true values is excellent. Near the time of maximum q, the calculated heat flux anticipates changes and starts to decrease two time steps too early. The abrupt decrease in the surface flux is followed quite well. The lower axis in Fig. 6 shows calculations for r = 4 and both Δt_{cal} = 0.25 ls and since there is negligible difference between the results for these two different calculational time steps. The r = 4 curve does not follow the sharp decrease in heat flux quite as well as the r = 3 calculation. For both curves shown in Fig. 6, the agreement of the calculated q values with the true values is excellent. This is particularly true in view of the large lag and damping of the measured temperatures and the large temperature variation of the thermal conductivity of material 1.

Figure 7 displays both the calculated and true temperatures for x = 0, 0.005 and 0.01 m. The dots are the calculated values and the continuous curves are the true values. The x = 0.005 and 0.01 m temperatures are used to estimate the surface conditions. It is quite remarkable that the surface temperature is calculated so accurately with the "measured" temperature rises at x = 0.005 and 0.01 m so small compared to the surface temperature. This result gives a graphic demonstration of the power of QUENCH1D to recover the surface temperature from very small errorless, interior temperature rises. Also note that the calculated surface temperature history shown in Fig. 7 is extremely accurate even though the thermal properties are varying greatly with temperature.

If realistic errors were introduced, a larger value of r such as 5 or 6 would be recommended.

This test case has unusually large changes in thermal conductivity with temperature (a factor of about 9). Even so, QUENCH1D produced accurate values. More examples are given in ref. 1 for the selection of number of future time steps r. In general, the closer is the temperature sensor to the heated surface, the better is the estimate for q and also the smaller is the required r value. Even at the heated surface, measurement errors are always present and thus the minimum recommended r is usually 2.

Test Case of a Cylindrical Body

The third test case involves a solid cylindrical body with unknown heat flux at the outer radius R = R_2. The temperature data is from an actual quench test. Figure 8 shows the the experimental data. The material is Inconel 700 and the thermal properties vary linearly with temperature. The cylinder is 0.5 inch in diameter (0.00635 m). The thermocouple is at the center (but it could have been any point in the body and more thermocouples could have been used).

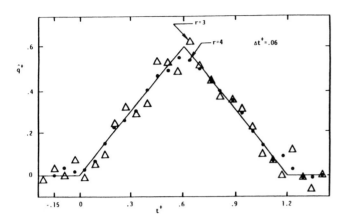

Fig. 5 Estimated heat fluxes for triangular heat flux test case with with added, normal uncorrelated random errors with $\sigma_Y = 0.0017$ C.

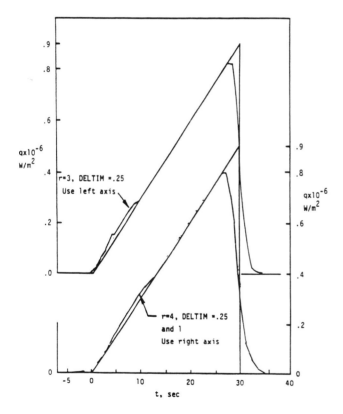

Fig. 6 Estimated heat fluxes for composite-body test case with temperature-variable thermal properties.

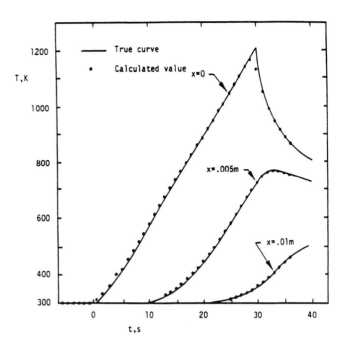

Fig. 7 Calculated surface temperatures for composite body test case with temperature-variable thermal properties.

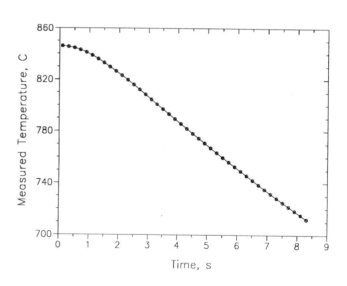

Fig. 8 Measured temperatures from actual quenching of a solid cylinder.

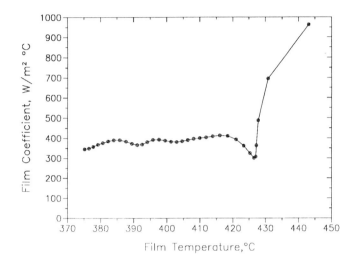

Fig. 9 Estimated film coefficient values for quenching of a solid cylinder.

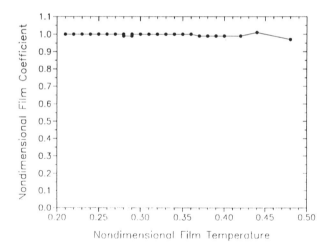

Fig. 10 Estimated heat transfer coefficient for end-quench of a solid cylinder with side heat losses.

The estimated film coefficient is plotted in Fig. 9 as a function of film temperature for a fluid temperature of 60°C. The number of future temperatures used is 6 because the time steps are relatively small (0.2 second) for a thermocouple 0.25 in (0.00635 m) from the heated surface for this relatively low conductivity material. The heat transfer coefficient could be made to have smoother results by using a larger number of future time steps for times > 3 s.

Test Case of Cylindrical Body with Small Side Heat Loss (Jominy End-Quench Test Case)

The case of small side heat transfer is illustrated for a solid cylinder suddenly exposed to a cooling fluid (of zero degrees) at the x = 0 end. The main heat transfer and the largest heat transfer coefficient is at the x = 0 end. The side heat transfer coefficient from the outer radius R = R_2 is relatively small (fin effect). The x = L end is insulated. This case simulates the Jominy end-quench test. An analytical solution for this case is

$$T(x^+, t^+) = 2 \sum_{n=1}^{\infty} \exp\left[-t^+(\beta_n^2 + m^2L^2)\right] A(\beta_n, B, x^+) \tag{5a}$$

$$A(\beta_n, B, x^+) = [C_n \sin\beta_n \cos((1-x^+)\beta_n)]/D_n \tag{5b}$$

$$C_n = \beta_n^2 + B^2, \quad D_n = (C_n + B)\beta_n \tag{5c,d}$$

where the eigenvalues, β_n, are calculated from

$$\beta_n \tan \beta_n = B, \quad B = hL/k \tag{6a,b}$$

The symbol h is the heat transfer coefficient at x = 0. The dimensionless group m^2L^2 is equal to

$$m^2L^2 = 2h_s L^2/(ka) \tag{7}$$

where a = cylinder radius and h_s is the side heat transfer coefficient. Temperature values were obtained from this equation at a series of dimensionless times, t^+, with time steps of 0.01 until $t^+ = 0.45$ and at $x^+ = 0.1$. By choosing dimensionless values, it is equivalent to h = 1, k = 1, α = 1, a = 1 and L = 1. The nondimensional side heat transfer coefficient, h_s, is assumed to be 0.25.

Figure 10 shows the the estimated heat transfer coefficient as a function of the surface temperature. Several regions of future times are used in this test case. The total time of the experiment was divided into three consecutive regions of 0 to 0.05, 0.05 to 0.2, and 0.2 to 0.45. The number of future time steps starts at one and increases to two and three. The number of future time steps is the smallest when the temperature is changing the most rapidly. An indication of the smallest time step is roughly

$$r\,(\alpha\Delta t/E^2) = 0.25, \qquad (8)$$

where r is the number of future time steps and E is the distance of the temperature sensor (closest to the surface) from the heated surface. In the present case with $\Delta t = 0.01$ and $E = 0.1$, $\alpha\Delta t/L^2 = 1$, which is a relatively easy case; $r = 1$ is satisfactory.

CONCLUSIONS

An inverse procedure is given for the analysis of quenching and heat treating experiments. The method uses the sequential-in-time function specification method for the estimation of the heat flux and/or the film coefficient by using transient temperature measurements at appropriate point(s) inside the body. Data from plates, cylinders, and spheres geometries with temperature-dependent thermal properties can be treated using the function specification method.

A computer program QUENCH1D based on the above method was developed and verified. A new and important feature is added to the function specification method to allow for variable number of future time steps over the course of the experiment. The time of the experiment can be divided into several subintervals and different value of the r parameter can be used for each subintervals. This feature allows for smooth curves with minimal biases of the estimated function.

The method is applied to several test cases including the analysis of simulated Jominy end-quench data. Results show that the function specification method with variable number of future time steps can provide fast and accurate analysis of quenching and heat treating data.

REFERENCES

1 - Beck, J.V., Blackwell,B. and St.Clair, Jr., C.R. "Inverse Heat Conduction: Ill-posed Problems", Wiley-Interscience, New York, (1985)

2 - Beck, J.V., Int. J. of Heat and Mass Transfer 13, 703-716 (1970)

3 - Beck, J.V., Litkouhi,B. and St. Clair, C.R., Numerical Heat Transfer 5, 170-178 (1982)

4 - Raynaud, M. and Beck,J.V. Journal of Heat Transfer 110, 30-37 (1988)

5 - Tikhonov, A.N. and Arsenin, V.Y. "Solutions of Ill-Posed Problems", V.H. Winston and Sons, Washington, DC. (1977)

6 - Alifanov, O.M., "Inverse Problems of Heat Exchange", Obratnye Zadachi Teploobmena Publishing House, Moscow (1988)

7 - Alifanov, O.M. and Rumyantsev, S.V. Journal of Engineering Physics, Vol. 34, No. 22, pp 223-226 (1978)

8 - Osman, A.M. and Beck, J.V. AIAA Journal of Thermophysics and Heat Transfer 3, 297-302 (1989)

9 - Osman, A.M. and Beck, J.V., "QUENCH2D: A General Computer Program for Two-Dimensional Inverse Heat Transfer Problems," Heat Transfer Group, Dept. of Mech. Eng., Michigan State University, Rept. MSU-ENGR-89-017 (1989)

10 - Osman, A.M. and Beck,J.V., "QUENCH2D: A General Computer Program for Two-Dimensional Inverse Heat Transfer Problems Involving an Internal Radiation Cavity," Heat Transfer Group, Dept. of Mech. Eng., Michigan State University, Rept. MSU-ENGR-90-005 (1990)

Table 1

Some calculated heat flux values for triangular heat flux case

		$\hat{q}^+$ values			
Time, t^+	Exact	r = 2	r = 3	r = 4	r = 5
.51	.51	.5099	.5096	.5037	.4893
.57	.57	.5690	.5557	.5345	.5101
.63	.57	.5815	.5573	.5324	.5058
.66	.51	.5059	.5159	.5016	.4799

Determination of Quench Heat-Transfer Coefficients Using Inverse Techniques

B. Hernandez-Morales, J.K. Brimacombe, and E.B. Hawbolt
The University of British Columbia
Vancouver, British Columbia, Canada

S.M. Gupta
Tree Island Industries Ltd.
Richmond, British Columbia, Canada

ABSTRACT

Heat-transfer coefficients in the quenching process have been determined by quenching stainless steel and mild steel specimens instrumented with thermocouples, in brine (3 % by wt NaCl), water, oil, and air, under controlled laboratory conditions that ensured one-dimensional heat transfer. From the experimentally measured thermal response, the relationship between the surface heat-transfer coefficient and surface temperature has been obtained by solving the heat-conduction problem inversely; the influence on this relationship, of important variables such as initial specimen temperature, quenchant temperature, type of quenchant and surface oxidation has been studied.

The inverse heat-conduction problem was solved numerically, following finite-difference schemes for one-dimensional heat flow in a flat plate having variable thermophysical properties. Two algorithms were implemented : 1) sequential matching of the thermocouple readings and 2) sequential function specification; both approaches give similar results with the latter being more computationally efficient, since it does not require iteration. Also, the sequential matching technique has stability restrictions when small time steps are used.

The results obtained in this work are reproducible to within ±18% and are in good agreement with previously reported results for similar experiments. The stability of the vapor film is increased when the initial specimen temperature is higher or the quench bath is hotter; on the other hand, agitation of the bath may reduce the vapor film stability.

QUENCHING IS A CRITICAL OPERATION in the production of many metal parts. Careful selection of processing variables will result in an end product with the desired microstructure (and, therefore, mechanical properties), free of cracks and distortion, and with the optimal distribution of residual stresses. More stringent specifications for heat treating complex alloys, together with goals of higher productivity (lower rejections due to cracks and distortion) and lower energy consumption, have prompted research toward the linking of final mechanical properties and processing conditions, such as initial part temperature, type of quenchant, quenchant temperature and bath agitation.

Several approaches have been developed to predict the performance of a given quenching process; they can be classified according to : i) measurements of quench bath quality, ii) factorial design, and iii) process modelling. Totten *et al.* [1] presented a comprehensive review of available tests applied to measure quench bath quality based on either its "hardening power" (hardenability, Jominy end quench) or its "cooling power" (GM quenchometer, hot wire, interval five-second, and cooling curve) and concluded that the latter approach, in particular analysis of the cooling curve, provides the best estimate. Factorial design provides a systematic method for conducting experiments but fails to give information on the processes that take place during quenching; nevertheless, this approach has been applied succesfully to the optimization of heat-treatment operations [2,3]. The process modelling approach takes advantage of the current availability of computing power to utilize concepts of transport phenomena and solid mechanics to calculate residual stresses and distortion in quenched steel parts which are the result of complex interactions among thermal, microstructural and mechanical phenomena taking place during rapid cooling.

In mathematical models of quenching processes [4,5,6], the thermal field in the interior of the body is computed as input for the calculation of both microstructural evolution and mechanical behaviour (thermal and transformation stresses); therefore, the need for an accurate determination of the thermal field as time progresses is apparent. A critical step in this process is the characterization of the boundary condition at the surface of the part, typically a Neumann condition. However, obtaining reliable measurements of surface temperature or heat-transfer coefficient directly during a quenching operation is very difficult. On the other hand, the temperature response at an interior location can be measured readily and the problem then is transformed into obtaining the surface heat flux that produced such a response; this is the inverse heat conduction problem (IHCP). It should be noted that, notwithstanding the complexities involved in solving the IHCP, robust techniques like sequential function specification are preferable to a purely theoretical approach based on coupled heat and fluid flow analyses of the quenching bath. The boiling which accompanies quenching is a very complex process involving hydrodynamics and several modes of heat transfer between the surface of the test piece, the bath and bubbles, as well as kinetics of bubble formation and growth [7,8,9], and essentially defies theoretical analysis.

Until recently, the most common approach to the solution of the IHCP relied on a "brute force" technique in which, at a given time, a guessed value of the heat-transfer coefficient was adopted as a starting point for succesive iterations of the heat-conduction equation; the iterations were stopped when a suitable convergence criterion, usually based on the difference between measured and calculated temperatures after two succesive iterations, was met and the process was then repeated sequentially. This procedure has the disadvantage of being extremely sensitive to meaurement errors.

In the present work, the inverse heat conduction problem during quenching of flat specimens has been solved by applying sequential matching and sequential function specification techniques.

Experimental

Apparatus. A schematic diagram of the apparatus is shown in Figure 1. An induction furnace was used to heat the quench sample to the initial quenching temperature. The furnace chamber was approximately 250 mm high × 200 mm long × 30 mm wide; the furnace could be opened at the bottom with remote handles and the atmosphere in the chamber could be controlled by supplying the inert gas (usually argon) through an inlet nozzle in the door of the furnace. A smooth vertical movement of the sample was ensured by installing a 150 mm long cylindrical brass guide just above the furnace top. A 1.1-m long, 13-mm OD steel pipe, fastened via a stainless steel bridge to the sample, passed through the brass guide and was connected to a steel cable that controlled the sample movement by means of a cable lock; a microswitch was mounted at the top of the sample tube to activate the recorder after the sample was dropped into the bath. A small piston with a long rod was attached to the bottom edge of the sample to centre it in the quench bath. When the sample was lowered into the bath, the piston slid into a fitted cylinder at the bottom of the tank, thus forcing all liquid out of the cylinder through a set of grooves; this absorbed the energy of the fall and provided a hold on the sample to avoid any oscillations. Stirring of the quenchant was accomplished by blowing cleaned air through a manifold bubbler at the bottom of the tank. Tap water, brine (3 w/o NaCl), oil (Houghton-Quench-G) and air were the quench media under study.

Quench Sample. The overall dimensions of the quench sample were approximately 200 mm-dia. × 20 mm thick (to ensure one-dimensional heat-transfer). The disks were made of stainless steel and mild steel plates machined to the shape shown in Figure 2. The quench sample had to be assembled from the three sections shown in the figure due to difficulties in machining very thin holes through a 20 mm thick disk. To prevent heat losses due to contact resistance, graphite was applied at the interfaces between sections. Once the three sections were aligned they were bolted as shown in the figure.

Type K (Chromel-Alumel) bare thermocouple wires of 0.31 mm-dia. by about 2.5 m long were welded to the bottom of two holes drilled at approximately 1.4 mm from the surface of the sample. The wires were then insulated by threading them through thin ceramic tube insulators of appropriate diameter and length. After each run the sample was sectioned to determine the exact distance of the thermocouple wires from the surface.

Procedure. Once assembled, the quench sample was mounted in the apparatus and the output leads were plugged into the thermocouple wires. The resistance of the thermocouple wires was measured while raising and lowering the sample. If any fluctuations in the resistance were observed, indicative of damaged thermocouples, they were repaired and retested. After placing the sample in the cold furnace, the thermocouple leads were connected to the chart recorder which was then calibrated. The thermocouple leads were removed to protect the recorder against damage due to induced currents in the thermocouple before the furnace was switched on. The temperature of the sample during heating up to the initial quenching temperature was then monitored with a potentiometer.

After heating to achieve a uniform temperature in the sample, the inert gas flow to the furnace was shut off, the thermocouple leads were connected to the recorder, the furnace doors were opened and the sample cable lock was released. As the sample dropped, it turned on the recorder by the microswitch mounted at the top of the sample tube. When the sample temperature was low enough, the sample was raised out of the quench medium which automatically shut off the recorder. Typically, it took 100 ms for the sample to reach the quench bath.

IHCP Algorithms

Because of instabilities in the solution introduced by the overspecification of boundary conditions at the thermocouple position, the IHCP is an ill-posed problem and, therefore, any solution suffers from inherent uncertainty. Even relatively small errors in the temperature measurements may result in large errors in the estimated surface temperature and heat flux when the "brute force" method is applied. In the following, the sequential matching and sequential function specification techniques are described; the geometry of the system, as well as the nomenclature adopted, are shown in Figure 3.

Sequential Matching. In this technique the whole domain was subdivided into two subdomains as shown in Figure 4. Heat conduction in the body is described by

$$\frac{\partial}{\partial x}\left(k\frac{\partial T}{\partial x}\right) = \frac{\partial}{\partial t}(\rho \, Cp \, T) \qquad x \in \Omega \qquad (1)$$

subject to the following boundary conditions :
Subdomain 1 (Ω_1)

B.C. 1 : $\quad \dfrac{\partial T}{\partial x} = 0$ $\qquad\qquad$ at $\quad x = 0$

B.C. 2a : $\quad T(x,t) = Y(t)$ $\qquad$ at $\quad x = x_1$

B.C. 2b : $\quad -k\left.\dfrac{\partial T}{\partial x}\right|_{\Omega_1} = -k\left.\dfrac{\partial T}{\partial x}\right|_{\Omega_2}$ $\qquad$ at $\quad x = x_1$

Subdomain 2 (Ω_2)

B.C. 1a : $\qquad T(x,t) = Y(t) \qquad\qquad$ at $\quad x = x_1$

B.C. 1b : $\qquad -k\dfrac{\partial T}{\partial x}\Big|_{\Omega_1} = -k\dfrac{\partial T}{\partial x}\Big|_{\Omega_2} \qquad$ at $\quad x = x_1$

B.C. 2 : $\qquad -k\dfrac{\partial T}{\partial x} = q(t) \qquad\qquad$ at $\quad x = L$

The heat flux at the surface, $q(t)$, is usually characterized as a combined convective and radiative flux :

$$q(t) = -\,\bar{h}(T_f - T)\big|_{x=0} \tag{2}$$

The entire domain ($\Omega = \Omega_1 + \Omega_2$) was divided into a grid of N equally-spaced nodes, with the sensor position located at node n. An implicit finite-difference method was applied to discretize the heat-conduction equation in both subdomains, resulting in a set of simultaneous algebraic equations of the form*:

$$a_i T_{i-1}^{j+1} + b_i T_i^{j+1} + c_i T_{i+1}^{j+1} = d_i \tag{3}$$

The inverse problem was solved as follows : the direct problem posed in Ω_1 was solved to obtain the temperature field in that subdomain. Then, the temperature at succesive points in Ω_2 was calculated as :

$$T_{i+1}^{j+1} = \frac{T_i^j - a_i T_{i-1}^{j+1} - b_i T_i^{j+1}}{c_i}, \quad n \le i \le N \tag{4}$$

The heat balance at node N (the active surface) results in :

$$a_N T_{N-1}^{j+1} + b_N T_N^{j+1} + c_N \hat{h} = T_N^j \tag{5}$$

Solving for the estimated heat-transfer coefficient gives, finally,

$$\hat{h} = \frac{T_N^j - a_N T_{N-1}^{j+1} - b_N T_N^{j+1}}{c_N} \tag{6}$$

Sequential Function Specification. In this second technique, the unknown surface heat flux, $q(t)$, at time t_M was calculated by using r additional future temperature measurements and assuming that $\hat{q}(t)$ (the estimated surface-heat flux) is constant during the time interval $t_M \le t \le t_{M+r-1}$ (see Figure 5). Time intervals between meaurements (Δt) are assumed to be constant. The solution strategy is based on minimizing the difference between calculated and measured temperatures with respect to q_M :

$$\frac{\partial S}{\partial q_M} = 0 \tag{7}$$

where S is defined, for a single thermocouple, by

$$S = \sum_{i=1}^{r} \left(Y^{M+i-1} - T^{M+i-1}\right)^2 \tag{8}$$

Substituting Eq. (8) in Eq. (7) results in :

$$2\sum_{i=1}^{r} \left(Y^{M+i-1} - T^{M+i-1}\right) X^{M+i-1} = 0 \tag{9}$$

---

*Throught the paper a subscript denotes the spatial coordinate and a superscript denotes the temporal coordinate when dealing with finite difference equations.

which is true only at $\hat{q}_M$. The quantity X, the sensitivity coefficient, is defined as :

$$X^{M+i-1} = \frac{\partial T^{M+i-1}}{\partial q_M} \tag{10}$$

In order to calculate T^{M+i-1} and its derivatives, it is assumed that T^{M+i-1} is a continuous function of q_i and, therefore, a Taylor series expansion about q_{M-1} results in

$$T^{M+i-1} = \overset{*}{T}{}^{M+i-1} + (q_M - \hat{q}_{M-1})\overset{*}{X}{}^{M+i-1} \tag{11}$$

where the asterix denotes that the functions T^{M+i-1} and X^{M+i-1} have been evaluated at time t_{M-1}.

Substituting Eq. (11) in Eq. (9) and solving for $\hat{q}_M$:

$$
\begin{aligned}
\hat{q}_M = \hat{q}_{M-1} \\
+ \ \frac{1}{\Delta_M}\sum_{i=1}^{r}\left(Y^{M+i-1} - \overset{*}{T}{}^{M+i-1}\right)\overset{*}{X}{}^{M+i-1}
\end{aligned} \tag{12}
$$

where

$$\Delta_M = \sum_{i=1}^{r}\left(\overset{*}{X}{}^{M+i-1}\right)^2 \tag{13}$$

An assumption is now made in order to linearize the problem, and avoid iteration : for the time interval $t_M \le t \le t_{M+r-1}$, the thermophysical properties at a given location are assumed to be constant and equal to their respective values at $t = t_{M-1}$. This assumption is justified if the properties at a given location do not change significantly from time to time, eventhough there might be a large temperature gradient in the part as a whole.

The heat-transfer coefficient can then be calculated as

$$\hat{h}^M = \frac{\hat{q}_M + \hat{q}_{M+1}}{2\left(T_f^M - \hat{T}_S{}^M\right)} \tag{14}$$

This algorithm has been coded in the computer program **CONTA** [11].

Numerical Calculations

The thermal response was recorded experimentally during each quench as a temperature (mV) *vs* time (s) plot. The plots were digitized and linear interpolation was applied to generate the data adopted as input for the computer programs.

For the sequential matching technique, data points were obtained at 10 °C intervals from the beginning of the quench; the time intervals used in the calculations, therefore, were fixed at the corresponding values. The domain was discretized into 20 nodes, making sure that an interior node was always positioned at the thermocouple location; the convergence criterion adopted was a difference in succesively calculated temperatures of 10^{-6} °C. The computer program **CONTA** requires the input data to be given at uniformly-spaced time intervals, Δt; satisfactory results were obtained for a value of $\Delta t = 0.0625$ s. The specimen was subdivided into two regions of 9.35 mm (14 nodes) and 1.4 mm (6 nodes), corresponding to Ω_1 and Ω_2 (Figure 4), respectively, and a value of $r = 2$ was adopted. Figure 6 shows the estimated h *vs* T_s obtained using sequential matching and sequential function specification to solve the IHCP for a stainless steel disk quenched in still water at 60 °C; good agreement is observed.

Results and Discussion

A typical plot of quench heat-transfer coefficient against surface temperature for a water quench has already been seen in Fig. 6. Thus the four commonly observed boiling zones [12] are evident : film boiling (above 600 °C); transition boiling (between 600 and 200 °C); nucleate boiling (from 200 to 120 °C) and pure convection boiling (below 120 °C).

In this work, a total of five quench variables were studied (see Table I); in each case, the temperatures that define the boundaries between the different modes of boiling were obtained in order to establish the influence of varying each of the quench variables. It should be noted that most of the research work on boiling phenomena has been performed under steady-state conditons. Peyayopanakul et al. [13] carried out transient experiments in order to determine the limits for which unsteady-state quenching results can be identified with those under steady-state. They worked with the circular end of copper cylinders of different lengths and found that a minimum value of Biot number of 0.9 was required. In several of the curves obtained in the present study, an inflexion point was observed early in the quench. This behaviour has been reported before (see, for example [14]) and corresponds to early transients due to the large temperature difference between the surface of the sample and the quench bath and should be ignored. The results obtained in this investigation are reproducible to within $\pm$ 18 % and are consistent with previously published work for equivalent cooling conditions [15,16].

Initial Sample Temperature. Figure 7 shows results for quench tests when two different initial sample temperatures were considered (842 °C and 891 °C). A late onset of nucleate boiling in the latter run (point A) suggests that a higher initial specimen temperature increases the stability of the vapor film during the transition and/or film boiling stages. The maximum difference in the h-values is 60 % at about 500 °C.

Lambert et al. [15] did not observe significant differences in the heat fluxes measured on a nickel specimen quenched in water from 850, 900, 925 and 950 °C. Mitsutsuka et al. [16] found that a higher initial specimen temperature tends to increase the stability of the film boiling phenomena, in agreement with the present findings.

Initial Bath Temperature. Figure 8 shows the h-values obtained by quenching in still water at three bath temperatures. For temperatures below 60 °C it is not possible to distinguish a transition region; for a bath temperature of 60 °C the start of the transition period occurs at approximately 575 °C (point A). The value of the maximum heat flux decreases as the bath temperature increases, due to a decrease in the difference between the boiling point of water and the bath temperature. Lambert and Economopoulos [15] and Mitsutsuka et al. [16] reported similar findings.

Agitation of the Bath. A comparison of heat-transfer coefficients obtained when the specimen was quenched in still water and stirred water is shown in Figure 9. Stirring of the bath reduces the stability of the vapor film by physically disturbing it and this is reflected in the graph which does not exhibit a region of stable film boiling. The cooling rates during transition boiling are significantly higher in the case of the stirred bath with a maximum difference in the estimated heat-transfer coefficient of 65 % at 600 °C. However slower

cooling rates were observed at temperatures where nucleate boiling was present in the still bath. An overall reduction in quench time, from 12.2 to 11.2 s, was observed when stirring was applied.

Surface Oxidation. Results of quench experiments involving plain carbon steel disks covered respectively with a layer of light oxide and heavy oxide are shown in Figure 10. The presence of the scale in the oxidized sample provides nucleation sites for the bubbles resulting in a higher temperature for the onset of nucleate boiling (point A). Since the thermal resistance of the heavy scale is now added to the heat transfer process, lower cooling rates are obtained. Loose oxide scale was observed on the specimen at the end of the quench in the oxidized case; it is thought that at the onset of nucleate boiling, water penetrates the thick scale, comes into contact with the hot sample surface, and boils, causing the oxide scale to loosen. Lambert and Economopoulos [15] suggested that the difference in shape observed between the two curves can be attributed to the fact that the vapour film adheres better to an oxidized surface.

Quench Medium. Figure 11 shows the effect of different quench media on the h vs T_s curve of stainless steel samples. The highest heat-transfer coefficient (highest cooling power) was obtained by quenching in brine, followed by water, oil, and air. This sequence agrees with published Grossman quench severity factors [17]. It can be observed that high values of the heat-transfer coefficient are obtained at temperatures under typical M_s values of hardenable steels when the specimen was quenched in water or brine; the consequence of this behavior is that distortion is highly possible when these quench media are used for quenching complex shapes. The maximum heat flux (corresponding to the maximum heat-transfer coefficient) and the start of non-boiling heat transfer is displaced to higher surface temperatures when quenching in oil relative to brine and water, due to its higher boiling point (the flash point temperature for Houghton-Quench oil is 355 °C). In the case of an air quench, cooling occurs by both radiation and natural convection, which are not as effective heat removal mechanisms as boiling; the results for the heat-transfer coefficient obtained in this study are within $\pm$ 12 % of theoretical values calculated for vertical plates in still air assuming an emissivity of 0.8 and a view factor of 1. Lambert and Greday [18] pointed out that, for brine, the presence of dissolved salt in the bath has a twofold effect on the heat transfer : it increases the thermal conductivity of water, and, therefore, increases the heat extraction power, and it modifies the adherence characteristics of the vapor film during film boiling.

Comparison with Previous Work. Results of selected runs were compared with previously published h vs T_s curves. It should be noted that quenching conditions were only *approximately* similar. In particular, the surface condition strongly influences the heat transfer.

Quenching in Still Air. Figure 12 compares the h-curve determined in this work to that obtained by Lambert and Economopoulos [15] for non-oxidized steel specimens quenched in still air under similar conditions. For most of the temperature range the results of both experiments are in good agreement. The maximum deviation in the two curves is at 550 °C, where the respective h-values are 68.6 and 54.4 W m^{-2} K^{-1}, or a difference of 24 %, which is in the order of the reproducibility of the results.

Quenching in Still Water. Figure 13 compares the *h*-curve determined in this work, with a stainless steel sample, to that obtained by Lambert *et al.* [15] using a nickel sample, for a still water quench under similar conditions. An earlier onset of nucleate boiling (point A) is observed in their results due to the lower initial temperature of the sample.

Quenching in Still Brine. Figure 14 compares the *h*-curve determined in this work using a stainless steel sample, to that obtained by Lambert and Greday [18] (nickel sample of cylindrical shape), for a still brine (3 w/o NaCl) solution quench under similar conditions. The maximum heat-transfer coefficient due to nucleate boiling is similar with the onset of nucleate boiling occurring at higher temperatures in this study. The maximum difference between the two curves occurs at 750 °C where their values are within 25 %. Below 210 °C the curve obtained in this work shows a slower reduction in *h*-value than that reported by Lambert and Economopoulos [18]

Quenching in Still Oil. Figure 15 compares the *h*-curve determined in this work with a stainless steel specimen, to that obtained by Stolz *et al.* [19] using a silver specimen of spherical shape, for a still oil quench. The properties of the two oils involved are given in Table II. The peaks in both curves occur at a common temperature of 485 °C, where their values are within 10 %. At higher temperatures, the heat-transfer coefficient measured in this study declines less rapidly through the transition boiling zone that that reported by Stolz *et al.* [19].

Summary and Conclusions

The effect of several quench variables on the heat-transfer coefficient was determined as a function of surface temperature for the quenching of disks of stainless steel and mild steel in water, brine, oil, and air. The inverse heat conduction problem of estimating surface heat-transfer coefficients and surface temperatures from the temperature response measured at a location inside the specimen was solved by applying sequential matching and sequential function specification techniques; the latter is robust, can be readily coded and produces good results even for small time increments (for which 'brute force' and sequential matching techniques become unstable).

The results obtained in this work are reproducible to within ± 18 % and are in reasonable agreement with previously published results. They can be summarized as follows :

- A higher initial specimen temperature increases the stability of the vapor film, thus prolonging the range of the stable film boiling and transition boiling stages.

- A hotter quench bath also increases the stability of the vapor film.

- Agitation of the bath may reduce the stability of the vapor film by introducing physical disturbances.

- Heavy oxide scale provides nucleation sites and reduces the heat extraction rate.

- In order of decreasing quenching efficiency the four media tested were brine (3 w/o NaCl), tap water, oil (Houghton-Quench-G) and air.

Acknowledgements

The authors are grateful to the Natural Sciences and Engineering Council of Canada for financial support of this study. The assistance of Prof. J. V. Beck (Michigan State University) in providing the original code **CONTA** is gratefully acknowledged. One of the authors (B. H-M.) is also grateful for the financial support provided by Universidad Nacional Autónoma de México.

List of Symbols

a_i	–	Coefficient in finite-difference eqns., dimensionless
b_i	–	coefficient in finite-difference eqns., dimensionless
c_i	–	coefficient in finite-difference eqns., dimensionless
Cp	–	heat Capacity, J kg^{-1} K^{-1}
D	–	rod diameter, mm
d_i	–	coefficient in finite-difference eqns., °C
$\bar{h}$	–	heat-transfer coefficient, W m^{-2} K^{-1}
$\hat{h}$	–	estimated heat-transfer coefficient, W m^{-2} K^{-1}
k	–	thermal conductivity, W m^{-1} K^{-1}
L	–	plate thickness, mm
n	–	node corresponding to thermocouple location, dimensionless
q	–	surface heat flux, W m^{-2}
$\hat{q}$	–	estimated surface heat flux, W m^{-2}
r	–	number of future time steps, dimensionless
S	–	least squares function, °C^2
t	–	time, s
T	–	temperature, °C
T_f	–	quench bath temperature, °C
T_o	–	initial sample temperature, °C
$\hat{T}_s$	–	estimated surface temperature, °C
x	–	axial coordiante, m
X	–	sensitivity coefficient, °C W^{-1} m^2
Y	–	measured temperature, °C
Ω	–	entire domain, m
Ω_1	–	subdomain 1, m
Ω_2	–	subdomain 2, m
ρ	–	density, kg m^{-3}
Δt	–	time step, s

References

[1] G.E. Totten, M.E. Dakins, and R.W. Heins. Cooling Curve Analysis of Synthetic Quenchants – A Historical Perspective, *J. Heat Treating*, 6, (2), 1988, pp. 87 – 95.

[2] T.E. Lim. Optimizing Heat Treatment with Factorial Design. *The Journal of The Minerals, Metals and Materials Society*, 41, (3), 1989, pp. 52 – 53.

[3] R.B. Frank and R.K. Mahidhara. Effect of Heat Treatment on Mechanical Properties and Microstructure of Alloy 901. In *Superalloys 1988*, S. Reichman, D.N. Duhl, G. Maurer, S. Antolovich, and C. Lund, Eds. The Metallurgical Society of AIME, Warrendale, PA, 1988. Pp. 23 – 32.

[4] F.M.B. Fernandez, S. Denis, and A. Simon. Mathematical Model Coupling Phase Transformation and Temperature Evolution During Quenching of Steels. *Mater. Sci. Technol.*, 1, (10), 1985, pp. 838 – 844.

[5] S. Sjöström. Interactions and Constitutive Models for Calculating Quench Stresses in Steel. *Mater. Sci. Technol.*, 1, (10), 1985, pp. 823 – 829.

[6] B. Hildenwall and T. Ericsson. Prediction of Residual Stresses in Case-Hardening Steels. In *Hardenability Concepts with Applications to Steel*, D.V. Doane and J.S. Kirkaldy, Eds., The Metallurgical Society of AIME, Warrendale, PA, 1978, pp. 579 – 606.

[7] V.K. Dhir. Nucleate and Transition Boiling Heat Transfer under Pool and External Flow Conditions. In *Heat Transfer 1990*, Vol. 1, G. Hetsroni, Ed., Hemisphere Publishing Comp., New York, 1990, pp. 129 – 155.

[8] H. Auracher. Transition Boiling. In *Heat Transfer 1990*, Vol. 1, G. Hetsroni, Ed., Hemisphere Publishing Comp., New York, 1990, pp. 69 –90.

[9] A. Sakurai. Film Boiling Heat Transfer. In *Heat Transfer 1990*, Vol. 1, G. Hetsroni, Ed., Hemisphere Publishing Comp., New York, 1990, pp. 129 – 155.

[10] J.V. Beck, B. Litkouhi and C.R. St. Clair, Jr. Efficient Sequential Solution of the Nonlinear Inverse Heat Conduction Problem, *Numerical Heat Transfer*, 5, 1982, pp. 275 – 286.

[11] J. V. Beck. User's Manual for CONTA – Program for Calculating Surface Heat Fluxes From Transient Temperatures Inside Solids. Sandia National Laboratory, Report SAND83-7134, Dec. 1983.

[12] F.P. Incropera and D. P. De Witt. **Fundamentals of Heat and Mass Transfer**. John Wiley & Sons, New York, 3rd. ed., 1990.

[13] W. Peyayopanakul and J.W. Westwater. Evaluation of the Unsteady-State Quenching Method for Determining Boilinf Curves. *Int. J. of Heat and Mass Transfer*, 21, 1978, pp. 1437 – 1445.

[14] D.Y.T. Lin and J.W. Westwater. Effect of Metal Thermal Properties on Boiling Curves Obtained by the Quenching Method. In *Heat Transfer 1982*, vol. 4, U. Grigull, E. Hahne, . Stephan and J. Straub, Eds., Hemisphere Publishing Comp., Washington, 1982, pp. 155 – 160.

[15] N. Lambert and M. Economopoulos. Measurement of the Heat-Transfer Coefficients in Metallurgical Processes. *JISI*, 208, (10), 1970, pp. 917 – 928.

[16] M. Mitsutsuka and K. Fukuda. The Transition Boiling and Characteristic Temperature in Cooling Curve During Water Quenching of Heated Metal. *Trans. ISIJ*, 16, (1), 1976, pp. 46 – 50.

[17] **ASM Handbook**, Vol. 4. ASM International, Warrendale, PA, 1991, p. 72.

[18] N. Lambert and T. Greday. *CNMR Metallurgical Report* 44, Sept. 1975, pp. 13 – 27.

[19] G. Stolz Jr., V. Paschkis, C.F. Bonilla and G. Acevedo. *JISI*, 197, 1959, pp. 116 – 123.

[20] J.V. Beck, B. Blackwell and Charles R. St. Clair, Jr. **Inverse Heat Conduction : Ill Posed Problems**. Wiley-Interscience, New York, 1985, p. 6.

Table I: Quench Variables Studied

Variable	Level			
Initial specimen temperature, °C	842		891	
Initial bath temperature, °C	20	40	60	
Bath agitation	yes		no	
Surface oxidation*	light oxide		heavy oxide	
Quench medium	Water	Brine	Oil	Air

* Mild steel specimens were used for these runs.

Table II: Physical Properties of Quenching Oils

Property	HQ-G (this work)	Slow Oil [19]
Flash Point, °C	180	182
Fire Point, °C	210	210
Viscosity at 37.8 °C, cP	105 – 115	103

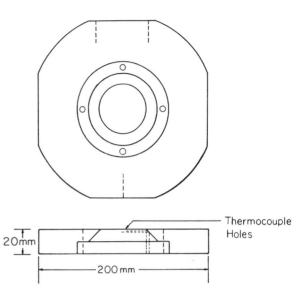

Figure 2: Configuration of disk quench sample showing thermocouple location.

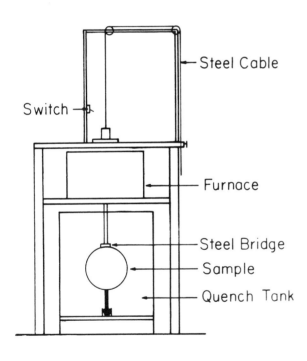

Figure 1: Schematic diagram of the apparatus used in this study.

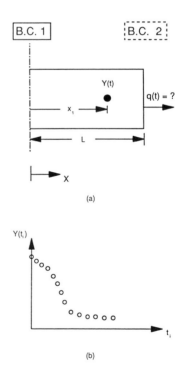

Figure 3: (a) Schematic representation of a single sensor IHCP, (b) discrete temperature measurements at position x_1.

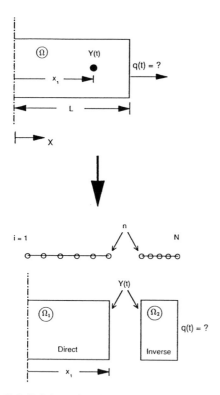

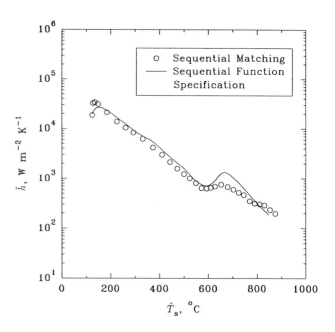

Figure 4: Subdivision of a single sensor IHCP into 2 subdomains (modified from [20]).

Figure 6: Estimated heat-transfer coefficient for quenching a stainless steel disk in still water at 60 °C as calculated by applying sequential matching and sequential function specification techniques.

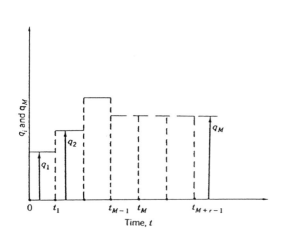

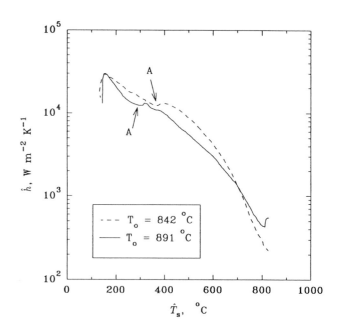

Figure 5: The temporary assumption of constant surface heat flux used in the function specification technique [10].

Figure 7: Estimated heat-transfer coefficient as function of estimated surface temperature for stainless steel disks quenched from 842 °C and 891 °C in still water at 20 °C.

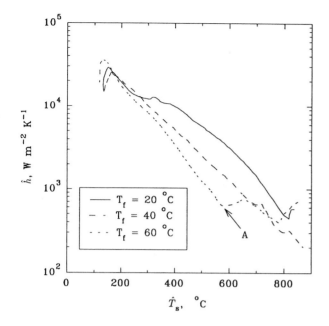

Figure 8: Estimated heat-transfer coefficient as function of estimated surface temperature for stainless steel disks quenched in still water at 20 °C, 40 °C and 60 °C.

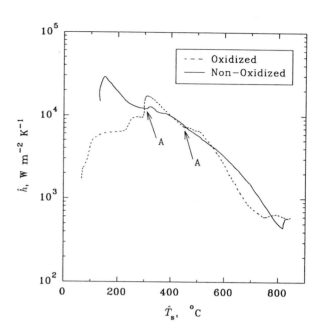

Figure 10: Estimated heat-transfer coefficient as function of estimated surface temperature for mild steel disks with and without surface oxidation quenched in still water at 20 °C.

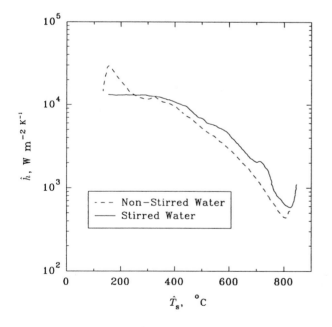

Figure 9: Estimated heat-transfer coefficient as function of estimated surface temperature for stainless steel disks quenched in still and stirred water at 20 °C.

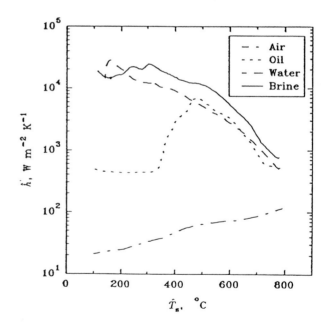

Figure 11: Estimated heat-transfer coefficient as function of estimated surface temperature for stainless steel disks quenched in water, brine, oil and air.

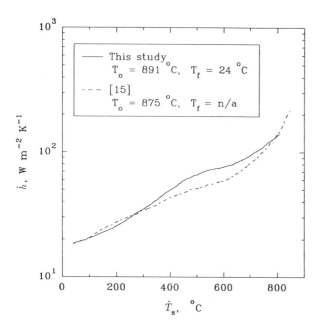

Figure 12: Comparison of estimated heat-transfer coefficients obtained in this study to those of Lambert and Economopoulos [15] as function of estimated surface temperature for mild steel samples quenched in air.

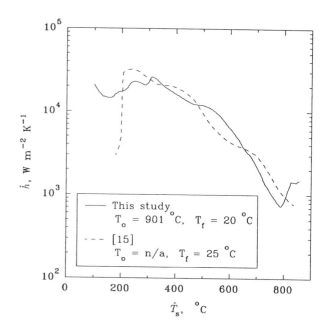

Figure 14: Comparison of estimated heat-transfer coefficients obtained in this study to those of Lambert and Greday [18] as a function of estimated surface temperature for stainless steel and nickel samples respectively, quenched in still brine (3 w/o).

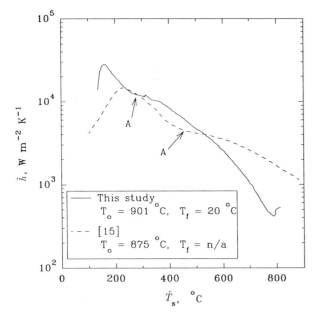

Figure 13: Comparison of estimated heat-transfer coefficients obtained in this study to those of Lambert and Economopoulos [15] as a function of estimated surface temperature for stainless steel and nickel samples respectively, quenched in still water.

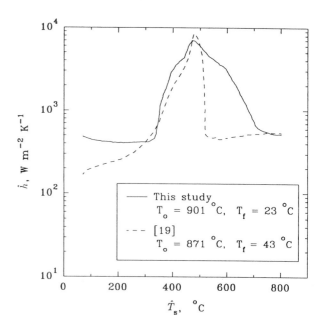

Figure 15: Comparison of estimated heat-transfer coefficients obtained in this study to those of Stolz *et al.* [19] as a function of estimated surface temperature for stainless steel and silver samples respectively quenched in still oil.

Variation in the Heat Transfer Coefficient Around Components of Different Shapes During Quenching

S. Segerberg and J. Bodin
Swedish Institute of Production Engineering Research
Göteborg, Sweden

ABSTRACT

Increasing use is being made of computers for calculation of effects such as temperature distribution, distortion and residual stresses in components after hardening. One of the input parameters used in such calculations is the heat transfer coefficient between the surface of the component and the quenchant. If the results are to be correct, it is necessary to know how this heat transfer coefficient varies from one part of the surface concerned to another. This article describes the results of measurements and calculations for cylinders and rings. It is apparent that the value of the heat transfer coefficient for upper, lower and vertical surfaces differs.

The article also describes the results of measurements and calculations of the heat transfer coefficient of the curved surface of Inconel 600 cylinders having diameters of 12.5 mm, 30 mm and 130 mm. It was found that the heat transfer coefficients at a position half-way along the cylinders do not differ greatly, despite the major differences in size.

COMPUTER PROGRAMS are available today for such applications as temperature curves or gradients, microstructure, hardness, distortions and residual stresses after hardening. A prerequisite for useful results is that there should be accurate mathematical models for calculating temperature curves, phase transformations and stresses in the interior of the parts. Another prerequisite is that the quality of input data, in the form of material data and heat transfer coefficients between the component and the quenchant, should be good. Modern computerised instrumentation permits relatively simple and detailed recording of quenching processes, from which the heat transfer coefficient can be calculated.

The different phases of cooling

The first phase of cooling of components, such as cylindrical test pieces, without oxidised surfaces or other surface coatings, in a hardening oil is the formation of a film of vapour around the test piece, see Figure 1a. At this stage, the heat radiated from the surface of the test piece is so high that the quenchant immediately adjacent to the surface is vaporised. The temperature falls most rapidly at corners and where sections are thin, with the result that radiation decreases and the thickness of the film of vapour is reduced. Eventually, the quenchant comes into contact with the surface and starts to boil, see Figure 1b. Boiling abstracts considerable heat from the surface, with the result that the temperature drops rapidly. As the vapour film collapses, boiling spreads upwards and downwards over the surface of the test piece until the two boiling fronts meet immediately above the centre of the test piece, see Figure 1c. When the surface temperature of the test piece has fallen below the boiling temperature interval of the quenchant, the boiling phase ceases and convection starts. Figure 1d shows boiling about to finish at the centre of the test piece, while convection has started on the upper and lower surfaces. When the temperature of the entire surface of the test piece has fallen below the boiling temperature interval of the quenchant, the only form of cooling still active is convection from all parts of the test piece, see Figure 1e. Figure 1 also shows how the overall cooling characteristic relates to the various stages of the quenching process.

At the same time, this means that the quenching process is dependent on where the temperature is measured on a test piece or component. In order to be able to compare cooling curves of different quenchants or different quenching conditions, it is important to use a standardised type of test probe when testing different quenchants. The test probe must be standardised in respect of shape and size, the position of thermocouple and the type of material, e.g. as set out in ISO/DIS Draft 9950 (2).

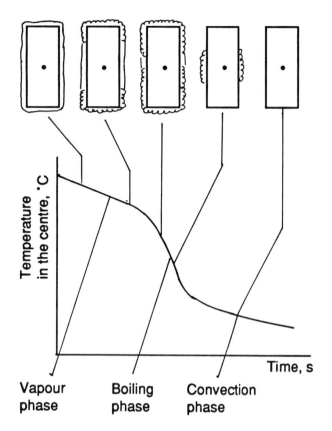

Fig. 1. Typical cooling curve, showing the three phases of quenching in relation to growth of the boiling and convection front around the test piece (1).

Cooling curves of long components

In the case of long components, the duration of the vapour film phase at a particular point on the surface depends on the distance between the point and the end surface (1), see Figure 2. When boiling starts from the end surface, the vapour phase changes to the boiling phase more quickly, the nearer the point is to the end of the component.

The boiling phase cannot, however, continue indefinitely. In the case of the particular hardening oil considered here, the vapour phase persists down to a temperature of about 650 °C. By then, the boiling phase has extended to about 35 mm from the end surface, which means that although the vapour film still covers most of the test piece as shown in Figure 2, it has become unstable, suddenly collapsing and allowing boiling to start over the rest of the surface of the test piece. The reason for this is that radiation from the cooling surface is no longer sufficient to vaporise the hardening oil, so that it comes into direct contact with the surface of the test piece and boiling starts spontaneously over the entire surface. Similar results have been previously described in the litterature (5).

Heat transfer coefficients of cylinders and rings

Even for such a simple geometry as that of a cylinder, the heat transfer coefficient varies over its curved surface and its plane end surfaces. The heat transfer coefficient was calculated by the "ivf quenchosoft" finite difference program, using an iterative calculation method, employing as input data the cooling curves as obtained from

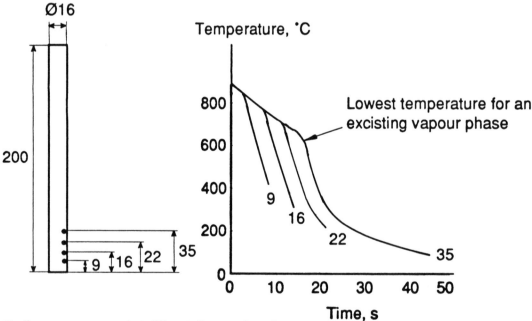

Fig. 2. Cooling curves measured at different distances from the lower end surface of the test piece.

thermocouple fitted immediately beneath the surface of the cylinder, see Figure 3. The calculations were based on the use of a one-dimensional model for calculation of the temperatures at points along the length of the cylinder, while its two-dimensional module was employed for calculation of temperatures at the corners and end surfaces.

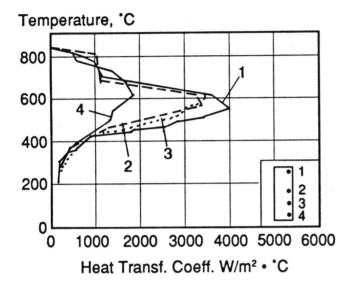

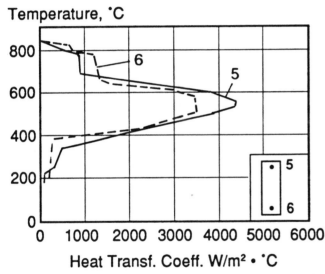

Fig. 3. Heat transfer coefficient of a cylinder, ø30 x 90 mm. Cooling in stationary oil at a temperature of 70 °C. Top: As measured along the curved surface and at the corners. Bottom: On the end surfaces.

The heat transfer coefficients at the two points on the curved surface are almost the same. The difference between the upper and lower corners is noticeable, but not unexpected when quenching in stationary oil. Oil closest to the surface of the test piece will flow upwards by natural convection as the oil heats up. This means that the upper corner of the test piece is not, in reality, beeing cooled by stationary oil, with the result that this major difference in heat

transfer coefficient arises. If forced circulation from below is employed, the difference in heat transfer coefficient can be reduced as a result of more uniform flow velocity over the surface.

The heat transfer coefficient at the upper surface of the cylinder is somewhat greater than on the lower surface. This, too, may also be due to natural convection, although there is also another mechanism involved, as will be discussed in the chapter four.

The geometry of rings is somewhat more complicated than that of cylinders. A ring that was quenched together with the cylinder above was fitted with five thermocouples: four immediately beneath the respective surfaces and one at the centre, as shown in, Figure 4. The heat transfer coefficients were calculated in the same way.

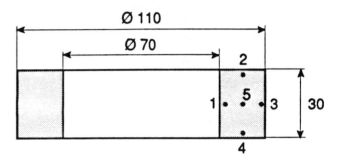

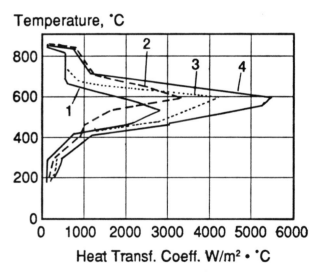

Fig. 4. The heat transfer coefficient of a ring. Cooling was carried out in stationary oil at a temperature of 70 °C with the ring in a horizontal position.

The magnitudes of the heat transfer coefficients are approximately the same as for the cylinder. The height of the ring is much less than that of the cylinder, which means that the effects of natural convection are less marked, which in turn reduces the value of heat transfer coefficient on the upper surface. Calculations employed the two-dimensional module. The same heat transfer coefficient was used for all surfaces for the first calculation of Point 1, when calculating the value for Point 2, the previously calculated value for Point 1 was used, while the values for the other surfaces were all the same. In this way, the most recently calculated values of the heat transfer coefficient were used for each subsequent calculation. It was necessary to perform this iterative process four times around the cross-section of the ring before stable values were obtained. A comparison of the measured and the calculated cooling curve for Point 5 indicated a maximum difference of only 12 °C, which occurred during the boiling phase when cooling is at its most rapid. The maximum difference during other parts of the quenching process was only 5 °C.

Heat transfer coefficients for horizontal and vertical surfaces

Another type of ring, flat and wide, was fitted with thermocouples on both the top and bottom, as well as on the inner diameter. It was quenched in a horizontal position, with the result that the heat transfer coefficient of the top was considerably higher than on the bottom, see Figure 5.

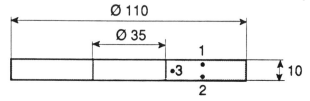

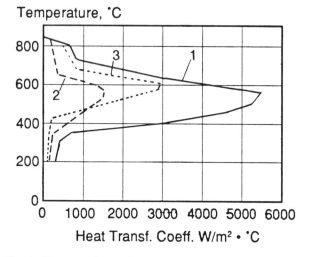

Fig. 5. Heat transfer coefficients for a flat, wide ring. Cooling in horizontal position in stationary oil at 70 °C.

The reason for the lower value of the heat transfer coefficient on the underside is probably to be found in the fact that the film of vapour formed there acts as an insulating layer. As the lower surface was horizontal, it is more difficult for the bubble of vapour to move away or disappear in any direction, so that it becomes relatively stable.

The heat transfer coefficient of the upper surface of the cylinder in Figure 3 was also somewhat higher than for other parts of the cylinder, while the opposite was the case for the ring in Figure 4. The reason for this difference may be that the ring was smaller than the cylinder, 20 mm instead of 30 mm, so that the vapour film had more difficulty in staying beneath the ring. Corresponding results have previously been described in the literature (4, 6), with the heat transfer coefficient for the upper surface having been higher than for the lower surface.

When the ring is quenched hanging in a vertical position, both sides are cooled equally, as shown in Figure 6. This means that the distortion is nowhere near as great as if the ring is quenched horizontally.

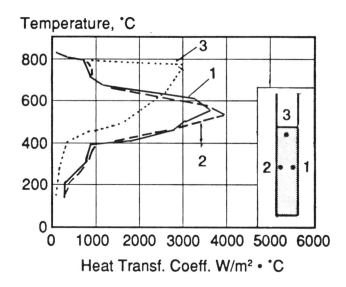

Fig. 6. Heat transfer coefficient for a flat, wide ring. The ring was cooled vertically in stationary oil at 70 °C.

If care is not taken to ensure that components which are to be quenched horizontally are actually horizontal, it is easy for the vapour bubble to rise away from the underside. This can lead to variations in the quenching process which will result in variations in the distortion.

The effect of heat transfer coefficient on test piece dimensions

In order to investigate whether the dimensions of cylinders are important in determining the value of the heat transfer coefficient, tests were performed with three cylinders of Inconel 600, a low-oxidation recistance alloy, having diameters of 12.5 mm, 30 mm and 130 mm. Temperatures were measured immediately beneath the surface of the two larger cylinders and at the centre of the smallest cylinder.

The diagram in Figure 7, top, shows the relative proportions of the test cylinders and their cooling curves. It should be noted that the vapour film phase is longer for the larger sizes. However, the duration of the vapour film phase for the largest test piece is not proportional to the length of the cylinder: instead, the film has become unstable and boiling has started spontaneously, as previously described in Chapter 2. The heat transfer coefficients were calculated from the cooling curves, see Figure 7, bottom. The results indicate very good agreement, which means that heat transfer coefficients calculated on the basis of test results obtained from test probe as recommended by IFHT (2) can clearly also be used for calculations of heat transfer coefficients for larger sizes.

Results of a corresponding comparison have previously been published (4) for cylinder diameters up to 30 mm. The report also describes tests made on silver cylinders. The results for the silver cylinders exhibited good agreement with the results for the 12.5 mm Inconel cylinder, despite the fact that the characteristics of silver differ very considerably from those of Inconel 600.

For measuring the cooling curves, see Figure 8, the "ivf quenchotest" equipment was used.

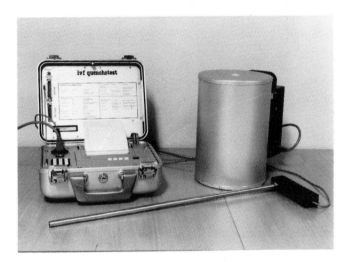

Fig. 8. Equipment used for plotting the cooling curves for the ordinary ø12.5 x 60 mm Inconel 600 test probe and for the larger test pieces.

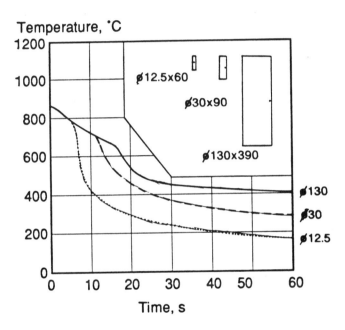

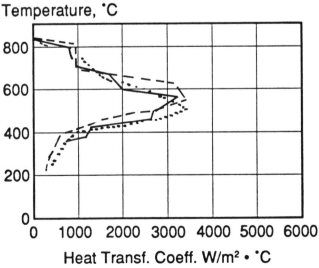

Fig. 7. Test piece diameters, temperature/time curves and heat transfer coefficients for cylinders of different diameters. Cooling in stationary oil at 70 °C.

The importance of heat transfer coefficient when calculating distortions

As described above, the heat transfer coefficient varies from one part of the surface of a component to another. It is therefore necessary to know how the heat transfer coefficient varies in order to be able correctly to calculate the temperature distribution in the interior of the part. This is needed, for example, when calculating distortion. Distortion of a ring have been calculated a) using the same value of heat transfer coefficient for all parts of the surface and b) using values which varied for the internal diameter of the ring (5), see Figure 9.

It can be seen from the figure that both the inner and the outer diameters increase when the heat transfer coefficient for the inner diameter is low. This may be due to the fact that the martensite conversion now occurs later at the inner diameter, allowing the martensite that has been formed on the outer diameter to expand and so increase the diameter. A more correct and accurate calculation requires knowledge of the respective values of the heat transfer coefficient for all surfaces.

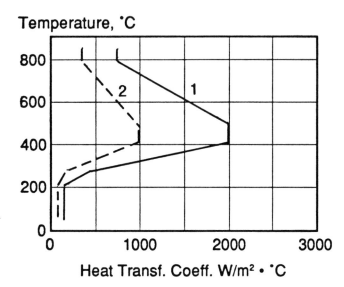

Temperature, °C

Heat Transf. Coeff. W/m² • °C

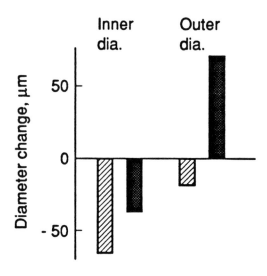

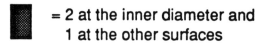

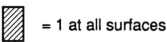

= 1 at all surfaces

= 2 at the inner diameter and 1 at the other surfaces

Fig. 9. Dimensional changes of inner and outer surfaces. Ring dimensions: thickness 10 mm, outer diameter 108 mm, inner diameter 36 mm.

Conclusions

In order to use computer programs for calculation of temperature distribution, hardness, distortion and residual stresses in components after hardning, it is necessary to have accurate input data for various parameters, including the heat transfer coefficient. However, the value of this coefficient varies from one part of a component to another, making it more difficult and time-consuming to define it.

Examples of significance of variations in the heat transfer coefficient for distortion of a ring have been described. Calculation of the heat transfer coefficient based on results from a small test probe can, with acceptable accuracy, be used for larger dimensions.

Acknowledgement

This work has been sponsored by the Swedish Board for Technical Development (NUTEK), the Nordic Fund for Industrial Development (NI) and five Swedish industrial companies, to which the authors express their gratitude.

References

1. S. Segerberg, *IVF-resultat 88605*, Sveriges Mekanförbund, Stockholm, (1988)

2. *Industrial Quenching Oils - Determination of Cooling Characteristics - Laboratory Test Method*. Draft international standard ISO/DIS 9950, International Organization for Standardization (submitted 1988)

3. N. Shimizu and I. Tamura, *Transaction ISII*, 16, (1976)

4. S. Segerberg, J. Bodin, in *Proc. 3rd Int. Seminar: Quenching and Carburising, Melbourne, 2-5 Sept. (1991)*, Institute of Metals & Materials Australasia Ltd. Parkville Vic 3052, Australia

5. A. Thuvander, *IM-2671*, The Swedish Institute for Metals Research, Stockholm, (1990)

6. R.A. Wallis, *Heat Treating*, Dec, (1989)

Proceedings of the First International Conference on Quenching & Control of Distortion, Chicago, Illinois, USA, 22-25 September 1992

The Influence of Heat Transfer on the Development of Stresses, Residual Stresses and Distortions in Martensitically Hardened SAE 1045 and SAE 4140

A. Majorek, B. Scholtes, H. Müller, and E. Macherauch
Institut für Werkstoffkunde I, Universität Karlsruhe (TH)
Karlsruhe, Germany

Abstract

In course of quenching during the martensitic hardening of steels, the cooling process is of great importance not only for the development and the distributions of temperatures, microstructures, deformations and stresses, but also for the residual stresses and distortions remaining after temperature balance. In such treatments the surface heat transfer conditions between steel parts and cooling media, which can readily be changed, are the most important factors influencing and controlling the generation of stress and distortion. For SAE 1045 and SAE 4140 steel cylinders martensitically hardened in oil or water, respectively, the consequences of heat transfer on the development of residual stresses and of changes in dimension and shape after quenching have been investigated systematically using FD-/FE-methods [1, 2, 3]. In this paper the results of numerical simulations of immersion cooling with selected cooling characteristics are described and the results compared with experimental observations. First the consequences of the consecutive stages of heat transfer comprising the vapor blanket stage, the nucleate boiling stage and the convective stage for the development of the residual stresses are outlined. Then, calculated results of immersion hardening processes are compared with experimental observations to find out whether or not effective local heat transfer coefficients must be considered to predict wetting kinematics as well as temperature and residual stress distributions correctly. In each case, selected examples are presented and discussed.

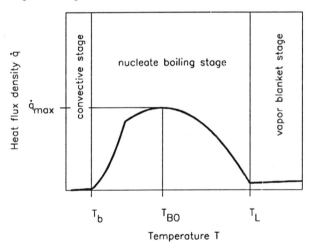

Figure 1: Heat flux density vs. temperature (schematically) for cooling of metallic parts in vaporizing liquids with cooling stages and characteristic points

THE SURFACE HEAT TRANSFER CONDITIONS during immersion cooling of metallic parts in vaporizing liquids, e.g. water or oil, are usually specified by T,t-curves measured at the center of a sample [4] or by temperature dependent heat flux densities [5] calculated with respect to experimentally determined cooling curves near the surface. If the initial temperature of the workpiece is well above the boiling point of the quenchant, successive stages of heat transfer will characterize the cooling process: vapor blanket stage, nucleate boiling stage and convective stage, see Figure 1.

After a very short initial liquid contact the workpiece is completely surrounded by a stable vapor film. The rate of heat removed from the workpiece at this stage is very low, because of the high thermal resistance of the vapor layer. The temperature, at which the stable film breaks down and bubbles start to detach from the surface of the workpiece, is called Leidenfrosttemperature T_L. Then, the heat flux density increases and reaches its maximum at the Burn-out-point. Below the temperature corresponding to the Burn-out-point T_{BO} the heat flux density decreases until the surface temperature of the specimen reaches the boiling temperature of the quenchant T_b. In the convective stage the heat transfer is determined by free or forced convection.

Calculating Procedure and Experimental Details

For studying the heat treating problems of interest, a special FD-/FE-program was used [1, 2, 3], which takes into account the temperature and phase dependencies of all relevant mechanical and thermophysical properties, including transformation plasticity effects. To approximate the real heat transfer conditions during martensitic hardening in vaporizing liquids temperature dependent as well as locally characterized heat transfer coefficients were used.

Since there are no reliable quantitative data available for describing the immersion cooling of cylindrical steel parts with phase transformation, experimentally determined [5] heat transfer coefficients h(T) of nickel cylinders (D = 30 mm) immersed in water and oil of room-temperature were app-

lied for the FE-calculations. They were linearly approximated within four temperature ranges: above the Leidenfrosttemperature T_L, between T_L and the temperature of maximum heat transfer coefficient T_{hmax}, between T_{hmax} and the boiling temperature T_b of the quenchant and below T_b. Figure 2 shows the experimentally determined as well as the linearly approximated h,T-curve for oil quenching. The characteristic values, T_L, T_{hmax} and h_{max}, were systematically varied for the calculations performed as indicated by arrows. For water quenching a qualitatively similar h,T-curve was used with the basic data $T_L = 700\ ^oC$, $T_{hmax} = 225\ ^oC$, $h_{max} = 0.015\ W/mm^2K$ and $T_b = 100\ ^oC$.

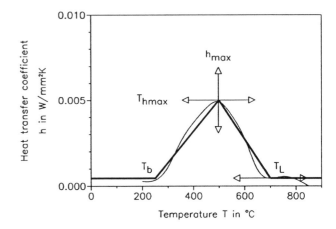

Figure 2: Heat transfer coefficient in dependence of surface temperature for oil quenching

Quenching experiments were carried out using SAE 1045 and SAE 4140 steel cylinders with a diameter to length ratio D/L = 20/60. They were immersed in unagitated water of room-temperature (20 ^oC) starting from an austenitizing temperature of 830 ^oC for SAE 1045 and 860 ^oC for SAE 4140, respectively.

Results and Discussion

Influence of the Leidenfrosttemperature on Residual Stresses. The Leidenfrosttemperature, at which the stable vapor film breaks down, is influenced by many factors, e.g. temperature of the cooling medium, material and shape of the sample, type and convection of the quenchant and surface state of the specimen [5]. An increasing Leidenfrosttemperature shortens the vapor blanket stage and increases the heat transfer during the nucleate boiling stage. However, the heat transfer conditions in the temperature range of martensitic transformation of steels, which are usually below 400 ^{o}C, are not markedly influenced. On the other hand, increasing Leidenfrosttemperatures suppress the diffusion-controlled ferritic/pearlitic transformation, since the heat transfer will be raised during the corresponding temperature range of the nucleate boiling stage.

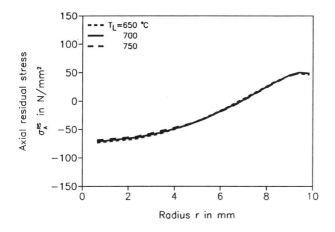

Figure 3: Axial residual stress vs. radius in the middle plane of a SAE 4140 cylinder with D = 20 mm quenched in oil for three different Leidenfrosttemperatures T_L

The influence of different Leidenfrosttemperatures and of the resulting heat transfer coefficients on the axial residual stress distribution in the middle plane of SAE 4140 steel cylinders with D = 20 mm, quenched from 860 ^{o}C in oil of room-temperature is shown in Figure 3. The calculations were carried out for the Leidenfrosttemperatures 650 ^{o}C, 700 ^{o}C and 750 ^{o}C and with $h_{max} = 0.005$ W/mm^{2}K and $T_{hmax} = 500$ ^{o}C. Cooling of the cylinder from all sides was assumed. The resulting axial residual stress distributions in the middle plane of the cylinder are

of the transformation type. Obviously in this particular case of full martensitic hardening changes of the Leidenfrosttemperature do not influence the calculated residual stress states.

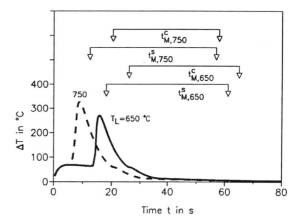

Figure 4: Temperature differences between core and surface in the middle plane of a SAE 4140 cylinder with D = 20 mm quenched in oil and time intervals of martensitic transformation at the surface t_M^S and in the core t_M^C calculated for two Leidenfrosttemperatures T_L

This happens, although the maximum temperature differences between core and surface as well as the martensitic transformations at the surface and in the core occur at different instants. Figure 4 shows for $T_L = 650$ ^{o}C and 750 ^{o}C the courses of the temperature difference between core and surface in the middle plane of the oil-quenched cylinders together with the martensitic transformation intervals at the surface t_M^S and in the core t_M^C. Besides of the variations in the maximum temperature differences between core and surface, temperatures and martensitic transformations develop nearly similar for both Leidenfrosttemperatures assumed. The courses are only temporally shifted as a consequence of the duration of the vapor blanket stage and the heat transfer conditions at the beginning of the nucleate boiling stage. In both cases the transformations at the surface start briefly after reaching the maximum temperature difference. At the beginning of the transformations in the core, which is about 8 seconds after transformation beginning at the surface, 58 Vol.-% austenite have transformed into martensite in the outer layers of the cylinder, respectively. In both cases the transformations in the core shortly end after termination of the transformations at the surface. Since at the instant, when martensitic transformations start, no large differences of the thermal stresses exist, the residual stress states of the full martensitically hardened SAE 4140 cylinders are comparable.

Similar results have been obtained for SAE 1045 and SAE 4140 steel cylinders with D = 10 and 20 mm quenched in oil and water, respectively.

Influence of Nucleate Boiling Stage on Residual Stresses. The nucleate boiling stage of a quenchant usually corresponds to the temperature range of the martensitic transformation of steels. Consequently, the heat transfer conditions during this cooling stage are of highest importance for the stresses and distortions in a martensitically hardened part.

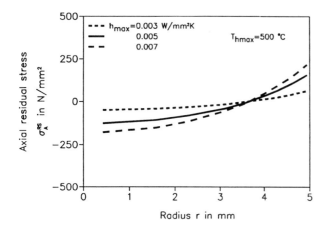

Figure 5: Axial residual stress vs. radius in the middle plane of a SAE 1045 cylinder with D = 10 mm quenched in oil for different maximum heat transfer coefficients h_{max}

In the following, the influence of heat transfer during the nucleate boiling stage on the development of axial residual stresses in SAE 1045 steel cylinders with a diameter of 10 mm quenched from 830 °C in oil of 20 °C is exemplarily demonstrated. The basic heat transfer coefficient h(T) was characterized by $T_L = 700$ °C, $T_{hmax} = 500$ °C and $h_{max} = 0.005$ W/mm²K. The boiling temperature of the oil was assumed to be $T_b = 250$ °C. For a constant T_L-value, calculations were performed for $T_{hmax} = 400$ and 600 °C and $h_{max} = 0.003$ and 0.007 W/mm²K. The heat transfer conditions during the vapor blanket stage and the convective stage were kept constant. Figures 5 and 6 show the calculated distributions of axial residual stresses in the middle plane of the cylinders.

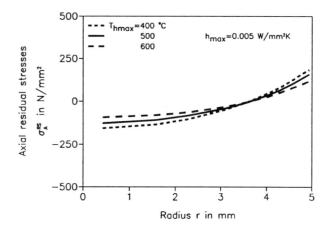

Figure 6: Axial residual stress vs. radius in the middle plane of a SAE 1045 cylinder with D = 10 mm quenched in oil for different temperatures T_{hmax}

All resulting residual stress distributions are of the transformation type with tensile stresses at the surface and compressive stresses in the core of the cylinder. Independently of the heat transfer conditions the axial residual stresses change their sign always at r = 3.7 mm. The influence of increasing h_{max}- and T_{hmax}-values on the resulting residual stress distributions is contrary. The middle plane stresses are homogenized with increasing temperatures T_{hmax} and decreasing coefficients h_{max}.

How $h_{max} = 0.007$ W/mm²K and $h_{max} = 0.003$ W/mm²K, for $T_L = 700$ °C and $T_{hmax} = 500$ °C, influence the time dependent formation of axial

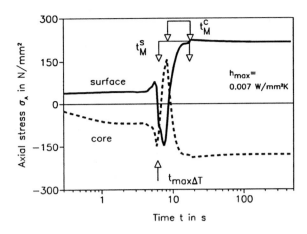

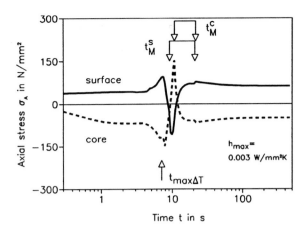

Figure 7: Formation of surface and core axial stresses in the middle plane of a SAE 1045 cylinder with D = 10 mm quenched in oil for $h_{max} = 0.007$ W/mm²K

Figure 8: Formation of surface and core axial stresses in the middle plane of a SAE 1045 cylinder with D = 10 mm quenched in oil for $h_{max} = 0.003$ W/mm²K

stresses is shown in Figures 7 and 8. In the diagrams the instants of maximum temperature difference between core and surface $t_{max\Delta T}$ and the martensitic transformation intervals at the surface t_M^S and in the core t_M^C are marked.

In both cases the same axial stresses develop until the film breaks down at the Leidenfrostpoint, because identical cooling conditions are assumed during the vapor blanket stage. The subsequent time dependent formation of stresses is similar, but different absolute stress values develop. The maximum temperature difference between core and surface occurs before the surface starts to transform. In each case martensitic transformations in the core start shortly after the beginning of transformations at the surface, but end almost at the same instant in the whole cross-section. Different amounts of residual stresses exist after temperature balance. The transformation of the cylinder cooled with the higher maximum heat transfer coefficient $h_{max} = 0.007$ W/mm²K is more inhomogeneous than that with $h_{max} = 0.003$ W/mm²K. This is due to the stronger gradient of the heat flux density $\partial \dot{q}/\partial T$ at the lower temperature range of the nucleate boiling stage. As a consequence, the amount of martensite at the surface formed at the beginning of transfor-

mation of the core is considerably different for the two heat transfer conditions. For cooling with $h_{max} = 0.007$ W/mm²K already 60 Vol.-% martensite are formed until the core starts to transform, in contrast to 45 Vol.-% martensite for cooling with the lower maximum heat transfer coefficient. The higher the amount of martensite transformed at the surface until the core reaches M_S, the more the transformation in the core dominates the last stage of the quenching process. As a consequence, the absolut values of the resulting residual stresses at the surface and in the core will become higher.

The same tendencies as discussed were obtained from model calculations with different temperatures T_{hmax}. With decreasing temperatures T_{hmax} the absolut values of residual stresses increase, because stronger temperature dependent gradients of the heat flux density during the temperature range of martensitic transformation occur.

Influence of Heat Transfer on Distortion. Besides of the knowledge of the residual stress state in the cylinders the generation of distortions during hardening is of great importance. The distortions are markedly influenced by small changes of the heat transfer conditions. This is shown in Figures 9 to 11 for martensitically hardened SAE 1045 steel cylinders with D = 20 mm quenched from 830 °C in water of 20 °C. The diagrams show the deformed cylinder quarters after temperature balance. Again the temperature dependent heat transfer coefficients were linearly approximated between the characteristic temperatures T_L, T_{hmax} and T_b. Different Leidenfrosttemperatures T_L and maximum heat transfer coefficients h_{max} were assumed.

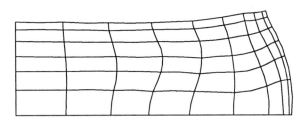

Figure 9: Fiftyfold enlarged deformations after quenching from 830 °C in water of 20 °C for T_L = 700 °C, T_{hmax} = 225 °C and h_{max} = 0.015 W/mm²K

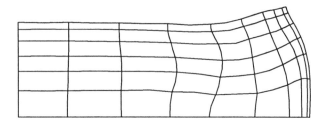

Figure 10: Fiftyfold enlarged deformations after quenching from 830 °C in water of 20 °C for T_L = 750 °C, T_{hmax} = 225 °C and h_{max} = 0.015 W/mm²K

After hardening the cylinders shown in Figures 9 and 10 take the shape of a spool. This is typical for water-quenching [6]. Near the end-face of the cylinder the changes in diameter are larger for calculations with T_L = 750 °C than for T_L = 700 °C. In addition, the cylinder with the shortener vapor blanket stage exhibits the larger changes in length over the whole cross-section. This can be seen comparing Figures 9 and 10.

Also, the reduction of the maximum heat transfer coefficient h_{max} in the nucleate boiling stage from 0.015 to 0.01 W/mm²K significantly affects the occuring changes in shape. Now, as demonstrated in Figure 11, the changes in diameter are nearly constant over the whole length of the cylinder. However, the changes in length are comparable to those calculated for the higher maximum heat transfer coefficients.

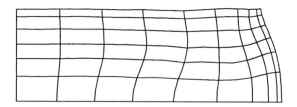

Figure 11: Fiftyfold enlarged deformations after quenching from 830 °C in water of 20 °C for T_L = 700 °C, T_{hmax} = 225 °C and h_{max} = 0.01 W/mm²K

Influence of the Effective Local Heat Transfer on Wetting Kinematics and Residual Stresses. In [7, 8], the wetting kinematics of steel cylinders quenched in vaporizing liquids from austenitizing temperature were studied in detail. In our investigation, SAE 1045 cylinders with 20 mm diameter and 60 mm length were used, which after final machining by grinding, had a surface roughness of 14 μm to 21 μm at the cylindrical surface and about 30 μm at the end-faces. The cylinders were induction heated within 5 minutes in argon atmosphere and austenitized for 3 minutes at 830 °C. Subsequently the samples were vertically immersed in unagitated water of 20 °C almost up to the top end-face. In addition to a visual and photographic registration of the wetting kinematics, the temperatures at the center of the specimen were registrated using compacted ceramic NiCr-Ni-thermocouples with an outside diameter of 0.5 mm.

Figure 12 schematically shows a typical wetting sequence during the immersion hardening process of a SAE 1045 cylinder. The unwetted regions, where the vapor blanket is stable, are unhatched. All the different stages of the cooling process are simultaneously present on the surface of the specimen. In the upper part of the cylinder the vapor blanket stage is effective, in the lower part drawn hatched the nucleate boiling and convective stage. Near the end-face, which has been first immersed, rewetting starts immediately.

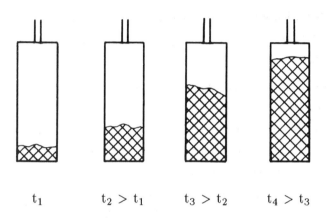

t_1 $t_2 > t_1$ $t_3 > t_2$ $t_4 > t_3$

Figure 12: Typical wetting stages of an immersed steel cylinder (schematically)

The change from vapor blanket stage to nucleate boiling stage is characterized by the rewetting time t_w [7] and the Leidenfrosttemperature T_L, respectively. t_w is the locally dependent time intervall between the stable vapor film break-down at the surface of the specimen and the first contact between sample and quenchant.

Figure 13 shows, that the rewetting time depends linearly on the distance from the lower end-face of the specimen. In the particular case investigated, the cylinder is completely rewetted after 12 seconds. Since similar experiments with SAE 4140 steel cylinders give the same results, the wetting kinematics seem to be independent of the steel type. The most essential influence on wetting is the geometry of the sample, especially the lower edge of

the cylinder, which immerses first and causes the break-down of the stable vapor film. Measurements using thermocouples placed near the surface of the cylinder at different distances from the lower face show, that at different locations different Leidenfrosttemperatures occur. T_L decreases with increasing distance from the lower end of the specimen. In the case of SAE 4140 steel cylinders T_L-values between 860 ^oC at the lower face and 750 ^oC in the middle plane of the cylinder were observed.

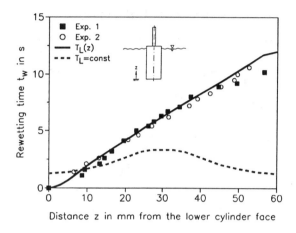

Figure 13: Experimentally and theoretically determined rewetting time in dependence of the distance from the lower cylinder face of a SAE 1045 cylinder with D = 20 mm immersed in water

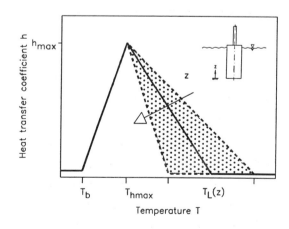

Figure 14: Effective local heat transfer coefficient (schematically)

Of course the non constant Leidenfrosttemperature influences the heat transfer of immersion cooled cylindrical parts and has to be taken into account for FE-calculations. Obviously, temperature dependent as well as effective local heat transfer coefficients have to be assumed, as schematically shown in Figure 14. For the calculations presented in Figures 13, 15 and 17, besides of $h_{max} = 0.015$ W/mm^{2}K and $T_{hmax} = 225\ ^oC$, Leidenfrosttemperatures were assumed to vary between 830 ^oC at the lower end-face and 630 ^oC at the upper one of the water-quenched SAE 1045 cylinder.

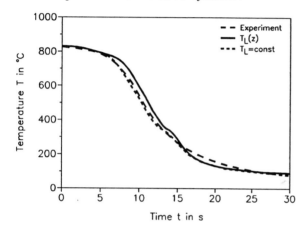

Figure 15: Experimentally and theoretically determined cooling curves for the center of a SAE 1045 cylinder with D = 20 mm immersed in water

In Figure 15 the experimentally determined T,t-curve in the center of the cylinder is shown. This curve agrees quite well with the calculated one. In addition, the result of a calculation with a constant Leidenfrosttemperature $T_L = 750\ ^oC$, T_{hmax} = 225 ^oC and $h_{max} = 0.01$ W/mm^{2}K is plotted. Also in this case, a good agreement with the experimental observation can be stated. However, in this latter case, the wetting kinematics are insufficiently described by the calculation. On the other hand, assuming effective local heat transfer coefficients leads to a quite good agreement between calculated and observed rewetting time. This can be seen from Figure 13, where the experimentally determined as well as the calculated values for $T_L =$

const. and $T_L(z)$ are plotted. The great discrepancies between experiment and calculation using constant T_L-values just at the upper and lower faces demonstrate without doubt, that the disturbing influences of the edges of the end-faces cannot be simulated with locally independent heat transfer coefficients.

The influence of effective local heat transfer coefficients on residual stresses can be seen comparing Figures 16 and 17, where the tangential residual stresses σ_T^{RS} are plotted in dependence of radius r and distance z from the lower end-face of the cylinder without (Figure 16) and with (Figure 17) consideration of local changes of heat transfer conditions.

Both calculations yield residual stresses of the transition type at the lower face (z = 0) and of the transformation type in the middle plane (z = 30 mm) of the cylinder. Remarkable differences occur near the upper face of the cylinder, where the residual stresses in both diagrams have opposite signs. In addition, the magnitudes of the residual stresses calculated with effective local heat transfer coefficients are larger all over the cylinder than in the case of locally independent heat transfer conditions. The results exemplarily shown for SAE 1045 steel cylinders with D/L = 20/60 immersed in unagitated water clearly show, that in addition to cooling curves measured at the center, detailed knowledge about the wetting kinematics is necessary to predict the development of temperatures and transformations as well as of residual stress states in a correct manner. Finally it should be pointed to the fact, that the influence of the effective local heat transfer coefficients especially on the development of residual stresses will be much more pronounced, when the developments of temperatures and transformations are more inhomogeneous than for the martensitically hardened cylinder investigated. This is valid in particular for immersion cooling in oil and/or of larger specimens.

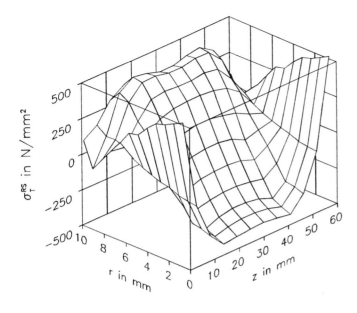

Figure 16: Tangential residual stresses of a SAE 1045 cylinder with D = 20 mm calculated without effective local heat transfer coefficients

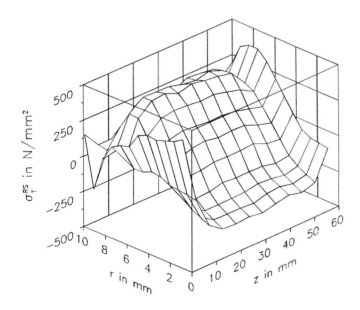

Figure 17: Tangential residual stresses of a SAE 1045 cylinder with D = 20 mm calculated with effective local heat transfer coefficients

Acknowledgement

The financial support of this research programme by the DFG (Deutsche Forschungsgemeinschaft, Bonn - Bad Godesberg) is gratefully acknowledged.

References

[1] Yu, H.-J., Wolfstieg, U., Macherauch, E., *Arch. Eisenhüttenwesen, Vol 49, pp 499-504 (1978)*

[2] Schröder, R., Scholtes, B., Macherauch, E., *HTM, Vol 39, pp 280-292 (1984)*

[3] Graja, P., *Dr.-Ing. Diss., Universität Karlsruhe (TH), (1987)*

[4] Rose, A., *Arch. Eisenhüttenwesen, Vol 13, pp 345-354 (1940)*

[5] Maaß, R., *Dr.-Ing. Diss., Techn. Universität Clausthal, (1988)*

[6] Schweizerische Fachgruppe für Wärmebehandlung, *Microtecnic, Vol 23, pp 303 and 410 (1969)*

[7] Tensi, H.M., Stitzelberger-Jakob, P., *Proc. of 6th Int. Conf. Heat Treatment of Met., Chicago, Ill., pp 171-176 (1988)*

[8] Tensi, H.M., Stitzelberger-Jakob, P., Stich, A., *HTM, Vol 45, pp 145-153 (1990)*

Heat Treatment, Microstructures, and Residual Stresses in Carburized Steels

G. Krauss
Colorado School of Mines
Golden, Colorado

ABSTRACT

This paper reviews the evolution of microstructure and residual stress during the heat treatment processing of gas carburized steels and relates these factors to fatigue resistance. The tempered martensite-retained austenite microstructure produced by quenching and tempering, and the prior microstructure produced by carburizing at temperatures where austenite is stable are described. Austenitizing affects grain growth, segregation of phosphorus, carbide formation, and surface oxidation. Factors which affect the development of surface residual stresses, i.e. quenching, refrigeration, oxide formation, tempering, shot peening, and strain-induced transformation of austenite are discussed. The microstructure and residual stresses are related to fatigue crack initiation mechanisms and fatigue performance.

INTRODUCTION

Carburized steels are widely used for shafts, gears, bearings, and other highly stressed machine parts. Most of these machine components are cyclically stressed and therefore resistance to fatigue fracture is a critical design factor. Resistance to rolling contact fatigue and/or bending fatigue may be required, depending on the application. This paper is restricted to bending fatigue performance.

Many processing and microstructural factors affect bending fatigue performance, and wide variations in endurance limits, ranging from 210 MPa to 1950 MPa have been reported in the literature [1]. Endurance limits in the mid and upper ranges of the reported values represent very good resistance to fatigue fracture, and are sensitive to microstructural factors such as the amounts and distribution of martensite and retained austenite and the development of favorable surface residual compressive stresses. The purpose of this paper is to review (1) the evolution of the processing-dependent microstructures and residual stresses in carburized steels, and (2) the relationship of microstructure and residual stress to fatigue performance of carburized steels.

PROCESSING AND MICROSTRUCTURE

PROCESSING CONSIDERATIONS - Carburizing of low-carbon steel may be accomplished by many techniques including gas, vacuum, plasma, salt bath and pack carburizing [2]. By far the most widely used process for large volume production is gas carburizing. In this process, the steel is exposed to a carrier gas atmosphere with gaseous hydrocarbons and carbon monoxide which decompose to introduce carbon into the surface of a steel [2,3]. The steel is held at a temperature at which the microstructure is austenitic, typically 930°C, and carbon diffuses into the interior of the steel. Typically, the carburizing is done in two stages: an initial stage during which the carbon potential is maintained at a level close to the solubility limit of carbon in austenite, at a carbon level which ranges from 1.0 to 1.2 pct C depending on temperature and alloy content of the steel, and a second step during which the carbon potential of the atmosphere is reduced to a level that will maintain surface carbon between 0.8 and 0.9 pct. In the second stage the excess case carbon diffuses further into the steel. The two-step process is therefore often referred to as boost-diffuse carburizing. Figure 1 shows a computed example of carbon gradients produced by two stage carburizing [2]. When the required case depth is achieved, the temperature is lowered to 850°C and the parts are quenched. Quenching from 850°C reduces distortion.

The austenite carbon gradients translate into hardness gradients after quenching. Figure 2 shows almost identical hardness gradients produced for the same steel, SAE 8719 containing

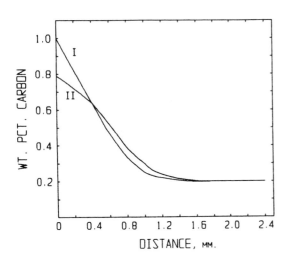

Fig. 1 - Simulated carbon profiles for two stage gas carburizing. Stage I: 927°C, 1.1 pct C potential, 3 hours. Stage II: 871°C, 0.8 pct C potential, 1.5 hours (ref. 2).

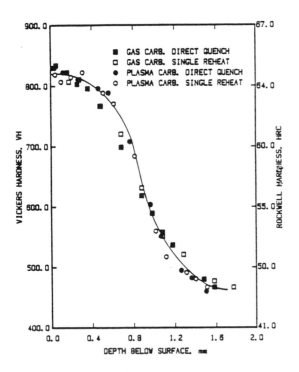

Fig. 2 - Hardness profiles from SAE 8719 steel carburized and hardened as shown (ref. 4).

1.06 pct Mn, 0.52 pct Cr, 0.50 pct Ni, and 0.17 pct Mo, carburized and hardened by four different schedules [4]. Case depths are usually defined as the distance from the surface to the point where hardness drops to a given level, typically HV 510 or HRC 50.

AUSTENITE MICROSTRUCTURE - The microstructure during carburizing consists of polycrystalline austenite. Since carburizing times are relatively long, for example 4 to 5 hours to produce a case depth of 1 mm at 930°C, there is a possibility of grain coarsening during carburizing. As a result carburizing steels are almost universally fine-grained, aluminum-killed steels in which aluminum nitride particles suppress grain growth. Figure 3 shows an example of the austenite grain structure produced by reheating a carburized low-carbon Cr-Ni-Mo steel. The specimen has been etched to bring out the austenitic grain boundaries [5]. Austenite grain size determines the size and distribution of the martensite which forms during quenching, and therefore plays an indirect but important role in fatigue resistance of carburized steels.

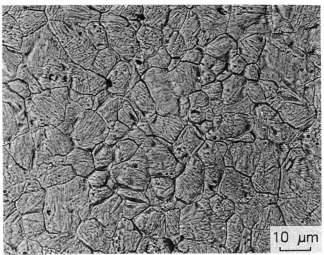

Fig. 3 - Prior austenite grain structure in SAE 8719 steel gas carburized and reheated to 850°C (ref. 4).

Another very important structural change which develops in the austenitic structure during carburizing is the segregation of phosphorus to austenite grain boundaries. The phosphorus concentrates in very thin layers, on the order of atomic dimensions, and can only be detected by Auger electron spectroscopy (AES). Such analysis has shown that phosphorus is present at austenitic grain boundaries in as-quenched steels [6,7]. Thus tempering is not required to cause the segregation.

During quenching, very small amounts of cementite form on the austenite grain boundaries in the high-carbon case, and the combination of phosphorus and cementite at the grain boundaries leads to a very high sensitivity to intergranular fracture [7,8,9]. Intergranular cracking, as discussed later, is a major cause of fatigue crack initiation in carburized steels.

Surface oxidation is another very important microstructural change introduced during gas carburizing when the steel is in the austenitic state [10,11]. All gas carburized atmospheres contain some partial pressure of oxygen due to the presence of CO_2 and H_2O [2], and this oxygen preferentially combines with some of the alloying elements present in the steel. Figure 4 shows an example of surface oxidation in a gas carburized SAE 8719 steel [4]. The surface oxidation has been found to develop in two zones: a shallow surface zone in which chromium-rich oxides form within austenite grains, and a deeper zone in which manganese-rich and silicon-rich oxides form along austenite grain boundaries.

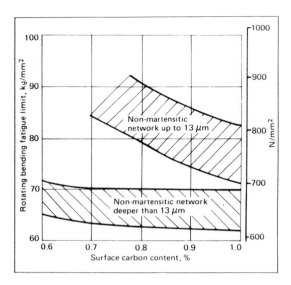

Fig. 5 - Effect of depth of non-martensitic networks and carbon content on fatigue of carburized steels (ref. 12).

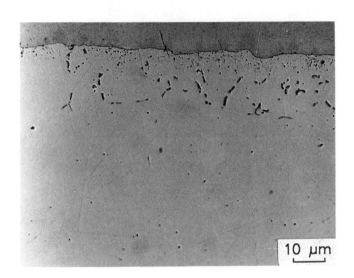

Fig. 4 - Surface oxidation in gas-carburized SAE 4820 steel (ref. 11).

The presence of oxides in the near surface zones of gas-carburized steels raises questions regarding the effect of oxides on fatigue fracture. There are two aspects which cause concern. One is the effect of alloying element depletion by the oxide formation to the point where hardenability is decreased and transformation products other than martensite form in surface layers. The other is the role that the oxides play in the initiation of fatigue cracks. The first aspect affects the residual stresses developed in the case and will be discussed later. The second aspect is related to the depth of the surface oxide zones. The depth of oxide penetration is diffusion-controlled and therefore dependent on the time of carburization. Figure 5 shows results of experiments which show that deeper non-martensite networks related to internal oxidation more severely degrade fatigue resistance than do shallow networks [12]. More recent work shows that very good fatigue resistance can be achieved with oxide penetration on the order of $10 \mu m$ when the surface microstructure associated with the oxidation remains martensitic [4,13].

Excess carbon, above the amount required for carbide formation, may be introduced into austenite during carburizing. If this excessive carbon content is not lowered during the diffusion stage of a carburizing cycle, massive carbides will form either during carburizing or during cooling [12]. Severe massive carbide formation has been associated with high temperature carburizing during which the carbon solubility in austenite in equilibrium with cementite is very high. The carbides form at specimen corners where lowering of carbon content by diffusion is geometrically constrained, and serve as preferred sites for fatigue crack initiation [14].

MARTENSITE-AUSTENITE-CARBIDE MICROSTRUCTURES - When the austenite with the surface carbon gradients introduced by carburizing is quenched, the austenite transforms to martensite. In the high-carbon case the martensite forms in a plate morphology, and because of the low M_s temperatures associated with the high carbon content, significant volume fractions of austenite are retained [15]. In the low-carbon core, martensite forms in a lath morphology with no resolvable austenite. As the carbon content drops from the surface to the core there is a transition from plate to lath martensite morphology, and the amount of retained austenite decreases continuously. Figures 6 and 7 shows examples of plate and lath martensite which have formed in the case and core regions of carburized specimens, respectively.

The formation of martensitic structures in carburized steels is a function of hardenability [16]. Alloy carburizing steels are selected to insure good case and core hardenability, but depending on alloying, quenching factors, and section size, varying amounts of nonmartensitic transformation products of austenite may form in the case and core.

Figure 6 shows the martensite typically formed in the case of steels directly quenched after carburizing. If carburized steels are reheated to a temperature below A_{cm}, i.e. at a temperature where austenite and cementite are stable, distributions of spheroidized carbide particles are developed in the case as shown in Fig. 8. The carbides tie up some of the carbon, and therefore M_s increases, producing less retained austenite in quenched specimens. Also the carbide particles effectively reduce grain boundary migration, and therefore, depending on their volume fraction and size, maintain a very fine austenite grain size.

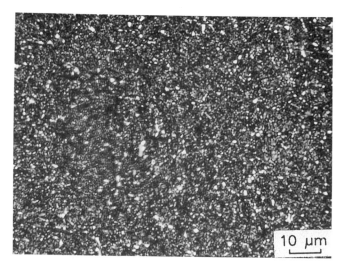

Fig. 8 - Spheroidized carbides (white) in matrix of martensite (dark) of carburized and reheated SAE 8620 steel (Ref. 14).

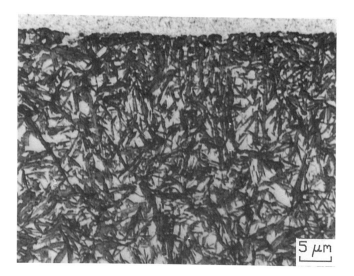

Fig. 6 - Plate martensite and retained austenite in case of carburized, quenched, and reheated SAE 8719 steel (ref. 4).

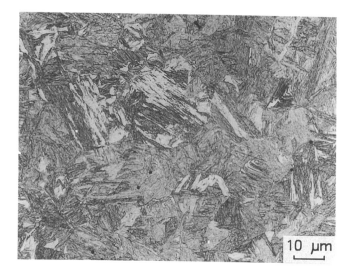

Fig. 7 - Lath martensite in core of carburized and reheated SAE 8719 steel (Ref. 4).

Almost all carburized steels are tempered, generally at temperatures between 150°C and 200°C. These low tempering temperatures preserve high strength and hardness. Residual stresses are reduced, fine transition carbides precipitate within martensite plates, and there is no change in austenite content from that retained in the as-quenched condition as a result of low-temperature tempering treatments [15].

RESIDUAL STRESSES

GENERAL CONSIDERATIONS - The chemical and transformation gradients in carburized steels produce surface residual compressive stresses. These stresses make it possible to readily manufacture high carbon martensitic structures which in quenched, through-hardened steels are difficult to produce because of surface tensile stresses and high sensitivity to quench cracking [15,17]. Also, the surface compressive stresses in carburized steels offset applied tensile bending stresses and thereby significantly increase fracture and fatigue resistance.

The compressive surface stresses are a result of the effect of temperature gradients established during quenching superimposed on M_s gradients associated with the carbon gradients introduced into the austenite by carburizing [18]. At some point during quenching, the temperature falls below the higher M_s temperature in the interior of the specimen, and the austenite begins to transform to martensite. Although the temperature is lower at the surface, the austenite there remains untransformed because of the low surface M_s. The interior formation of martensite causes an associated volume expansion which at high temperatures can be readily accommodated by the surrounding austenite. Eventually during the quenching process, the temperature at the surface falls below the M_s and high carbon martensite forms. The interior martensite, now at a lower temperature and significantly stronger than austenite, resists the expansion of the high-carbon surface martensite and thereby places it in compression. Concomitantly the interior martensite is placed in tension.

Figure 9 shows a schematic diagram of the residual stress profiles that typically form in carburized and hardened steels [19]. The compressive residual stress peaks at some distance from the surface and the surface residual stresses are balanced by interior tensile residual stresses. Also shown schematically are ae carbon profile responsible for the transformation gradients on quenching and a retained austenite profile. Larger amounts of austenite transformation would be expected to increase compressive stresses according to the scenario outlined above, but there is a limit to the benefits of austenite transformation, especially if accomplished by refrigeration treatments. The latter is discussed in a later section.

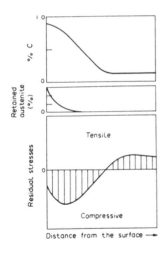

Fig. 9 - Schematic diagram of residual surface stresses in carburized steels (Ref. 19).

Parrish and Harper [19] report the results of a literature survey which included the evaluation of 70 residual stress curves for carburized steels. The specimens had case depths around 1 mm or less, the core carbon contents were between 0.15 and 0.20 pct, and the parts were oil quenched and tempered between 150 and 180°C. Figure 10 shows the range of residual stress profiles which are typical of gas carburizing accomplished according to the conditions outlined above. As shown, the compressive residual stress zone encompasses most of the case.

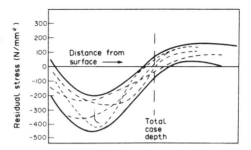

Fig. 10 - Ranges of residual stresses measured in 70 carburized steels (Ref. 19).

MEASUREMENT AND MODELING OF RESIDUAL STRESSES - Residual stress profiles are routinely measured by x-ray diffraction analysis [20]. The residual stresses cause changes in the interplanar spacings of the lattice planes in the martensite and austenite of carburized steel, and the changes in interplanar spacing are used to calculate strains which in turn are used to calculate stresses by use of Poisson's ratio and the Young's modulus [21]. Because of the limited depth of penetration of x-rays, residual stress profiles must be obtained by serial examination of subsurface layers exposed by electrolytic or chemical polishing.

The development of residual stresses depends on complex interactions between specimen size and geometry, heat flow through the steel, heat transfer associated with quenching, the transformation of austenite as a function of chemical composition, position and time as determined by cooling rates and the composition gradients established by carburizing, and the temperature-dependent mechanical properties and plastic flow characteristics of the mixtures of austenite, martensite, and other phases which form as a carburized specimen is quenched. All of these phenomena are now being incorporated into computer models which predict residual stresses and distortion of carburized steels, and although a major short coming is the unavailability of accurate high-temperature constitutive flow stresses for austenite, martensite and other phases, the models make valuable predictions of residual stress as a function of variations in processing parameters [22-26].

Table 1 lists a series of residual stress parameters calculated by Ericsson et al. for cylinders of various sizes of a low-carbon steel carburized to three different case depths [23]. Calculated and measured values agree well, and with increasing bar diameter, case depth decreases because of increased formation of nonmartensitic products. However, the calculations show that the maximum compressive stresses and the position of the maximum increase with increasing bar diameter. This finding is related to the greater amount of nonmartensitic transformation products in the core of the heavier sections. The latter

TABLE 1 - Compilation of Characteristic Depths and Residual Stresses
For Case Hardened Cylinders (Ref. 23)

Diam (mm)	Case depth DC (mm) calc	exp	σ_{max} MPa calc	Depth to σ_{max} mm calc	Depth to σ_{max} divided by DC calc	Depth to Start Mart mm calc	Depth to 0-Crossing mm calc	Core Hard HVI exp
10	0.67	0.67	-450	0.06	0.10	1.0	0.83	360
17	0.66	0.63	-520	0.15	0.23	0.9	0.87	320
30		--	-540	0.15	--		0.86	--
10	0.98	1.04	-420	0.06	0.06	1.7	1.18	360
17	0.95	0.94	-480	0.25	0.27	1.3	1.40	320
30	0.90	0.83	-520	0.45	0.50	1.1	1.41	280
10		--	-340	0.25			1.68	--
17	1.89	1.74	-415	0.60	0.32	1.3	2.10	380
30	1.71	1.49	-480	0.90	0.53	1.1	2.35	260

situation is explained by a greater difference
in the case transformation strain and the
average transformation strain, leading to the
predicted increased compression residual stress.

RESIDUAL STRESSES AND SURFACE OXIDATION -
In an other study, Hildenwall and Ericcson [24]
calculate the tensile stresses produced by
excessive surface oxidation produced by gas
carburizing. If the oxidation is severe enough,
sufficient alloying elements are removed from
solid solution and the hardenability of the
surface may be decreased to the point where
pearlite forms instead of martensite. As a
result the surface transforms at high
temperature early in the quenching cycle and
when the unoxidized case austenite eventually
transforms to martensite, the resulting
expansion places the surface in tension. The
surface residual tensile stresses then adversely
affect fatigue behavior [11,12,27].

RESIDUAL STRESSES AND SHOT PEENING -
Systematic measurements of residual stress
profiles provide valuable information concerning
the performance of carburized steels. Sometimes
a direct correlation can be made to fatigue
performance if parallel fatigue testing is
performed. In other cases if high values of
residual compressive stresses as measured,
improved fatigue performance can be inferred.
For example, shot peening is commonly applied to
carburized steels in order to increase fatigue
performance. Figure 11 shows significant
increases in residual surface compressive
stresses due to increased shot peening velocity,
and Fig. 12 showing the improved fatigue
performance produced by the shot peening [21].
Kim et al. also measured the effects of shot

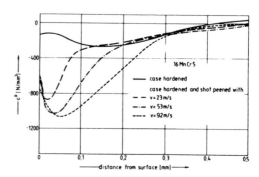

Fig. 11 - Effect of shot peening at velocities
marked on residual stresses.

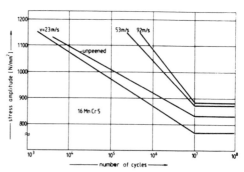

Fig. 12 - Effect of shot peening on fatigue of
carburized steel (Ref. 21).

peening on residual stress and showed a
beneficial effect on fatigue performance [28].

QUENCHING AND TEMPERING EFFECTS ON
RESIDUAL STRESS - Shea [29,30] has
systematically measured the effect of quenchant

temperature and tempering on the residual stresses, microstructure and properties of carburized steels. Figure 13 and 14 show the effect of quenchant temperature and tempering on the surface compressive stresses of SAE 4130, containing 1.0 Cr and 0.20 pct Mo, and SAE 1526 containing 1.25 pct Mn, respectively. The compressive residual stresses increase with increasing quenchant temperature from 50°C typical of conventional warm oil quenches to 270°C which is classified as a hot oil quench. Tempering at 165°C in all cases reduced the surface compressive stresses. The high residual stresses in the hot oil quench of the high-hardenability 4130 steel were attributed to a reduced temperature gradient which shifted the transformation start temperature toward the core, causing substantial core transformation prior to the surface transformation. The lower compressive stresses of the hot oil quenched low-hardenability SAE 1526 steel were attributed to lower transformation strains associated with bainite rather than martensite.

A study by Stickels and Mack [31] also evaluated the effect of tempering on surface residual stress formation in carburized steels. They examined carburized Mo-Cr-V secondary hardening steels and a carburized SAE 8620 steel, a commonly used low-alloy carburizing steel. Figures 15 and 16 show residual stress profiles after tempering at 150°C and 300°C. At 300°C, most of the retained austenite has transformed and the martensitic structure is decreasing in volume as the tetragonality of the martensite is relieved by carbide formation.

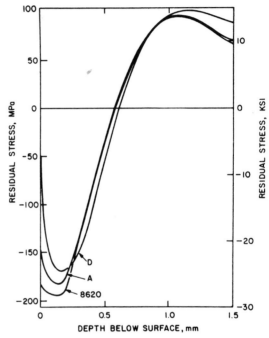

Fig. 15 - Residual stresses in gas carburized steels quenched and tempered at 150°C. Steels A and D are secondary hardening steels (Ref. 31).

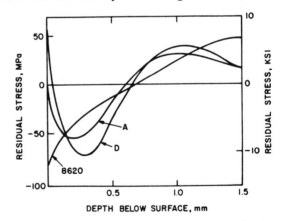

Fig. 16 - Residual stresses in gas carburized steels tempered at 300°C. Steels A and D are secondary hardening steels (Ref. 31).

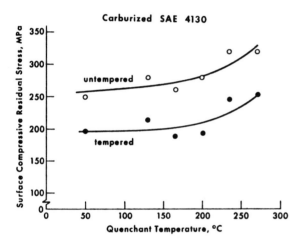

Fig. 13 - Effect of quenchant temperature on residual stress in carburized SAE 4130 steel (Ref. 29).

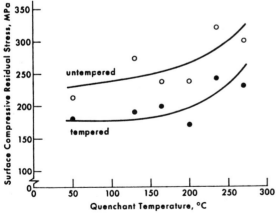

Fig. 14 - Effect of quenchant temperature on residual stress in carburized 1526 steel (Ref. 29).

RESIDUAL STRESS AND REFRIGERATION - Refrigeration or sub-zero cooling treatments are sometimes applied to carburized steels. The lower temperatures cause some of the retained

187

austenite to transform to martensite, and thereby increase case hardness especially if excessive amounts of austenite were present at room temperature. Reduction in retained austenite content also increases dimensional stability by minimizing the volume expansion associated with thermally strain-induced transformation of austenite to martensite during tempering or in service. However, a number of investigators show that refrigeration treatments are detrimental to fatigue resistance [28,32,33].

The reason for the decreased fatigue performance has been shown to be associated with localized residual stresses developed by refrigeration treatments. Although compressive stresses continue to develop in the martensite of the case, Kim et al. [28] have measured high tensile stresses in the austenite of sub-zero cooled specimens. These tensile stresses then act in concept with applied bending tensile stressed to cause crack initiation at susceptible microstructural features during cyclic loading. If refrigeration is used, the carburized parts should be tempered before and after sub-zero cooling [19].

STRAIN-INDUCED AUSTENITE TRANSFORMATION AND RESIDUAL STRESS - The deformation-induced transformation of austenite to martensite is well known in stainless and other highly alloyed steels [34]. The relatively large amounts of retained austenite, typically between 20 and 30 pct, in the near surface case regions of direct-quenched carburized steels also makes deformation-induced austenite transformation a factor in the fatigue performance of carburized steels. Zaccone et al. [35], in a study of the bending fatigue of notched specimens of a series of high carbon steels with microstructures similar to those in the case of carburized steels, show that significant amounts of austenite transform in the plastic zone around a growing fatigue crack. The strain-induced austenite transformation was shown to be beneficial to low-cycle, high strain fatigue resistance. Specimens with more austenite showed longer fatigue life than specimens with lower amounts of retained austenite. The beneficial effects of the strain-induced austenite transformation were attributed to the associated increased strain hardening and generation of increased compressive stresses which arrested and slowed fatigue crack propagation.

Shea, in a study of impact properties of carburized steels, has measured the changes in surface residual stresses which develop as a function of bending strain in carburized steels [36]. Figures 17 and 18 show results for carburized 4120 and 3310 steels, respectively. In the as-quenched and tempered specimens, containing greater than 20 pct retained austenite after an incubation strain, compressive stresses increase significantly with strain. These increases in compressive stress

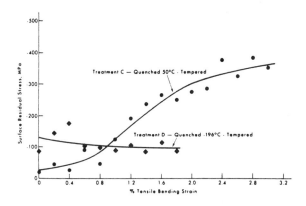

Fig. 17 - Effect of tensile bending strain on residual stress in carburized 4620 steel (Ref. 36).

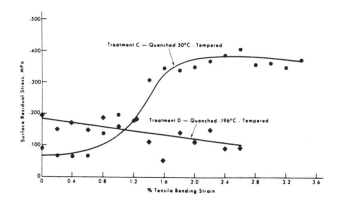

Fig. 18 - Effect of tensile bending strain on residual stress of carburized 3310 steel (Ref. 36).

correlated directly with strain-induced austenite transformation. Specimens quenched in liquid nitrogen, with much reduced retained austenite contents, about 10 pct when strained showed a slight decrease on compressive stress. Higher fracture resistance was associated with the strain-induced martensite formation in the microstructures with the larger amounts of retained austenite.

FATIGUE MECHANISMS AND PERFORMANCE

Fatigue cracks in carburized steels subjected to cyclic bending stresses may be nucleated at a variety of features: surface discontinuities such as machining grooves and pits, inclusions, massive carbides, and prior austenite grain boundaries. Fatigue crack growth is limited because of the relatively low case fracture toughness, 20 to 27 MPam$^{1/2}$, of the high carbon martensitic microstructure [37]. Once the critical crack size is achieved, unstable crack propagation proceeds through the balance of the case and into the core.

Figure 19 shows the macroscopic fracture cross section through a bending fatigue specimen of a carburized Type 8219 steel [13]. The case fracture region has a granular appearance and is delineated from the core fracture which has a ductile, fibrous appearance. On this scale the fatigue crack initiation and propagation areas are too small to be resolved.

Fig. 19 - Macrograph of fatigue fracture cross section of carburized Type 8219 steel. Granular fracture of case and ductile fracture of core is shown (Ref. 13).

Figure 20 and 21 show the fatigue crack initiation and growth areas at higher magnifications. These fractographs were taken from a high carbon steel with a plate martensite-retained austenite microstructure similar to that in the near surface case region of carburized steels. Carburized steels subjected to bending fatigue show exactly the same type of fatigue crack topography [4,13,35]. The arrows in Figs. 20 and 21 point to intergranular fracture initiation sites. These first cracks are arrested, and slow transgranular fatigue crack growth proceeds. When the critical crack size is attained, unstable crack propagation occurs by intergranular fracture as best shown in Fig. 20. The latter intergranular fracture makes up the case fracture shown in the macrograph of Fig. 19.

The above sequence of fatigue fracture is typical of that observed in properly carburized specimens, i.e. those without excessive surface oxidation or massive carbides, with smooth surfaces. The intergranular crack initiation and intergranular unstable case crack propagation are caused by the combination of phosphorus segregation and cementite formation at austenite grain boundaries as described earlier. The intergranular crack initiation extends only a few grains, and is arrested, perhaps because of the generation of favorable residual compressive stresses by strain-induced retained austenite transformation [35,36].

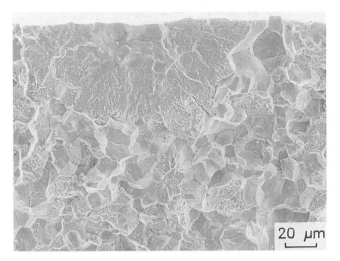

Fig. 20 - Fatigue crack initiation (arrow), transgranular growth zone, and overload intergranular zone in carburized case of direct quenched SAE 8719 steel (Ref. 4).

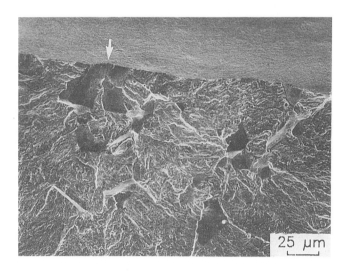

Fig. 21 - Intergranular fatigue crack initiation (arrow) and transgranular crack propagation zone in a simulated carburized case microstructure in an 0.85 pct C steel containing 1.3 pct Cr (Ref. 35).

Erven has shown in a series of 8 alloy steels that the intergranular initiation mode of fatigue cracking is associated with endurance limits between 1070 MPa and 1260 MPa [13]. This range of fatigue resistance appears to be the range associated with well-prepared surfaces and small ranges in residual stresses, austenitic grain size, alloy content, and retained austenite. Rough surfaces, massive carbides, excessive surface oxidation, and high densities of inclusions would all lower fatigue performance.

A second mechanism of fatigue cracking in carburized steels is associated with transgranular crack initiation as shown in Fig. 22. These cracks indicate at surface discontinuities and apparently depend on slip mechanisms of fatigue crack initiation [38]. The endurance limits associated with this type of initiation are much higher, 1400 to 1950 MPa, than those associated with microstructures sensitive to intergranular cracking [4,39,40]. Carburized microstructures which have these high levels of fatigue resistance have very fine austenitic grain sizes and relatively low retained austenite contents, microstructures which may be produced by reheating carburized specimens to below A_{cm} [4,39].

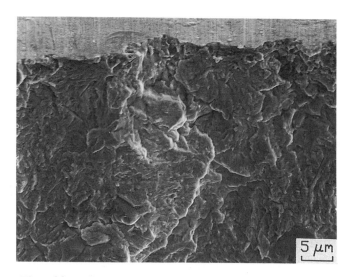

Fig. 22 - Transgranular fatigue crack initiation and propagation in carburized and reheated SAE 8719 steel (Ref. 24).

ACKNOWLEDGEMENTS

I thank Ms. Mimi Martin for her assistance with the preparation of this manuscript, Scott Hyde, and Steve Tua for assistance with the figures. This paper was first presented in slightly modified form in the International Seminar Quenching and Carburizing, Melbourne, Australia, September, 1991, sponsored by the International Federation for Heat Treatment Surface Engineering and the Institute of Metals and Materials, Australia. Research on fatigue of carburized steels at the Colorado School of Mines is supported by the Advanced Steel Processing and Products Research Center.

REFERENCES

1. R.E. Cohen, P.J. Haagensen, D.K. Matlock, and G. Krauss, SAE Technical Paper No. 910140, SAE, Warrendale, PA, 1991.

2. C.A. Stickels and C.M. Mack, in Carburizing: Processing and Performance (ed. G. Krauss), ASM INTERNATIONAL, Materials Park, OH, 1989, 1.

3. Case Hardening of Steel (ed. H.E. Boyer), ASM INTERNATIONAL, Materials Park, OH, 1987.

4. J.L. Pacheco and G. Krauss, J. Heat Treating, 1989, 7, No. 2, 77.

5. A. Brewer, K.A. Erven, and G. Krauss, to be published in Materials Characterization.

6. O. Ohtani and C.J. McMahon, Jr., Acta Metallurgica, 1975, 23, 337.

7. T. Ando and G. Krauss, Metall. Trans. A, 1981, 12A, 1283.

8. G. Krauss, Metall. Trans. A, 1978, 9A, 1527.

9. H.K. Obermeyer and G. Krauss, J. Heat Treating, 1980, 1, No. 3, 31.

10. R. Chatterjee-Fischer, Metall. Trans. A, 1978, 9A, 1553.

11. C. Van Thyne and G. Krauss, in Carburizing: Processing and Performance (ed. G. Krauss), ASM INTERNATIONAL, Materials Park, OH, 1989, 333.

12. G. Parrish, The Influence of Microstructure on the Properties of Case-Carburized Components, ASM, Materials Park, OH, 1980.

13. K.A. Erven, D.K. Matlock, and G. Krauss, to be published in Proceedings of First ASM Heat Treatment and Surface Engineering in Europe, Amsterdam, 22-24 May 1991.

14. K.D. Jones and G. Krauss, in Heat Treatment '79, The Metals Society, London, 1979, 188.

15. G. Krauss, Steels: Heat Treatment and Processing Principles, ASM INTERNATIONAL, Materials Park, OH, 1990.

16. U. Wyss, Harterei Technisehe Mitteilungen, 1988, 43, 27.

17. L.J. Ebert, Metall. Trans. A, 1978, 9A, 1537.

18. D.P. Koistinen, Trans. ASM, 1958, 50, 227.

19. G. Parrish and G.S. Harper, Production Gas Carburizing, Pergamon Press, New York, 1985.

20.	*SAE Handbook Supplement: Residual Stress Measurement by X-Ray Diffraction*, SAE J784a, SAE, Warrendale, PA, 1971, 19.

21.	G. Scholtes and E. Macherauch, in *Case-Hardened Steels: Microstructural and Residual Stress Effects* (ed. by D.E. Diesburg), TMS, Warrendale, PA, 1984, 141.

22.	B. Hildenwall and T. Ericssen, in *Hardenability Concepts with Applications to Steel* (eds. D.V. Doane and J.S. Kirkaldy), TMS, Warrendale, PA, 1978, 579.

23.	T. Ericsson, S. Sjostrom, M. Knuuttila, and B. Hildenwall, in *Case-Hardened Steels: Microstructural and Residual Stress Effects* (ed. D.E. Diesburg), TMS, Warrendale, PA, 1984, 113.

24.	B. Hildenwall and T. Ericsson, J. Heat Treating, 1980, 1, No. 3, 3.

25.	J.A. Burnett, in *Residual Stress for Designers and Metallurgists* (ed. L.J. Vande Walle), ASM, Materials Park, OH, 1981, 51.

26.	M. Henriksen, D.B. Larson, and C.J. Van Tyne, submitted to ASME Journal of Engineering Materials and Technology.

27.	S. Gunnarson, Metal Treatment and Drop Forging, 1963, 30, 219.

28.	C. Kim, D.E. Diesburg, and R.M. Buck, J. Heat Treatment, 1981, 2, No. 1, 43.

29.	M.M. Shea, J. Heat Treating, 1980, 1, No. 4, 29.

30.	M.M. Shea, J. Heat Treating, 1983, 3, No. 1, 38.

31.	C.A. Stickels and C.M. Mack, J. Heat Treating, 1986, 4, No. 3, 223.

32.	K.D. Jones and G. Krauss, J. Heat Treating, 1979, 1, No. 1, 64.

33.	M.A. Panhans and R.A. Fournelle, J. Heat Treating, 1981, 2, No. 1, 55.

34.	G.B. Olson and M. Cohen, Metall. Trans. A, 1975, 6A, 791.

35.	M.A. Zaccone, J.B. Kelley, and G. Krauss, in *Carburizing: Processing and Performance* (ed. G. Krauss), ASM INTERNATIONAL, Materials Park, OH, 1989, 249.

36.	M.M. Shea, SAE Technical Paper No. 780772, SAE, Warrendale, PA, 1978.

37.	G. Krauss, in *Case-Hardened Steels: Microstructural and Residual Stress Effects* (ed. D.E. Diesburg), TMS, Warrendale, PA, 1984, 33.

38.	*Fatigue and Microstructure* (ed. M. Meshii), ASM, Metals Park, OH, 1979.

39.	C.A. Apple and G. Krauss, Metall. Trans., 1973, r, 1195.

40.	L. Magnusson and T. Ericsson, in Heat Treatment '79, The Metals Society, 1980, 202.

Residual Stresses and Their Measurement

C.O. Ruud
Penn State University
University Park, Pennsylvania

ABSTRACT

It has long been recognized that the rapid temperature changes and temperature gradients inherent in quenching, as well as the abrupt volume changes that occur during phase transformation induce high residual stresses and severe stress gradients in heat treated steels.

Residual stresses are the inevitable consequence of any manufacturing process which transforms the shape or changes the properties of materials. These stresses have been given many labels, including self, internal, welding, quenching, forming, fabrication, bulk, and in-situ stresses. They are usually considered as being represented by a statistical average of stress-induced elastic strain across a finite length, usually spanning several metallic grains, i.e., on the order of millimeters or fractions of millimeters. These are called residual macrostresses to distinguish them from microstresses, which are considered to be stresses causing interatomic strain variation across single metallic grains.

Residual Stresses, both macro and micro, are usually formed when portions of a component undergo nonuniform, permanent dimensional change, i.e., either elastic or plastic strain. Plastic strain may be induced by large temperature gradients, as by casting, welding, quenching, or rapid heating, as well as by forming processes. Elastic strain is often induced by volume changes induced by phase transformation. The contribution of residual stresses to fracture, fatigue failure, and stress corrosion cracking of metallic components has long been recognized. However, the cost, difficulties, and limited accuracy of the mechanical methods (stress relief techniques) and, to a large extent, the x-ray diffraction techniques, have discouraged their study for all but the most critical and/or frequently occurring problems.

This talk will describe the practical methods and mention the proposed methods of residual stress measurement. Applications of the methods and their limitation will be discussed and examples of residual stresses induced by thermal treating will be cited.

INTRODUCTION

Definition and Origin: Residual Stresses are inevitably introduced into a material in the process of shaping it into a useful object. The simplest definition of residual stresses is that they are those stresses remaining in a solid in the absence of external loading and thermal gradients, excluding gravity. These stresses have also been called internal, welding, forming, casting, quenching, etc. stresses. Residual stresses are caused by several mechanisms inherent in materials processing. As described by Littmann [1], the origins may be viewed as thermal, chemical, or mechanical.

The thermal origins are due to plastic deformation which from non-uniform thermal expansion or contraction will produce residual stresses in the absence of phase transformation. An example is the drastic quenching of steel from a temperature below the critical range to produce compressive surface stresses, or the water quench process used in solution heat treating of aluminum alloys.

The chemical origins are due to volume changes from chemical reactions, precipitation, or phase transformations which produce residual stresses in various ways. Nitriding produces compressive stress in the diffusion region because of expansion of the lattice and precipitation of nitrides. Carburizing causes similar growth but volume changes from austenite transformation during cooling in combination with stresses due to thermal gradients during the quench determine the final residual stress distribution [1].

Finally, the mechanical origins are due to non-uniform plastic deformation which can result in undesirable residual surface stresses or may be used to impose favorable residual stress patterns. Examples are surface rolling or shot peening of fillets and seats for press fits on shafts, autofrettage of gun tubes or pressure vessels to produce favorable stresses. Unwanted tensile stresses from machining, cold straightening, etc. represent the opposite situation.

When thermal, mechanical and transformation effects operate in combination, as in the grinding of hardened steel and other metal removal processes, e.g. EDM or laser hole drilling, extremely complex patterns of residual stress may result [1].

Stress Versus Strain: The mechanisms associated with the aforementioned origins of residual stresses for the most part operate on a microstructural or crystallographic basis. Yet usually, residual stresses are investigated and reported on a macro scale, that is to say those stresses averaged over several millimeters or centimeters. Nevertheless, these macro stresses are composed of the summed and averaged microstresses.

However, in most, if not all methods of stress measurement, strains are really measured, and stresses are derived from the strains using elastic constants.

Thus, it is appropriate to define macro and micro strains. Macrostrain is usually considered the mean strain over any finite gage length of measurement large in comparison with grain size. Macrostrain can be measured by several methods, including electrical resistance strain gages and mechanical or optical extensometers. Elastic macrostrain can be measured by x-ray diffraction. Microstrain is the strain over a gage length comparable to a few grains or fractions of grains. These are the strains being averaged by the macrostress measurement. Microstrain is not measurable with existing techniques. Variance of the microstrain distribution can, however, be measured by x-ray diffraction.

Importance of Residual Stresses: The relevance for the measurement, prediction and control of residual stresses is based upon their effect on the serviceability of manufactured components. Residual stresses may induce premature failure through, i.e., causing cracks which results in stress risers, reducing fatigue strength, inducing stress corrosion or hydrogen cracking, and causing distortion. On the other hand, compressive residual stresses may improve the fatigue, stress corrosion, hydrogen embittlement, etc. properties of a component. Thus, the investigation and subsequent understanding of residual stresses on a macro and micro scale is vitally important to the quality and reliability of many components.

MEASUREMENT

As mentioned previously residual stresses are always present and important to varying degrees. However, they are seldom, if ever, measured directly. What is measured is the elastic strain caused by the residual macro and micro-stresses. This is so for all of the measurement methodologies mentioned subsequently.

Ultrasonic and Magnetic: There has been a great deal of investigation and instrumentation development towards applying ultrasonic and magnetic nondestructive testing methods to the measurement of residual stresses [3]. It is a fact that when a crystalline material is placed under stress, the velocity of the ultrasound changes, and with suitable instrumentation and techniques, the change in velocity can be measured. However, the problem comes in establishing the velocity of ultrasound in the unstressed material. Since microstructural features such as grain size, shape, orientation and second phases, as well as composition and plastic deformation all effect the velocity of the ultrasound, a specimen which is exactly the same as the component in which residual stress is to be measured, but without stress, must be used to obtain the unstressed velocity. This condition is virtually impossible to meet in a practical situation. Another problem with the application of ultrasound is that residual stresses are usually very inhomogeneous; changing tens of MPa over distances of tens of microns. However, the nature of the propagation of ultrasound and instrumentation for measurement of its velocity results in spacial resolution on the order of tens of millimeters to tens of centimeters, i.e., three to four orders of magnitude too coarse [3]. Thus, except for a few instances, ultrasonic techniques have not found wide application towards the measurement of residual micro or macro stresses.

With regard to magnetic methods, a technique called Barkhausen Noise Analysis (BNA), has been tested, and instrumentation marketed, for residual macro-stress measurement [4]. However, the BNA signal is as sensitive to slight changes in material chemical composition, phase composition, dislocation density, microstructure, etc., as it is to residual macro-stress. Thus, BNA has not found broad application for the quantitative measurement of residual macro stress, nor for micro stress [3].

Mechanical: Mechanical methods for the measurement of residual macro stresses have been used longer than any of the other methods. There are many techniques used for the mechanical measurement of residual macrostresses, but for the most part they all consist of removal of stressed material and the measurement of the resultant strain change. The most popular technique presently in practical use is that of hole drilling. Here a specially designed set of electrical resistance strain gages are used to measure the strain change caused by removing stressed material by hole drilling [3]. This technique has many limitations and is not suitable where high stresses and stress gradients exist.

X-Ray Diffraction: When a metal or ceramic polycrystalline material is placed under stress, the elastic strains in the material are manifest in the crystal lattice of the individual grains. The stress applied externally or residual within the material, when below its yield strength, is taken up by interatomic strain. The x-ray diffraction techniques are capable of actually measuring the interatomic spacings, which are indicative of the macro-strain undergone by the specimen. Stress values are obtained from these elastic strains in the crystals by knowing the effective elastic constants of the material and assuming that stress is proportional to strain, a reasonable assumption for homogeneous, nearly isotropic materials as are most metals and alloys of practical concern.

The important parameters involved in the mechanics of x-ray diffraction are described by the Bragg relation,

$$n\lambda = 2d \sin \theta$$

where

n = 1,2,... i.e., any integer; it is the order of reflection, and for stress work is usually unity.

λ = wavelength of the diffracted radiation

d = spacing of the reflecting planes of atoms

θ = Bragg angle.

This relationship is illustrated in Figure 1.

In all applications where the determination of d is required, either λ or θ is known. In most practical applications, including stress analysis, λ is a constant and is known, and 2θ is measured. In a powder where particles show no preferred orientation, an x-ray beam only a few square millimeters in crosssection is diffracted as a cone of x-ray beams by a large number of randomly oriented crystallites (see Figure 2). The cone's semi-apex angle ($180°-2\theta$) is not infinitely sharp, in that a plane perpendicular to the cone's axis would show that the diffracted energy is distributed over a few tenths to several degrees of 2θ (see Figure 3). Therefore, the angular distribution of the diffracted beam intensity must be ascertained as a function of 2θ. In other words, the diffracted x-ray intensity must be measured at several angular positions or a continuum of positions, in order to determine the mean diffraction angle 2θ.

The basic equation relating x-ray diffraction principles to stress-strain relations can be rewritten as

$$\varepsilon_{\phi\psi} = \left(\frac{1+v}{E}\right)\sigma_{\phi}\sin^2\psi - \frac{v}{E}(\sigma_1 + \sigma_2) \qquad (1)$$

where v and E = elastic constants

σ_{ϕ} = stress in the plane of the surface of the specimen at an angle of ø with a principal stress direction in the specimen surface.

ψ = angle between the surface normal and the normal to the crystallographic planes from which an x-ray peak is diffracted.

$\varepsilon_{\phi\psi}$ = strain in the direction defined by the angles ø and ψ

σ_1 and σ_2 = principal stresses in the surface plane of the specimen.

This may be rewritten as

$$\sigma_{\phi} = \frac{d_{\phi\psi}-d_1}{d_1}\left(\frac{E}{1+v}\right)\frac{1}{\sin^2\psi} \qquad (2)$$

where $d_{\phi\psi}$ = interatomic planar spacing for those crystal planes for which the normal is defined by the angles ø and ψ.

d_1 = interatomic planer spacing for crystal planes parallel to the specimen surface.

It would be noted that only those planes within a selected d-space range will be measured due to the narrow selection of Bragg angle imposed by the stress measurement arrangement.

Now since the difference between $d_{\phi\psi}$ and d_1 is small, then from Bragg's Law $\Delta d/d = -\cot\theta\left(\dfrac{\Delta 2\theta}{2}\right)$ and Equation 2 may be rewritten as

$$\sigma_\phi = (2\theta_1 - 2\theta_\psi)\frac{\cot\theta_1}{2}\left(\frac{E}{1+\nu}\right)\frac{1}{\sin^2\psi}\left(\frac{\pi}{180}\right) \qquad (3)$$

Equation (3) is the basic equation relating macro stress to the measured Bragg angles. An important assumption in this equation is that the material measured is under plane stress. There are several variations to this equation for use to different techniques of application, but they all assume the plane stress condition. There are also techniques which do not need to restrict the conditions to plane stress, these are sometimes referred to as triaxial stress measurement techniques [4].

It is important to recognize that the foregoing description of the x-ray diffraction stress measurement method is for the measurement of macro stress. However, information about the degree of micro strain in the volume of material interigated by the x-ray beam may be obtained from the same data as used for macro stress determination. This is because the greater the range of micro strain about the average macro stress, the longer is the two-theta distribution of the x-ray intensity. Thus, as micro strain increases, the breadth of the peak shown in Figure 3 increases, and from this peak breadth, quantitative information about the micro strain can be derived [6].

Variation in micro strains are caused from several sources which cause inhomogeneous strain within each grain or among the grains over which the macro stress is being determined. Variations within grains are caused by second phases, plastic deformation (dislocations tend to be more numerous near the grain boundaries and each dislocation contains a tension and compression region), and variation in stiffness of neighboring grains (i.e. the matrix). Variations among grains are caused by orientation differences of neighboring grains and macro-stress gradients.

EXAMPLES

The following paragraphs describe a few examples of residual stresses induced by quenching.

Solution Heat Treatment: Solution heat treatment involves the thermal soaking at a high enough temperature that the alloying elements of a metal are placed in solid solution. The metal is then cooled rapidly to obtain a super saturated solid solution at room temperature. Upon aging, either at room or an elevated temperature, some of the alloying elements precipitate to form coherent second places that produce hardening of the matrix metal. Aluminum alloys are quenched from near 530°C in cold water to produce the super saturated

solid solution. This process rapidly shrinks the surface, while the center of the material is still hot, and this inhomogeneous cooling causes a combination of compressive and tensile plastic deformations from the interior to the surface. The result is compressive stresses at the surface and tensile stresses at the center of the cross-section. The magnitude of the stresses depend upon the thickness of the cross-section as well as the quenching efficiency of the water. Stresses at the surface of -150 MPa and in the center of +100 MPa are not uncommon. However, aging (precipitation) tends to ameliorate the magnitude of the stresses slightly.

An example of quenching stresses is that of an 50 cm diameter 7075-T6 forged aluminum ring with a cross-section of about 15 cm^2. The residual stresses on these forgings varied markedly due to inhomogeneous quenching conditions on the surface of the rings. An example was a ring that, measured at four azimuthal positions, i.e., 0, 20, 45, and 90 degrees, and showed stresses in the tangential direction of -126, -154, -56 and -119 MPa, respectively. The precision of stress measurement was ± 20 MPa. These stresses tended to distort the rings to the degree that they could not be installed.

Phase Transformations: Any phase transformation that causes a change in volume will produce or modify the state of residual stress (RS) in a material. Since many transformations involve heating and quenching, the combined effects of thermal gradients and the transformation influence the resulting RS. Examples are carburizing and hardening, case hardening by flame or induction heating of the surface followed by quenching, nitriding and laser glazing. In these processes, the objective is to produce a beneficial and controlled level of compressive RS in the surface. However, in some manufacturing processes, RS can be created in a way that is not controlled. Examples of such processes are: machining by chip cutting tools, abrasive grinding, electro-discharge machining (EDM), laser hole drilling, or any other surface forming or finishing process that cause local plastic deformation or intense heating of the metal surface.

While any phase transformation that produces a volume change can produce or modify residual stresses, the transformations involved in the quench hardening and tempering of steel provides an illustration of such effects. Furthermore, RS from the austenite to martensite transformation and the tempering reactions in medium and high carbon steel are deliberately utilized to create and control RS in case hardening by flame or induction hardening in the former, and by carburizing and quench hardening in the latter. However, adverse RS can be created by machining with chip-cutting tools, by abrasive wheel grinding, or surface melting as in EDM.

The grinding of case-carburized and hardened steel will be used to illustrate both effects, controlled beneficial RS in case hardened steel and detrimental RS resulting from inadequate control of grinding.

Controlled RS in Case-Carburized Steel: Figure 4 is a plot of RS vs. depth for a carburized and hardened part. The level of surface compressive residual stress is controlled by the surface carbon content, the carbon gradient (case depth), quenching conditions, hardenability of the case and core material, and tempering after hardening. The optimum microstructure and RS for a given part will depend on its intended function. In rolling contact bearings or gear teeth, for example, a controlled level of retained austenite (RA) is beneficial, but the level of surface compressive stress and hardness will be lower than if the part is refrigerated before tempering to reduce the level of RA, all other things being equal.

Figure 4 also illustrates the detrimental effect of abusive or uncontrolled grinding on the surface RS and microstructure of hardened high carbon steel (7). The surface temperatures generated in the grinding of hardened steel and the resulting transformations have been described by Littmann and Wulff (8). Figure 4 is an example of detrimental RS from grinding at the root of a gear tooth which contributed to a premature failure of the tooth by bending fatigue.

Additional examples of detrimental RS due to the grinding of steel can be found in (2).

CONCLUSIONS

1. Residual stresses often show high gradients over the surface and in depth which are nearly impossible to measure by any method except x-ray diffraction.
2. Residual stresses are not uniform.
3. Many stress measurements are required to understand the existing pattern sufficiently to evaluate its influence on performance or other observed effects such as distortion.

REFERENCES

1. Littmann, W.E., "Measurement and Significance of Residual Macrostress in Steel", SAE 793A, Proceedings of the Automatic Eng. Cong., Detroit, MI, January 13-17 (1964).
2. Littmann, W.E., "Control of Residual Stress in Metal Surfaces", Proc. of the CIRP, 1303-1326, ASTME, September 25-28 (1967).
3. Ruud,. C.O., "A Review of Nondestructive Methods for Residual Stress Measurement", J. of Metals, 33(6), July (1981).
4. Ruud, C.O. and P.C. Chen, "Application of an Advanced XRD Instrument for Surface Stress Tensor Measurement on Steel Sheets", Exp. Mech., 25(3), 245-250, September (1985).
5. Cullity, B.D., Elements of X-Ray Diffraction, 2nd Ed, 285-292, Addison-Wesley (1978).
6. Delhez, R., Th.H. de Keijser, and E.J. Mittemeijer, "Role of X-Ray Diffraction Analysis in Surface Engineering: Investigation of Microstructure of Nitrided Iron and Steels", Surf. Eng., 3(4), 331-341 (1987).
7. Littmann, W.E., "The Influence of Grinding on Workpiece Quality", American Society of Tool and Manufacturing Engineers, MR 67-593, 1976.
8. Littmann, W.E. and Wulff, J., "The Influence of Grinding on the Structure of Hardened Steel", Transactions of the American Society for Metals, Vol. 47, 1955.

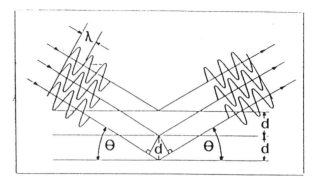

Figure 1. Illustration of the Bragg relation.

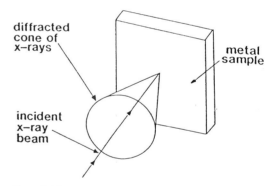

Figure 2. Diffracted cone of x-rays from a polycrystalline metal.

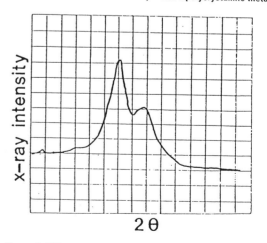

Figure 3. Diffracted x-ray intensity distribution plotted against 2θ for a $K\alpha_{1,2}$ doublet.

Grinding Injury on a Carburized Gear
Residual Stress vs Depth

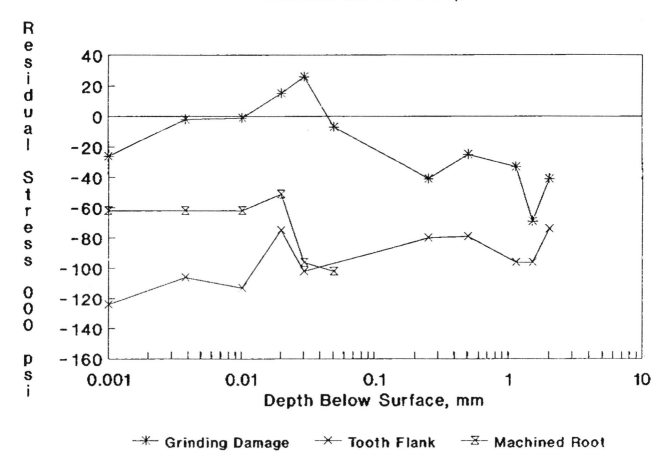

FIGURE 4

Reduction of Distortion Caused by Heat Treatment on Automobile Gear Steels

T. Fukuzumi
Mitsubishi Steel Mfg. Co. Ltd.
Tokyo, Japan

T. Yoshimura
Yoshimura Technical Office Inc.
Tokyo, Japan

ABSTRACT

It is hard to neglect the influence of Aluminium and Nitrogen contained in the steel for obtaining the aiming hardness within H-band of Jominy testing on gear steel of auto use (SAE8620, 4120 etc).
The commercial composition of steel can be controlled more severely by using the ladle furnace and it can be controlled within the underdescribed ranges for the respective elements:

C: +/-0.01%, Si: +/-0.02%, Mn: +/-0.02%, Cr: +/-0.02%, Mo: +/-0.01%, Ni: +/-0.01%

However, on steel making process of these steels, Aluminium is used for Deoxidizing and for obtaining fine Austenite grain, and generally steel of EAF contains relatively high Nitrogen content. First, it was examined the influence of Aluminium and Nitrogen against hardness by Jominy test after controlling the contents of C, Si, Mn, Cr, Mo and Ni within min. range. Second, it was measured the distortion caused by Carburizing with Navy C-type test pieces.
It has been found that it could minimize the distortion caused by heattreatment as well as the variation of Jominy hardness by making the A/N contents ratio in steels within 1-2.5 range.

1. Introduction

In end of 1970s', Japanese auto makers were suffered from the elimination of noise, especially, gear noise caused from their distortion after the heat treatment. So they expected and desired the steel makers to supply "low and stable distortion" steel.
In order to reply to these demand, Mitsubishi Steel Tokyo Works installed ladle refining equipment in 1976. Since that year, we have been producing the precisely controlled "clean" and "low and stable distortion" steel. This report is the experimental results to reduce and stabirize the distortion more.

Fig. 1 shows the effect of Al content on the gap distortion of SAE 8620H steels as measured with Navy C type specimens before and after the carburization. We have experienced on many occasions that Al content of steel has a large effect on the quenching distortion or the hardenability as showing in the examples.

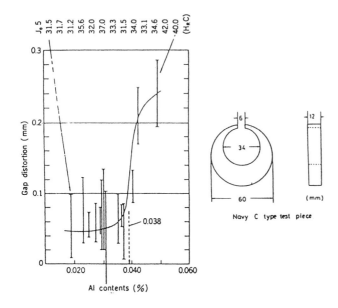

Fig. 1 : Effect of Al content on gap distortion of SAE8620H steels.

Both Aluminium and Boron belong to the same group in the periodic table of elements, so Aluminium is expected to behave similary to Boron. In other words, since the effect of Boron in steel is affected by the Nitrogen content of steel, it would also be desirable to discuss the effect of Aluminium under the combination of Aluminium and Nitrogen.

The following is a report of the investigation on the effect of Al-N on the hardenability and quenching distortion behavior.

2. Experimental Results

2.1 Effect of Al-N on the hardenability.
The important factors usually affecting the hardenability of steel are believed to be the alloying elements and grain size. Although the effects of the alloying elements; C, Si, Mn, Cr, Mo etc. and the grain size of Austenite have been studied by many investigators including M.A.Grossman, no standarized conclusion

has been reached on the effect of Al, that is to say, multiplying factor, many differents views being proposed by each investigators. On the other hand, few studies on N have been reported, due to the difficulty in controlling the content of N. Especially, any study has scarcely been reported on the effect of Al-N on the hardenability.

Fig.2 shows the effect of Al-N on the hardenability for various compositions of the amounts of Al and N in SCM420H steel while its principal alloying elements; C, Si, Mn, P, S, Ni and Mo, were kept almost constant.

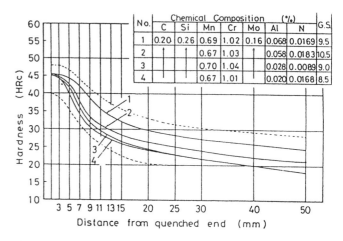

Fig. 2 : Effect of Al-N on the hardenability.

The grain size of the steels remained ASTM NO. 8.5-10.5, and this much of difference affects the hardness by HRC 2.0 at the most; thus the large effect of Al-N on the hardenability is clearly shown.

2.2 Form of Al and N in steel.
Before discussing the effect of Al and N, let us examine the form of Al and N in the steel.
Fig.3 shows the curves of solubility product of [Al]γ, [N]γ in Austenite for the reaction [Al]+[N] =AlN at 1200℃ and 925℃ respectively.

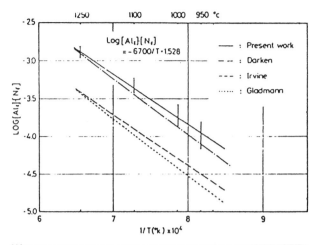

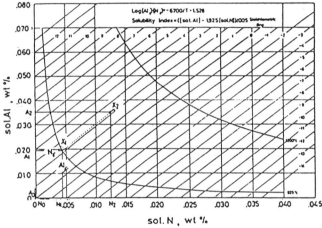

Fig. 3 : Curves of the solibility product of [Al] and [N] in Austenite and concept of Sol.Index.

Many equations concerning the Al-N solubility product have been reported, however, we experimentally prepared the our own equation Log[Al]γ[N]γ =-6700/T +1.528 shown with existings in Fig.3. The temperature of 1200 ℃ and 925℃ were selected because the ordinary temperature for hot working such as forging and rolling is almost 1200℃ and the temperature for normalizing or carburization is about 925 ℃ respectively.

In Fig.3, the inclined line starting origin indicates the stoichiometric ratio of Al to N in AlN (Al/N=1.9253) and other lines parallel to the stoichio-

metric line with equal spaces therebetween indicate 0.005 wt% of Al which is the minimum adjustable amount of Al when refining steels.

We named these regions separated by these parallel lines "Solubility Index (Sol.Index)."

The Sol.Index is decided by the following equation. That is, Sol.Index=([Al%--1.9253[N]%/0.005.

Explanation on how to read this figure is given as follows: Let us assume that the total contents of Al and N are 0.030 wt% and 0.0125 wt% respectively, that is, Point X2 in the figure. When a steel having composition X2 is heated to 1200 ℃, all of the Al and N are dissolved in to the Austenite as solid solution. By cooling the steel then to 925 ℃ and holding at that temperature, AlN is precipitated along Line X2X1 that is parallel to the stoichiometric line. And then the amounts of [Al] γ and [N]γ that canbedissolved in the Austenite as solid solution are A0A1 and N0N1, respectively, while those of [Al] AlN that can be precipitated as fine Aln particles are A₁A₂ andN1N2. In other words, on the line parallelto the stoichiometric line in the Al-Ndiagram, the amounts [Al]γ and [N]$\gamma$are the same although the amount of the precipitated [AlN] particles changes. And it is well-known that this precipitated [AlN] inhibit the grain growth of Austenite. From these point of view based on theAl-N diagram, let us now examine the effects of [Al]γ and [N]γ on the hardenability and the quenching distortion.

2.3 Relationships between amount of [Al]γ, [N]γ, [AlN] and Sol.Index and the hardenability.
Fig.4 shows the measured hardenability (Jominy Curves) of specimens A-E having various combinations of Al-N amounts,

and N, and their amounts calculated from the Al-N diagram. The Sol.Index are also shown in the figure which are calculated by above mentioned method.

NO.	Chemical Composition (%)							A.G.S. NO.	Calculated (%)		
	C	Si	Mn	Cr	Mo	Al	N		$[Al]_\gamma$	$[N]_\gamma$	AlN
A	.19	.25	.71	1.00	.15	.016	.0036	4.6	.014	.004	0
B	.20	.26	.70	1.02	↑	.029	.0149	8.5	.009	.007	.023
C	.21	.26	.71	1.02		.025	.0228	8.2	.004	.015	.023
D	.19	.25	.72	1.01		.065	.0036	6.5	.057	.001	.008
E	.23	.26	.65	1.00		.056	.0101	8.2	.035	.002	.025

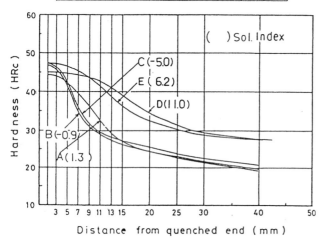

Fig. 4 : Example of relationship between $[Al]_\gamma$, $[N]_\gamma$, Sol. Index and Hardenability

The results showing a large difference in the hardenability in spite of the chemical composition being almost the same, reveals that $[Al]_\gamma$, contributes largely on the hardenability and then that the Austenite grain size is also decided by the amount of [AlN] to some extent.

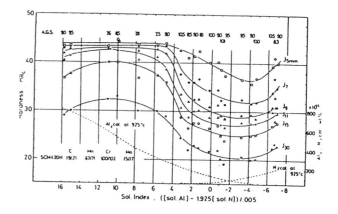

Fig. 5 : Relationship between amount of $[Al]_\gamma$, $[N]_\gamma$, Sol. Index and Jominy hardness.

Fig. 5 shows the relationships between Jominy hardness of SCM420H steels in various depth from quenched end and Sol. Index.
From this, it is clear that the low $[Al]_\gamma$, and $[N]_\gamma$, region are low and stable in the hardenability. It is also shown that the content of $[Al]_\gamma$, contributes greatly to the hardenability as compared with that of $[N]_\gamma$.

2.4 Relationship between Sol.Index and Gap distortion in Navy C type specimen.

Fig. 6 shows a summary of the gap distortion of Navy C type specimens before and after the carburization for SCM420H steel containing C:0.20 ± 0.01, Mn:0.69 ± 0.02, Cr:1.02± 0.02, and Mo:0.16± 0.01wt% against Sol.Index. The result shows that in order to reduce the gap distortion, 0.015-0.025 wt%N is desirable for the steels belonging to Sol. Index (-1)-(-4) having 0.030 wt% Al content.

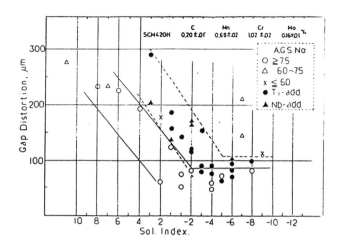

Fig. 6 : Relationships between Sol. Index and gap distortion in Navy C type specimen.

The above results explained concern the case where nitride-forming element is Al only, but there case where other element such as Nb and Ti are added as nitride-forming elements for inhibiting grain growth. Therefore, the amount of the distortion in Navy C type specimens of steels, of similiar chemical composition to that of steels except contents of Nb or Ti, was plotted in Fig.6 agains Sol.Index.

An interesting phenomenon is observed by comparison of the both steel groups. That is, the low and stable quenching distortion region of the steels where merely Al-N are contained, lies in the Sol.Index (-1)-(-5), but if other Nitride-forming element is Sol.Index (-5)-(-8) in Fig.6 for the same Al content but with a higher N content. This is probably due to an increase in $[Al]\gamma$, as a result of Nb or Ti fixing N in steel, that increase the quenching distortion. In other words, it is also estimated that $[Al]\gamma$, has a large influence on the quenching distortion.

3. Summary

A summary of the results of the examination on the effect of Al and N on the hardenability and quenching distortion is as follows:

(1) Al and N in steel have a large influence on the hardenability and the quenching distortion of steel. Their behavior can be grasped by $[Al]\gamma$, and $[N]\gamma$ which dissolve into the Austenite, especially the $[Al]\gamma$, as large effect on the hardenability and the quenching distortion.

(2) Under addition of Nitride-forming element such as Nb and Ti besides Al, they also form carbonitrides, this consequently results in an increase in $[Al]\gamma$, Therefore, it is necessary in this case to increase the amount of $[N]$ over that of Al-N steel somewhat for decrease $[Al]\gamma$, in order that the quenching distortion in lowered and stablized.

Metallo-Thermo-Mechanical Simulation of Quenching Process—Theory and Implementation of Computer Code "Hearts"

T. Inoue and D-Y. Ju
Kyoto University
Kyoto, Japan

K. Arimoto
CRC Research Institute Inc.
Osaka, Japan

ABSTRACT

A finite element code "HEARTS" (Version 2.0) for 3-D simulation of heat treatment process developed by the authors based on the metallo-thermo-mechanical theory is presented. Coupled equations of heat conduction, inelastic stresses and kinetics of phase transformations are derived as well as the diffusion equation during carburization, followed by finite element formulation. The program is available for 2-D and 3-D simulation of various heat treatment processes, such as quenching, tempering and so on. The system is used in the CAE environment of a pre- and post-processor over "I-DEAS". The architecture of the code consisted of over 200 subroutines described by approximately 35,000 steps are briefly stated. Some examples of simulated results of carburized quenching processes for cylindrical rod, ring and gear wheel are also presented.

INTRODUCTION

In the processes incorporating phase transformation, such as quenching, welding, casting and so on, the change of material structures is necessary to be included in the analysis of temperature and stresses since the structural change due to phase transformation is strongly coupled with the fields of temperature and stresses, which is termed as *metallo-thermo-mechanical coupling* by the authors (1-6), and the mode of such coupling is schematically illustrated in Fig.1 (2).

When the temperature distribution in a material (and in some cases the heating and cooling rate) is unequal, thermal stress ① is caused in the solid, and the temperature-dependent phase transformation ② leads to alternations in structure such as austenite-pearlite and austenite-martensite transformation. Local dilatation due to phase changes in the solid brings about transformation stress ③ (7) and also the effect of transformation plasticity (8) interrupts the stress/strain field of the body.

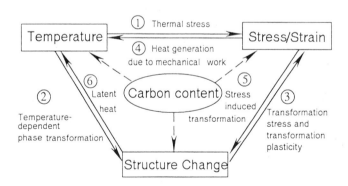

Fig. 1. Metallo-thermo-mechanical coupling during processes involving phase transformation and the effect of carbon content

Contrary to these effects which are taken into account in ordinal stress analysis, arrows facing the opposite direction indicate different coupling modes. Part of the mechanical work done by existing stress in the material converts into heat generation ④, so disturbing the temperature. The acceleration of phase transformation by stress or strain, called stress- (or strain-) induced transformation ⑤(9), has been discussed, and is sometimes applied to improve the mechanical properties of metallic materials. Arrow number ⑥corresponds to the latent heat generation due to phase transformation which affects the temperature distribution. In addition to such effect of coupling, content of carbon, which diffuses during the carburization process, is considered to cause some influences on the fields as shown by broken lines.

In the first part of the paper, governing equations for temperature, stress and strain, and metallic structures in addition to diffused carbon are summarized relevant to simulating the process of carburized quenching based on the theory of *metallo−thermo−mechanics* mentioned above.

A lot of finite element codes are available for uncoupled calculation of inelastic thermal stresses. However, no program has been developed which is capable to simulate such coupled metallo-thermo-mechanical behaviour. This project motivates to develop a code "HEARTS" [HEAt tReaTment Simulation program] to be used for engineers(10). Brief introduction of the structure and the implementation strategy of the developed code are made in the next section. The program of version 2.0 is available for 3-D simulation in addition to the 2-D analysis equipped in version 1.0 including axisymmetric and generalized plane strain problems. The system architecture of the program is briefly stated which is utilized in the CAE environment by adopting pre/post processor, I-DEAS, being combined with the solver.

The final part of the paper deals with simulated results of quenching process of rod, ring, and gear wheel, where carbon content after carburization and temperature, structure change and stress distribution including residual stress are presented.

GOVERNING EQUATIONS

The detail of introducing the governing equations in the framework of thermodynamics capable of describing such three fields of metallic structures, temperature and stress/strain are already reported elsewhere (3-6). Here, the fundamental equations are summarized in the following:

Diffusion Equation of Carbon Sintered carbon content C during carburization process is easily determined by solving the unsteady diffusion equation

$$\frac{\partial C}{\partial t} = D \frac{\partial^2 C}{\partial x_i \partial x_i} , \qquad (1)$$

with diffusion coefficient

$$D = exp[a(C - b)] , \qquad (2)$$

under the initial and boundary conditions.

Constitutive Equation Total strain rate $\dot{\varepsilon}_{ij}$ is assumed to be divided into elastic, plastic, thermal strain rates and those by structural dilatation due to phase transformation and transformation plasticity such that

$$\dot{\varepsilon}_{ij} = \dot{\varepsilon}_{ij}^e + \dot{\varepsilon}_{ij}^p + \dot{\varepsilon}_{ij}^T + \dot{\varepsilon}_{ij}^m + \dot{\varepsilon}_{ij}^{tp} . \qquad (3)$$

Here, elastic and thermal strains are normally expressed as

$$\varepsilon_{ij}^e = \frac{1 + \nu}{E} \sigma_{ij} - \frac{\nu}{E} \sigma_{kk} \delta_{ij} , \qquad (4)$$

and

$$\varepsilon_{ij}^T = \alpha(T - T_0)\delta_{ij} , \qquad (5)$$

with Young's modulus E, Poisson's ratio ν and thermal expansion coefficient α.

Strain rates due to structural dilatation and transformation plasticity depending on the I-th constituent read

$$\dot{\varepsilon}_{ij}^m = \sum_{I=1}^N \beta_I \dot{\xi}_I \delta_{ij} \equiv \sum \beta_I \dot{\xi}_I \delta_{ij} , \qquad (6)$$

and

$$\dot{\varepsilon}_{ij}^{tp} = 3/2 \sum K_I h(\xi_I) s_{ij} , \qquad (7)$$

with

$$h(\xi_I) = 2(1 - \xi_I) \qquad (8)$$

where s_{ij} is deviatoric stress, β stands for the dilatation due to structural change, and K_I is the intensity of transformation plasticity. ξ_I in the above equation is the volume fraction of the I-th constituent to be given later.

The plastic strain rate is reduced to the form when employing temperature dependent materials parameters

$$\dot{\varepsilon}_{ij}^p = \Lambda \frac{\partial F}{\partial \sigma_{ij}}$$

$$= \hat{G}(\frac{\partial F}{\partial \sigma_{kl}} \dot{\sigma}_{kl} + \frac{\partial F}{\partial T} \dot{T} + \sum \frac{\partial F}{\partial \xi_I} \dot{\xi}_I) \frac{\partial F}{\partial \sigma_{ij}} , \qquad (9)$$

with a temperature dependent yield function

$$F = F(\sigma_{ij}, \varepsilon^p, \kappa, T, \xi_I), \qquad (10)$$

where

$$\frac{1}{\hat{G}} = -\left(\frac{\partial F}{\partial \varepsilon_{mn}^p} + \frac{\partial F}{\partial \kappa}\sigma_{mn}\right)\frac{\partial F}{\partial \sigma_{mn}} \quad . \qquad (11)$$

Either isotropic or kinematic hardening type of yield function F is available to be used in the code.

Kinetics of Phase Transformation   A material point undergoing quenching is assumed to be composed of three kinds of constituents. One is austenite, which initiates above Ac temperature. The second is pearlite, bainite and others which are induced by diffusion mechanism, and the third is martensite. When we denote the volume fraction of these three kinds of structures respectively as ξ_A, ξ_P and ξ_M, or $\xi_I(I = 1, 2, 3)$, the equation

$$\xi_A + \xi_P + \xi_M = 1 \ , \qquad (12)$$

holds.

An assumption is made that the material parameters χ are described by the mixture law (11)

$$\chi = \sum \chi_I \xi_I \quad , \qquad (13)$$

in terms of the parameter χ_I of the I-th constituent.

By modifying the Johnson-Mehl relation (12), the volume fraction of phases induced by diffusion mechanism is expressed as

$$\xi_P = 1 - \exp(-V_e) \ , \qquad (14)$$

$$V_e = \int_0^t \bar{f}(T, \sigma_{ij})(t - \tau)^3 d\tau \ . \qquad (15)$$

Here, the function in the form

$$\bar{f}(T, \sigma_{ij}) = \exp(C\sigma_m)f(T) \qquad (16)$$

is used, since the time-temperature-transformation (TTT) diagram under the applied mean stress σ_m in logarithmic scale deviates from one without stress which is represented by the function $f(T)$.

From thermodynamic consideration, the formula for martensitic reaction from austenite is assumed to be controlled by the modified Magee's rule (13)

$$\xi_M = 1 - \exp[\varphi(T - M_s) - \psi(\sigma_{ij})] \ , \qquad (17)$$

with

$$\psi(\sigma_{ij}) = A\sigma_m + BJ_2^{1/2} \ , \qquad (18)$$

where M_s is the martensite-start temperature under vanishing stress and J_2 means the second invariant of deviatoric stress s_{ij}. The parameters A and B can be identified if we have the data on the dependence of the martensitic transformation on applied stress.

Heat Conduction Equation   The local energy balance is usually given in the form

$$\rho\dot{e} = \sigma_{ij}\dot{\varepsilon}_{ij} - \frac{\partial h_i}{\partial x_i} \ , \qquad (19)$$

in terms of internal energy $e = g + T\eta + \sigma_{ij}\varepsilon_{ij}/\rho$, with the stress power $\sigma_{ij}\dot{\varepsilon}_{ij}$, where ρ and h_i are the density and heat flux, respectively. Introducing the expressions for specific heat $c = T\partial\eta/\partial T$, Eq.(19) is reduced to a heat conduction equation

$$\rho c\dot{T} - \frac{\partial}{\partial x_i}(k\frac{\partial T}{\partial x_i}) - \sigma_{ij}\dot{\varepsilon}_{ij}{}^p + \sum \rho_I l_I \dot{\xi}_I = 0 \qquad (20)$$

where k and l_I denote the coefficient of heat conduction and the latent heat produced by the progressive I-th constituent.

FINITE ELEMENT SCHEME AND METHOD OF NUMERICAL CALCULATION

Finite element scheme is applied to the fundamental equations (1)-(20), and a new version 2.0 of "HEARTS" approximately with 35,000 steps consisted of 200 subroutines in several levels is coded by FORTRAN 77 for three dimensional problem as well as two dimensional and axisymmetrical problems (plane stress and strain problems including that of generalized plane strain for stress analysis), which were available in the version 1.0. The 2-D and 3-D isoparametric elements with variable-number-nodes are selected from an element library.

A skyline scheme and modified or full Newton-Raphson method are employed to solve these nonlinear equations in each time step (14). In order to treat unsteady heat conduction equation depending on time, a numerical time integration scheme 'step-by-step time integration method' is introduced, while an incremental method is used for deformation and stress analysis.

The formulated finite element equation system considering the coupling between increment of nodal

207

displacement $\{\Delta u\}$ and temperature $\{T\}$ as well as phase transformation ξ_I can be expressed in the forms;

$$[P]\{\dot{T}\} + [H]\{T\} = \{Q(\xi_I, \sigma_{ij})\} \ , \qquad (21)$$

$$[K(\mathbf{u})]\{\Delta u\} = \{\Delta F(T, \xi_I)\} \ . \qquad (22)$$

Here, matrices $[P]$, $[H]$ and $[K]$ represent the matrices of heat capacity, heat conduction and stiffness, respectively, and the vectors $\{Q\}$ and $\{\Delta F\}$ are the terms of heat flux, increments of thermal load.

Data listed in Tables 1 and 2 are introduced for initial and boundary conditions, respectively, in which h and γ represent coefficients of heat transfer and heat radiation on the cooling boundary with a unit normal n_i, and T_w is the temperature of coolant.

Table 1. Initial conditions

Initial temperature	$T = T^0$
Initial displacement	$u_i = u_i^0$

Table 2. Boundary conditions

Heat transfer.	$-k(\partial T / \partial x_i)n_i = h(T - T_w)$
Heat radiation	$-k(\partial T / \partial x_i)n_i = \gamma(T^4 - T^4{}_w)$
Fixed temp.	$T = T_f$
Imposed disp.	$u_i = \bar{u}_i$

ARCHITECTURE OF "HEARTS"

The heat treatment simulation program "HEARTS" is utilized in the CAE circumstance as illustrated in Fig.2, being combined with the solver, and pre/post processor such as I-DEAS, or other popularly used processors, and the interface. The data necessary for the simulation is generated by the pre-processor, is output in the form of intermediate file. The data in the file is transferred into the data file for control and initial-boundary conditions as well as the file for the element and node data, while the material data file is constructed separately.

The solver of "HEARTS" is divided mainly into four parts corresponding to the equations, and they can be connected by the user's requirement what kind of solutions, coupled or uncoupled, to be solved.

The output of the numerical results calculated by the solver are transferred into the files for post-treatment, list image and final results. The data for

post-treatment is again stored in the intermediate file through the interface to convert into the final data for post-processor, and several kinds of illustration are available by the user's requirement as shown in Fig. 3.

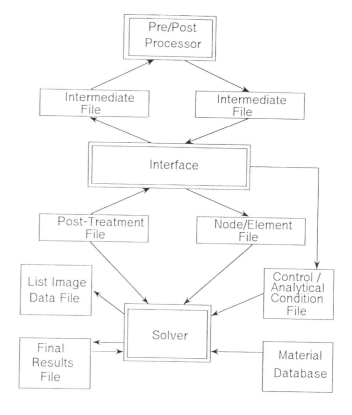

Fig.2. System architecture of "HEARTS".

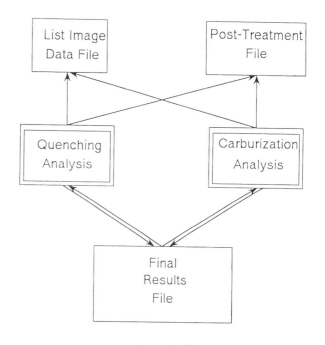

Fig.3. Architecture of result files.

208

DATA BASE FOR SOME STEELS

To complete the "HEARTS" system, it is necessary to provide some material parameters and boundary conditions. Typical two kinds of steels are, in this stage, formally propounded: They are 0.45%C carbon steel (S45C) and 3.5 % Ni-0.75 % Cr steel (SNC815), and successive effort is undergoing to accumulate the data for other materials. Such material data as well as data of some boundary conditions, say heat transfer or radiation coefficients, are filed in data base as the material data file and the boundary condition data file.

EXAMPLES OF SIMULATED RESULTS

Simulated results of some quenching process are presented in this paper by use of the code "HEARTS". The first example is the carburized quenching of a long cylinder of a case-hardening steel containing molybdenium.

Figure 4 shows the result of calculated carbon content after carburization for a long cylinder, while the circles represent measured data by X-ray diffraction method. The cylinder was water-quenched from 800 $^{\circ}C$, and temperature variation on the surface is plotted in Fig.5. Solid line in the figure shows the result when considering the effect of transformation plasticity, while the broken line is without the effect. Volume fractions of retained austenite ξ_A, martensite ξ_M and pearlite ξ_P are shown in Fig. 6 with some measured data. Figure 7 depicts the distribution of residual stresses with some measured data for axial stress, and comparison of the residual stress σ_z with and without the effect of the transformation plasticity is shown in Fig.8.

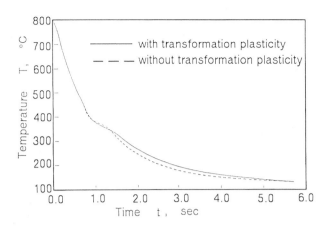

Fig.5. Temperature variation — cylinder.

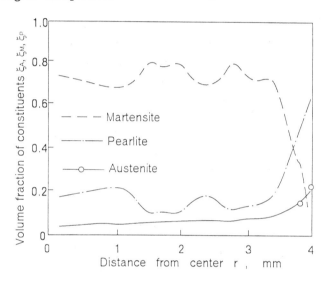

Fig.6. Distribution of volume fraction ξ_A, ξ_P and ξ_M —cylinder.

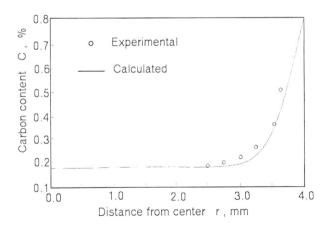

Fig.4. Distribution of carbon content.

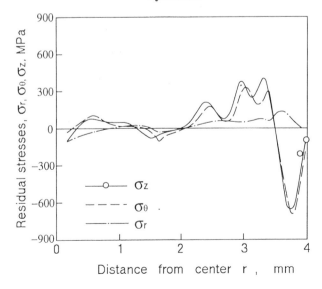

Fig.7. Distribution of residual stresses —cylinder.

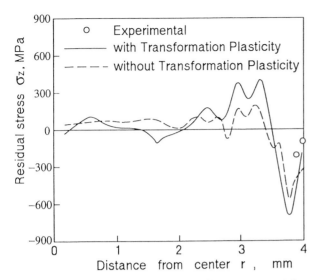

Fig.8. Comparison of the residual stress σ_z with and without the transformation plasticity – cylinder

A ring shaped body as illustrated in Fig. 9 was also quenched into water as a model of a gear wheel. Figures 10 (a) and (b) respectively represent examples of the distributions of temperature and martensite in the upper half of the section. The changes in deformation mode with progressing time are represented in Figs.11 (a),(b) and (c).

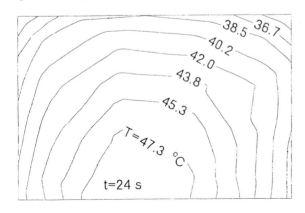

Fig.9. A ring model for quenching process.

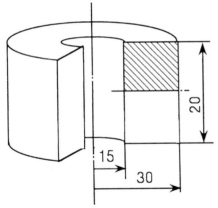

(a) Temperature distribution at t=24 s.

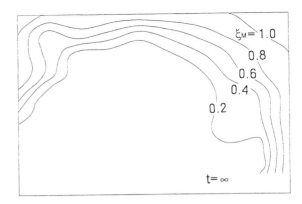

(b) Volume fraction of martensite after quenching
Fig.10. Distribution of temperature and martensite
—ring.

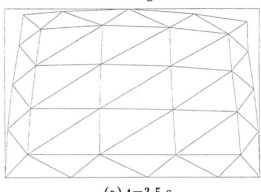

(a) t=3.5 s

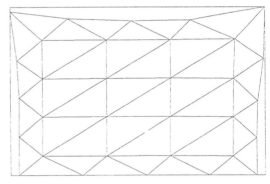

(b) t=15.0 s

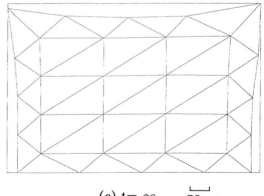

(c) t= ∞   75 μm

Fig.11. Change in deformation of a section – ring

Final example is the 3-D simulation of a gear tooth. Figure 12(a) depicts the distribution of carbon content after carburization process, and the temperature and volume fraction of induced martensite are shown in Figs.(b) and (c), respectively. Simulated residual axial and radial stresses are illustrated in Fig.13.

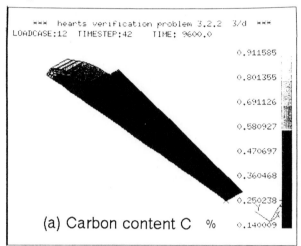

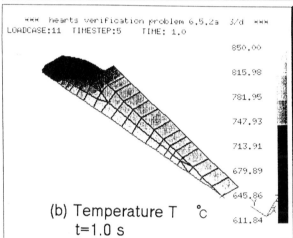

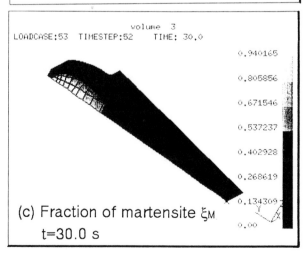

Fig.12. Distribution of carbon, temperature and martensite — gear tooth.

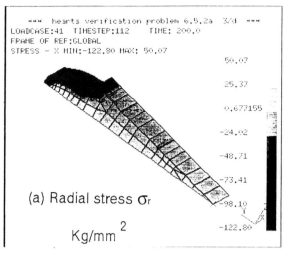

(a) Radial stress σ_r

Kg/mm^2

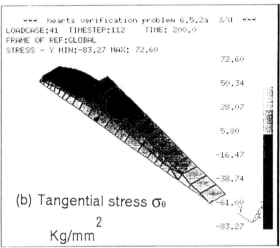

(b) Tangential stress σ_θ

Kg/mm^2

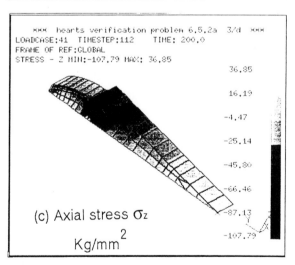

(c) Axial stress σ_z

Kg/mm^2

Fig.13. Distribution of residual stresses — gear tooth.

CONCLUDING REMARKS

A developed theory of *metallo−thermo−mechanics* relevant to analyze some engineering processes incorporating phase transformation is summarized. The strategy and architecture of a finite element code "HEARTS" for the simulation of carburization and quenching processes are stated based on the theory.

Implementation of the code were carried out by a EWS, and some examples of the results of carburization and quenching are presented.

The code "HEARTS" is under further development: The functions to simulate other kinds of processes with phase transformation such as welding (15) and casting (16-18) are scheduled to be included in addition to heat treatment processes.

ACKNOWLEDGMENT

The authors express their gratitude to Dr. H. Yaguchi, Iron and Steel Research Laboratories, Kobe Steel Co., for his cooperation to provide some experimental data for the cylinder.

REFERENCES

1- Inoue, T., S. Nagaki, T. Kishino and M. Monkawa, Ing.-Arch.,50, 315-327 (1981)

2- Inoue, T. and Z. G. Wang, Mat. Sci. Tech.,1, 845-850 (1985)

3- Inoue, T., Berg- und Hutten. Mon., 132, 63-71 (1987)

4- Inoue, T., "Thermal Stresses", 3, (R.B. Hetnarski, ed.), Elsevier Science Publishers, B.V., 192-278 (1989)

5- Inoue, T., T. Yamaguchi and Z. G. Wang, Mat. Sci. Tech., 1, 872-876 (1985)

6- Inoue, T., Z. G. Wang and K. Miyao, Proc. 32nd Japan Cong. Mat. Res., 21-26 (1989)

7- Inoue, T. and B. Raniecki, J. Mech. Phys. Solids, Vol.26, 187-212(1978)

8- Denis, S., A. Simon and G. Beck, Trans. ISIJ, 22, 504-513 (1982)

9- Greenwood, G. W. and R. H. Johnson, Proc. Roy. Soc. London, 283A, 404-422 (1965)

10- Inoue, T., K. Arimoto, and D.Y. Ju, Proc. 3rd Int. Conf. Residual Stresses, (to appear)

11- Bowen, R.M., "Continuum Physics", Vol.3, (A.C. Eringen ed.), Academic Press, New York, 2-129 (1976)

12- Johnson, W.A. and R. F. Mehl, Trans. AIME, 135, 416-458 (1939)

13- Magee, C.L., "Nucleation of Martensite", ASM, New York, (1968)

14- Bathe, K.J., "Finite Element Procedures in Engineering Analysis", Prentice-Hall, New Jersey, (1982)

15- Wang, Z. G., and T. Inoue, Mat. Sci. Tech., 1, 899-903 (1985)

16- Inoue, T. and D.Y. Ju, Int. J. Plasticity, 8, 161-183(1992)

17- Inoue, T. and D. Y. Ju, Advances in Continuum Mechanics,(ed. O. Bruller, et.al,) Springer-Verlag, Berlin, 389-406(1991)

18- Inoue, T. and D. Y. Ju, J. Thermal Stresses, 15, 109-128(1992)

Proceedings of the First International Conference on Quenching & Control of Distortion, Chicago, Illinois, USA, 22-25 September 1992

Modeling Distortion and Residual Stress in Carburized Steels

M. Henriksen
University of North Dakota
Grand Forks, North Dakota

D.B. Larson and C.J. Van Tyne
Colorado School of Mines
Golden, Colorado

Abstract

This presentation appraises the accuracy with which distortion and residual stresses can be predicted using finite element techniques and presently available mechanical property data. The authors used a model which incorporated thermal and composition dependent elastic-plastic material properties. The model predicted the distortion of thin metallic strips of SAE 4023 and SAE 4620 steels having various thicknesses, and being carburized on one side. The specimen configuration was chosen because of its amplification of distortions caused by transformation induced strains. Predicted distortions were compared with experimental data, collected from samples which had been carburized. Predicted distortions did not agree with experiments. A parametric study showed that the distortions, and hence the residual stresses, were quite sensitive to variations in constitutive parameters and carbon contents. A discussion of constitutive modeling requirements for more accurate analyses is included.

Introduction

Motivation for this Work. This work was originally motivated by a need to explain the distortion patterns produced by a carburizing process, used in gear manufacturing. The sponsors further expressed a general desire for having available an accurate and reliable method to predict residual stress and geometric distortions of carburized and quenched steel products. An accurate modeling tool that predicts deformation and residual stress states should enhance the design process and lead to a reduction in the degree of empiricism inherent in the design process. A reliable modeling capability should also lead to product quality improvement and the enhancement of reliability and manufacturing economy.

Mechanisms of Distortion. Permanent distortion in carburized and quenched steels is generated by those phase transformation strains which occur in the quench process. These strains induce stresses which are often sufficiently severe that the hot yield strength of the alloy is exceeded. Occasionally, the tensile strength is exceeded and the result is quench cracking. Local plasticity will invariably produce macroscopic dimensional change; the result is commonly referred to as quench induced distortion.

Three distinct distortion patterns are observed during heat treatment of helical gears. "Unwinding" is a phenomenon in which the helix angle is decreased and the tooth faces tend to align with the axis of the gear. The involute profile may also change. This causes misalignment which produces excessive noise and wear. Finally, the gear blank may, because of asymmetry, distort, producing an effect referred to by gear manufacturers as "potato chipping." Traditionally, these problems have been encountered in manufacturing and remediation has generally been an expensive trial and error procedure.

During the carburizing process, steel is heated above the A_3 temperature and exposed to a carbon rich atmosphere, increasing the concentration of carbon adjacent to exposed surfaces. At the end of the carburizing process, the steel is quenched to the ambient temperature, generally through a complex sequence of steps. In the quenching process, austenite is transformed to several decomposition phases which may include ferrite, pearlite, carbide, bainite, and martensite. The decomposition process is accompanied by swelling as the decomposition products have a lower density than the austenite.

The increased carbon concentration at the surface lowers the martensite start temperature (M_s); thus transformation of the case austenite occurs at a lower temperature. In small cross-sections, where thermal differentials are minimal, the higher surface carbon content causes a delay in the transformation of case austenite relative to the core austenite. This time lag causes transformation dilatational strains to occur progressively, from the core to the outer case. The strains are sufficiently high, generally exceeding yield, to give rise to progressive plastic deformation of the system.

While other mechanisms such as differential thermal cooling also produce residual stress in large and relatively stiff components, permanent deformation of gears are believed to be the by-product of transformation driven mechanisms.

Previous Work. Considerable effort has been expended in modeling and analysis of quenched and carburized steels. Burnett (1) gave a rather complete account of work up to 1977. In his dissertation, Burnett did a transient analysis of a cylindrical specimen, subjected to an oil- and water quench for SAE 86XX steel; the analysis employed the von Mises yield criterion, an associated flow rule and isotropic hardening. Residual stress data were obtained indirectly through measurements of lattice strains, using x-ray diffraction techniques. No attempts were made to measure macroscopic distortions. Burnett only considered the austenite to martensite transformation process.

Swedish researchers (2,3) have assembled much useful quantitative data about products of diffusion controlled transformations in steels. Hildenwall (2) has presented in a useful analytical form methods for determining the relative amounts of transformation products (ferrite, pearlite, bainite and cementite) that occur as a function of time, using Avrami's equation. Hildenwall's measurements of mechanical properties of various constituent phases are particularly noteworthy. These data allow the dilatational responses of steels to be modeled as time dependent. Hence, the prediction of residual stress for products other than martensite is possible. However, while Hildenwall's analyses agreed with his experimental data, the range of materials to which Hildenwall and Sjostrom's data can be applied has not been established.

Some commercial finite element packages contain sophisticated constitutive elastic-plastic models. Among these, ABAQUS, includes a theoretically rigorous set of models. Dougherty (4) used ABAQUS to model a heat treatment process which included a double quench. A visco-plastic model allowed the inclusion of time-dependent high temperature effects.

Current Work. With a continuing increase in computing power and the steady decline of computational cost and with the development of "user-friendly" stress analysis software, there is currently much emphasis on modeling and simulation of commercial processes. There is also an unfortunate tendency for finite element users to extend the commercial codes well beyond the limitations imposed by the scarcity of constitutive data. The work described herein was partly done to ascertain whether residual stress states could be modeled reliably, using currently available material data.

It is probable that many proprietary experiments have focused on the analytical prediction of residual stress, using finite element methods. Unfortunately, the measurement of residual stress is difficult since the methods that are available are invasive and modify the material state. Thus quantitative experimental verification is at best difficult, at worst impossible. The authors are not aware of any conclusive data which validate residual stress prediction by finite element methods.

In the present work, Larson and his co-workers (5) devised a test in which distortion was the dependent parameter. The test specimens resembled an Almen strip, commonly used for measurement of residual stress induced by shot peening. The strip, being 50 mm long, 10 mm wide and having a thickness ranging from 1.295 mm to 3.175 mm, was copper plated on one side as a diffusion barrier, carburized, and subsequently quenched. Like the Almen strip, the residual stresses caused the post-carburized strips to deform substantially. Thus, the strips provided a set of very sensitive gauges which indicated the state of strain in the post-heat treated material. The strip was of such a size that the thermal transformation was too rapid to allow significant diffusion. Therefore, the transformation products were essentially martensite. Fig. 1 shows the experimental strips, deformed by the heat treatment and quenching operation.

A finite element code that incorporated a constitutive model which could represent the transient thermal and mechanical behavior was assembled. Constitutive data, generally available in Hildenwall's dissertation (2), were incorporated in the analysis. The carburized specimen proved effective as a sensitive experimental medium.

In the following sections, the finite element model is described briefly; the finite element discussion does not stress those aspects of continuum mechanics that went into the the design of the analysis code; the section concentrates on material properties needed for a realistic portrayal of the physical system, described by the model. A more complete description, including a detailed derivation, is available elsewhere (6). The finite element results are included in a section in which experimental and modeling data are compared. The paper concludes with a discussion of needed future work.

The Finite Element Model

The computer codes employed in this project were assembled from existing software modules. The finite element method was used to simulate both the cooling and the carburizing processes. The stress analysis code employed 8-node isoparametric elements. The program allowed analysis of elasticity and plasticity with isotropic hardening. The code was validated against 12 transient thermal elastic-plastic benchmark problems, each with a known closed form solution. Larson (5) detailed the verification process. The code was also verified for large finite element systems which included several thousand elements.

The authors had two principal reasons for using a customized code, assembled from existing components, rather than a commercial code. First, the code needed to be

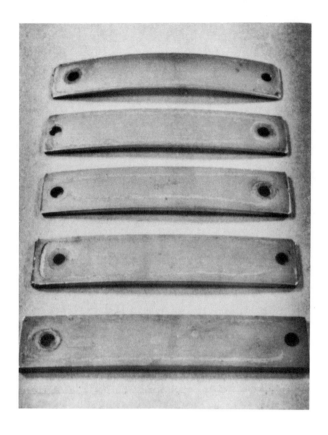

Figure 1. Carburized and Deformed Test Specimens

present, the carbon content and temperature. Where a mixture of phases exists, a rule of mixtures is invoked. Thus, if $P_m(T)$ and $V_m(T)$ are the properties and volume fractions of martensite and $P_A(T)$ and $V_A(T)$ are the properties and fractions of austenite, then the composite mixture is computed as

$$P(T) = V_m(T)\ P_m(T) + V_A(T)\ P_A(T) \qquad 1$$

Kinetics. The thin sample configuration shown in Fig. 1 was selected because it was presumed that the rapid cooling rates would induce the martensite transformation rather than those producing pearlite and bainite. Hence, the volume fraction of martensite was computed using a relationship proposed by Marburger and Koistinen (7),

$$V_m(T) = 1 - \exp[-b(M_s - T)] \qquad 2$$

where V_m is the volume fraction martensite, M_s is the martensite start temperature and b is a rate constant. The M_s temperature may be computed from the well known empirical equations, readily available in the literature. In this work, the empirical data published by Andrews (8) was used.

Mechanical Properties. From an analytical perspective, **Young's modulus** of elasticity may be considered independent of carbon content. Although they exhibit similar trends with increasing temperatures as evidenced by Fig. 2, the martensitic and austenitic phases have different moduli. Hence, the austenite-martensite transformation process will induce discontinuous changes in elastic moduli. However, the elastic modulus is primarily sensitive to temperature as is evident in Fig. 2.

flexible in its ability to model thermo-mechanically complex materials. It was particularly important to be able to modify the constitutive laws to include new approaches, including viscoplastic behavior. Secondly, the code needed to be sufficiently extensible to allow for subsequent incorporation of coupled thermo-mechanical constitutive laws.

Details of the computer codes used in this analysis were outlined by the authors elsewhere (6,7). The finite element software incorporated methods which are commonly used in elastic-plastic formulations. The material was considered as homogeneous and isotropic; all strains were assumed to be infinitesimal; stresses were expressed as Cauchy stresses. The von Mises yield theory of failure was considered as the yield criterion. Plastic flow was assumed to be adequately described by the Prandtl-Reuss flow rule. The computer code included an algorithm which enforced the appropriate consistency constraints.

Material Characteristics

This section describes the kinetic and mechanical properties that were used in the analyses. The discussion of each property set includes a review of its dependence on the phases

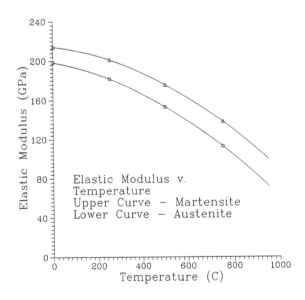

Figure 2. Elastic Modulus v. Temperature

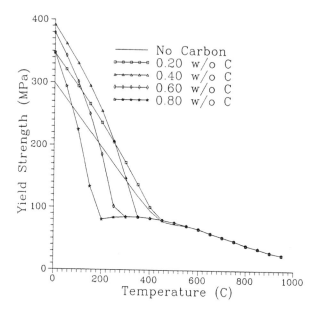

Figure 3. Yield Strength v. Temperature

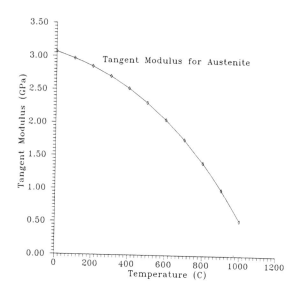

Figure 4. Tangent Modulus v. Temperature

The **yield strength**, Y, is strongly dependent on the specific phases present as well as the temperature. The yield strength of martensite decreases rapidly with increasing temperature. The decrease is less pronounced in the austenitic phase. Fig. 3 depicts the yield strength of low alloy steels with various carbon contents. These data were originally reported by Hildenwall (2). Note the pronounced change in slope at the martensite start temperature.

Yield strength is also significantly affected by the strain state of the material through work-hardening effects. Mathematically, this may be stated as

$$Y = Y(\epsilon_p, T) \qquad 3$$

where the first variable, ϵ_p is the total equivalent plastic strain, referred to as the effective strain in many finite element user manuals, and T is temperature.

Plastic deformation depends on the **work-hardening** modulus (or the plastic modulus) of the material. The plastic tangent modulus is shown in Fig. 4; this analysis incorporated Hildenwall's empirical data (2). Again, a strong dependence on temperature is seen while the effect of carbon content is minor.

The Prandtl-Reuss **flow rule** was used to predict the mechanical response temperature and phase induced strains. The yield surface, F, may be described using the following symbolic form:

$$F(\tau, \epsilon_p, T) = J_2(\tau) - \frac{1}{3}[Y(\epsilon_p, T)]^2 \qquad 4$$

where τ represents the Cauchy stress components and J_2 is the second deviatoric stress invariant. The gradient of the yield surface mathematically represents the potential for plastic flow. Thus, if the consistency condition is reviewed mathematically (5,6), it becomes clear that either an increment in applied load, an increment in plastic strain or an increment in temperature may trigger additional plastic flow. This may be illustrated by the following equation:

$$\Delta F = \frac{\partial F}{\partial \epsilon_p}\Delta \epsilon_p + \frac{\partial F}{\partial \tau}\Delta \tau + \frac{\partial F}{\partial T}\Delta T = 0 \qquad 5$$

During the quenching process, the incremental total strain that develops in the system, $\Delta\epsilon$, will consist of four components:

$$\Delta\epsilon = \Delta\epsilon_e + \Delta\epsilon_p + \Delta\epsilon_V + \Delta\epsilon_T \qquad 6$$

where $\Delta\epsilon_e$ is a vector containing the components of the incremental elastic strains and $\Delta\epsilon_p$ is a vector containing the incremental plastic strains. These two sets of strains develop as a response to the incremental thermal strains, denoted symbolically as $\Delta\epsilon_T$, and the phase transformation induced incremental strain components, $\Delta\epsilon_V$. The elastic and plastic strain components are computed from the plastic flow rule, incorporated in the finite element program. Thus, the first two sets of strain components (elastic and plastic) may be considered dependent parameters in the analysis while the two sets of dilatational strain components (thermal and phase induced dilatation) could be viewed as the independent parameters.

The incremental thermal strain components, $\Delta\varepsilon_T$, were assumed to be linearly dependent on the carbon content and a quadratic function of the temperature. Hildenwall's empirical equations were used to define the coefficient of thermal expansion for each of the component phases in the iron-carbon system.

The incremental strains, induced by phase transformation, $\Delta\varepsilon_V$, were defined in the finite element code using dilatometric information, also contained in Hildenwall's dissertation (2). The data were included in the finite element code in tabular form and applied in the analysis through appropriate interpolation.

Analytical Results

The five specimens depicted in Fig. 1 were plated on one side and subjected to a commercial carburizing treatment at a gear manufacturing plant. The resulting carbon profiles, as measured by Larson (5), are depicted in Fig. 5 as a function of their normalized position within the specimen. Microhardness traverses across the thickness of each of the five specimens were converted to carbon concentrations using appropriate conversion factors. Statistical curve fitting subsequently allowed the carbon profiles to be described in the computer analysis.

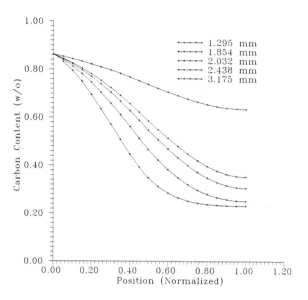

Figure 6. Carbon Content (Computed)

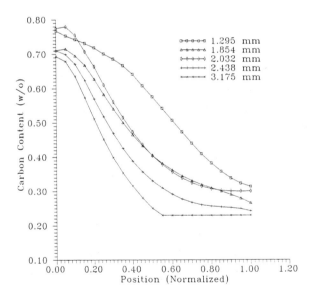

Figure 5. Carbon Content (Experimental)

One set of analyses was completed using the measured carbon profiles. Reference (6) reports on these results. The second series of analyses, published herein, used carbon profiles computed from first principles. The carbon traverses,

computed from first principles, are depicted in Figure 5.

The finite element grid used in the sequence presented herein modeled one half of the Almen strip, using a plane strain representation. The code employed 8-node isoparametric elements, with 12 and 10 elements representing the thickness for the two series of runs respectively.

Fig. 7 depicts the axial response of the specimens. On Fig. 7 and Fig. 8, the temperature increases along the abscissa; hence, the time domain spans from right to left. An inspection reveals that the axial response effectively duplicates the dilatometric curve for a specimen which has a carbon content equivalent to the average in each specimen. The authors made no effort to measure the axial distortion of the experimental.

The transverse response is depicted in Fig. 8. Note that virtually no transverse deformation takes place until the M_s temperature has been reached. When the M_s temperature is reached on the low carbon side, the austenite-martensite transformation results in a significant volumetric expansion on the low carbon side; this volumetric expansion causes the tip of the specimen to deflect away from the low carbon side (denoted as the negative direction on Fig. 8). When the M_s temperature is reached on the high carbon side, however, the deformation trend is reversed. As the high carbon side of the specimen transforms, it will, because of greater carbon content, cause greater distortion in the opposing direction.

Fig. 9 depicts the transverse deformations that were

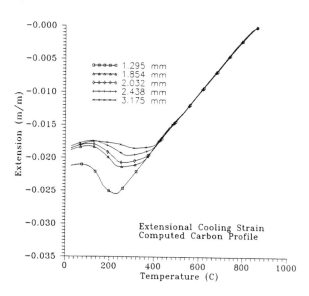

Figure 7. Extensional Deformation (Computed)

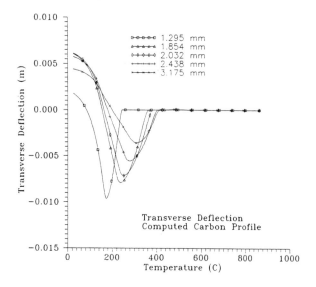

Figure 8. Transverse Deformation (Computed)

predicted by the two sets of numerical simulations. Fig. 9 also contains the experimental data obtained by Larson (5). For the transverse results, the agreement between experiment and analysis is poor. The lack of correlation for the transverse deflections is the subject of discussion in the following section.

Discussion and Conclusions

Discussion. The specimen configuration was selected because of the ease with which transverse deflections could be measured; it was also selected because of its sensitivity to minor variations in residual stress and its ability to magnify dilatational effects.

The data depicted by Fig. 9 show that the analytical results are very sensitive to minor variations in carbon contents (compare Fig. 5 and Fig. 6). This obviously suggests that very substantial differences between theory and practice should be expected if the carbon content is not established with great accuracy.

Sensitivity analyses also have shown (6) that a 20 percent variation in the M_s temperature caused a significant change in the transverse deflection (39 percent) in one instance. Similarly, a 20 percent change in the elastic modulus throughout the temperature range affected the transverse deflections by as much as 30 percent. Similar modifications in yield strength, however, did not produce sizeable changes in the transverse deflections.

The authors are unable to definitively explain the apparent differences between the trends exhibited by experiment and analysis for the thin specimen in Fig. 9. It may be that the properties published by Hildenwall and chosen for these analyses do not adequately represent the behavior of the gear blank alloys. However, the choice of the use of Hildenwall's data, in the judgment of the investigators, is justified based on chemical composition.

Conclusions. Several conclusions are inescapable. The elastic-plastic transient analysis provided exact solutions for a set of idealized thermo-mechanically complex benchmarks. These transient benchmarks included 12 test cases, all of which correlated well with theory. Hence, we are convinced that the finite element code faithfully executes the constitutive models as intended.

Other investigators (1,2,3) have found good correlation between analytical and experimental data. However, in all these reported analyses, a relatively large non-carburized core dominated local plastic deformation. The residual stresses predicted in these experiments were considered to be in a plausible range. However, since plasticity analysis by necessity includes the consistency condition referenced earlier, numerical results in finite element plasticity will always be of a physically realistic magnitude so long as the materials properties are appropriate.

In our view, constitutive parameters such as yield strength, elastic moduli, and work hardening characteristics must be established with greater certainty for each phase present if confidence is to be placed in the prediction of transient phenomena as discussed herein. Many assumptions were inherent in this work as well. They included assumptions that the martensite transformation was solely dependent on temperature and carbon content. That is in our view

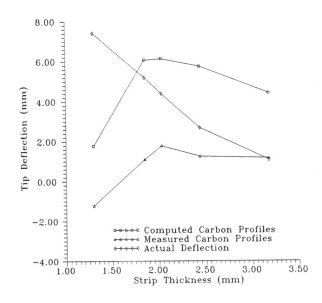

Figure 9. Transverse Deformation Comparison

probably not a defensible assumption, considering that the transformation is driven by shear. The rule of mixtures may also be over-simplification.

We will conclude that in the absence of better constitutive data, the results which may be derived from finite element analyses of metallurgical transformations should primarily be considered of value from a phenomenological perspective. We do not believe that great confidence in the quantitative results from such analyses are warranted at this time.

Acknowledgement

This work was performed under the sponsorship of the Advanced Steel Processing and Products Research Center of the Colorado School of Mines. The support of the Center is gratefully acknowledged.

References

1- Burnett, J., "Evaluation of Elastic-Plastic Stresses in Quenched Carburized Cylinders by the Finite Element Method, " Ph.D. Dissertation, University of Akron, Akron, OH (1977).

2- Hildenwall, B., "Prediction of the Residual Stresses Created During Quenching - Especially the Quench Response in Carburized Steels," Linkoping Studies in Science and Technology, Ph.D. Dissertation (39), Linkoping, Sweden, (1979).

3- Sjostrom, S., "The Calculation of Quench Stresses in Steel," Linkoping Studies in Science and Technology, Ph.D. Dissertation (84), Linkoping, Sweden, (1982).

4- Dougherty, J., "Thermal Elastic Visco-Plastic Finite Element Analysis of Heat Treatment Processes for Carburized Components," M.S. Thesis, University of Akron, Akron, OH (1987).

5- Larson, Donald B., "Finite Element Analysis of Residual Stress and Distortion in Forged and Carburized Gear Steels," M.S. Thesis T-3932, Colorado School of Mines, Golden, CO (1990).

6- M. Henriksen, D. B. Larson, C. J. Van Tyne, ASME, J. Engr. Materials and Technology, (to Appear 1992).

7- Koistinen, D., Marburger, R., Acta Metallurgica, 7, 59-60 (1959).

8- Andrews, K., J. Iron and Steel Institute, 203, 721-727 (1965).

Proceedings of the First International Conference on Quenching & Control of Distortion, Chicago, Illinois, USA, 22-25 September 1992

A Fundamental Based Microstructural Model for the Optimization of Heat Treatment Processes

B. Buchmayr
Graz University of Technology
Graz, Austria

J.S. Kirkaldy
McMaster University
Hamilton, Ontario, Canada

Abstract

A coupled finite element model for the calculation of the temperature field, microstructural evolution and mechanical response (residual stresses and distortion) will be presented. The microstructural model is based on fundamentals of thermodynamics and kinetics taking into account alloying and synergistic effects. Industrial objectives of this program are the assurance of reduced residual stresses or distortion, avoidance of crack formation, achievement of a specified microstructure or hardenability for optimum service behaviour. Due to the fundamental approach, the program can be applied to manifold problems. The widespread capability is shown with respect to the optimization of cooling procedures for heavy heat-resistant turbine casings and with respect to weldability aspects.

HEAT TREATMENT OF LOW ALLOY STEELS is an economical way to produce components with reliable service properties. Both the chemical composition and the kind of heat treatment contribute to the determination of the material properties. The objectives of controlling the heat treatment can be classified according to
- reduction of residual stresses,
- reduction of distortion,
- avoidance of crack initiation,
- achievement of sufficient hardenability,
- microstructural control to meet improved properties, such as creep strength, wear resistance and toughness, and
- achievement of a specific hardness distribution.

To achieve these particular aims the proper coupling of the calculation parts pertaining to temperature, microstructure, and stress must be effected.

In the past, a number of conferences and articles have been devoted to the modeling of quenching in order to predict the microstructure and the residual stresses [1-4]. The state of the art about computer applications for heat treatments has been described in [5,6]. These have provided a better understanding of the processes with which to model the transformation and thermomechanical behaviour of low alloy steels during the quenching process.

Global Model Structure

In [7], the coupled model components to describe the transient temperature field, the transformation behaviour, the microstructural development and the corresponding mechnical response have been reported.

The main parts (see Figure 1) of the software are:
- thermodynamics based prediction of the multicomponent phase diagram corresponding to the chemical steel composition
- analytical prediction of the isothermal transformation diagram
- automatic mesh generation program for the finite element (FE) analysis
- FE calculation of the transient temperature field taking into account microstructural effects such as latent heat
- prediction of the microstructural evolution during quenching incorporating the possibility for intermediate isothermal heat treatments
- FE calculation of the stress/strain response after temperature and microstructure prediction, taking into account dilatation, transformation plasticity, and creep
- prediction of the hardness distribution as a function of microstructure and cooling conditions
- several postprocessing routines for graphical presentation of the results

The program is designed to accomondate any 2-dim. or axisymmetric geometry as well as time-dependent boundary conditions. The material properties are considered as a function of the individual phases and temperature. Stress effects on the transformation behaviour have been neglected as well as deformation heat. By these assumptions the residual stress calculation can be made separately.

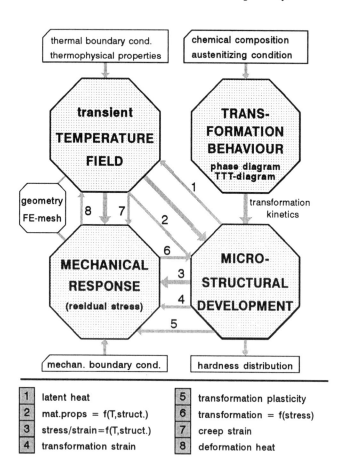

Fig. 1 - Global model structure to predict the microstructure and residual stresses in quenched components.

calibrated to the USS Atlas of Transformation Diagrams, were used as a preprocessing part of the microstructural calculations. A schematic overview of the essential components for the thermodynamical prediction of the phase diagram (Fe-C-X) of the multicomponent system and for the kinetics of diffusional transformations, which gives the time-temperature transformation (TTT) diagram, is given in Figure 2. Input data for these calculations are the chemical composition and the austenite grain size. The data on the transformation kinetics (times for transformation of distinct phase amounts) are stored for later use to calculate the transient microstructural changes during cooling. It should be noted that the model is only valid for low-alloy steels.

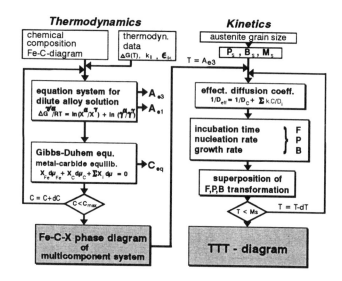

Fig. 2 - Schematic overview of calculation of the transformation behaviour of low-alloy steels.

Calculation of transient temperature field. The first step in any case of microstructural problems is to describe the temperature field as a function of geometry and time under given boundary conditions. Based on the fundamental Fourier differential equation, the applied algorithm may be an analytical solution, the finite difference method or the finite element method. From the modelling point of view, it is useful to begin with the simplest case and with one-dimensional considerations. In many cases, substantial information and an improved understanding may be obtained by these simplifications. On the other hand, practical problems often require the use of more sophisticated approaches, although this does not mean a priori, that they are more accurate. The microstructural model, described here, can be applied to any of the aforementioned methods, as it will be shown later.

Prediction of the transformation behaviour. The hardenability concepts developed by Kirkaldy and coworkers [8-11],

Prediction of the microstructural evolution. A coupled computation of the transient temperature field and the microstructural evolution gives the microstructure in any location of a component as a function of time and temperature, respectively. For the transfer from isothermal conditions (TTT) to continous cooling conditions (CCT), the consumption of incubation period and time to transform a particular phase amount is calculated for each temperature step. This procedure is similar to the integration of the Avrami equation, as used by various colleagues [6]. As a result the start temperatures for the diffusional transformations and the isolines for a particular phase amount can be determined for every point. The lowering of the Ms-temperature caused by carbon enrichment of austenit during diffusional transformation is also taken into account. A flow chart for this coupled computation is shown in Figure 3.

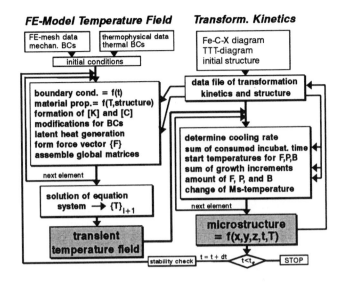

FE-Model Temperature Field

Transform. Kinetics

Fig. 3 - Flow chart of the prediction of the microstructural development during cooling.

Prediction of the mechanical response. In a next step, an elasto-plasto-mechanical approach [12] is used to describe the stress/strain response during cooling and to predict distortion and residual stresses. A flow chart of this routine is given in Figure 4.

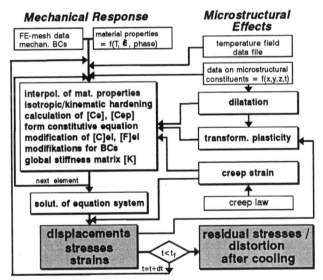

Mechanical Response

Microstructural Effects

Fig. 4 - Simplified flow chart to predict the mechanical response during cooling and prediction of residual stresses.

Prediction of Hardenability Effects. In [7], the model has been applied for the description of the hardenability of a quenched and tempered steel (type 5140). For verification reasons, a jominy test was simulated. Figure 5 shows the calculated TTT diagram (a), the CCT diagram for the Jominy cooling curves (b) and the resulting Jominy hardness curve (c). Regarding the mechanical response, the distortion of the sample during quenching (see Figure 6) and the residual stresses after cooling were predicted.

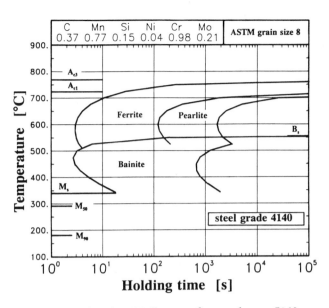

Fig. 5a - Calculated TTT-diagram for steel type 5140.

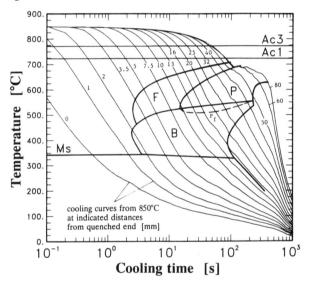

Fig. 5b - Calculated Jominy cooling curves and corresponding CCT-diagram for steel type 5140.

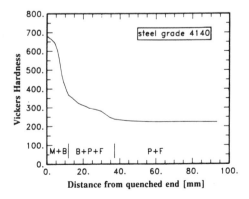

Fig. 5c - Predicted Jominy hardness curve for steel type 5140 based on thermodynamical and kinetic calculations.

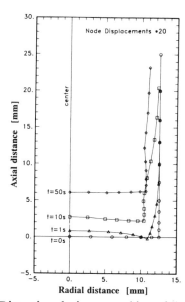

Fig. 6 - Distortion during quenching of the Jominy sample (displacements are shown in 20 x magnification).

Optimization of the cooling technology for cast turbine casings

Steam turbine casings are usually made out of heat resistant 1%CrMoV cast steels. To achieve optimal creep strength and toughness properties, the microstructure should contain mainly upper bainite, i.e. martensite and more than 15% ferrite should be avoided [13]. The reasons for these microstructural specifications are: a martensitic structure at the surface reduces the creep strength and leads to notch-embrittlement. More than 15% ferrite in the centre of thick-walled components results in a remarkable reduction of long-time creep rupture strength. Further it is known, that by means of Mn- and/or Ni-addition the hardenability (shift of ferrite transformation to longer times) can be improved. Nevertheless, there is only a small range of suitable quenching severities to fullfill the aforementioned requirements. Regarding a complete hardening of the cross section, circulated air cooling can only be used for a thickness less than about 150mm, for larger wall thicknesses (casings with up to 50 tons) oil quenching is usually applied.

In the early seventies, VOEST ALPINE Stahl Linz developed an alternative cooling technique based on intermediate water/air cooling [14,15]. Besides of excellent mechanical properties obtainable, this technology also protects the environment. In a recent research project, further computer-aided optimization were performed in order to adjust the cooling sequence for various casing geometries and to achieve a better reliability of the quenching procedure. The mathematical model takes also into account the actual chemical composition and the wall thickness. The first screen of the program and the main menu is given in Table I.

Table I. Input parameters and main menu of the program.

```
TEMPERATURE AND MICROSTRUCTURAL PREDICTION

MAIN INPUT FILE              :VATURB.DAT
Problem description          :VA turbine casing
Node points data file        :W200.KNP
TTT-diagram data file        :GSKIN.DAT
Material                     :GS-17CrMoV 5 11
  Chemical Analysis in wt%   :
C   Mn   Si   Ni   Cr   Mo   Cu   V    Nb
0.16 0.70 0.38 0.22 1.19 0.96 0.02 0.23 0.00
  IIW-carbon equivalent      :0.67
  transformation temp.°C     : Ae3  Ae1  Bs   Ms
                             : 876  752  565  424
  austenitizing temperature °C : 950
  austenitizing time h       : 8
  ASTM-grain size            : 7
material properties file     :JOMMAT.DAT
heat transfer data file      :WD_HTP.DAT
time schedule file           :DIPPEN.ZTP
process name                 :water dipping
File of results              :VATURB.ZTV

MAIN MENU
> Do calculation
  Show TTT-diagram
  Show CCT-diagram
  Show temperature
  Show structure
  Show main results
  Program end
```

A one-dimensional finite difference method was used for the temperature prediction, taking into account temperature-dependent material properties and thermophysical variables. The heat transfer coefficient for water cooling with no agitation was chosen as shown in Figure 7. There is a low heat transfer at high temperatures due to a stable steam film, a pronounced transfer between 100 and 200°C due to unstable bubble formation (boiling) and a conventional heat transfer in the liquid phase.

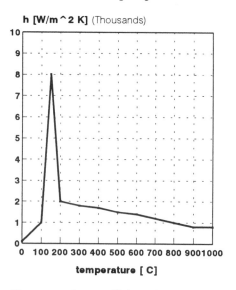

Fig. 7 - Heat transfer coefficient for water cooling as a function of surface temperature.

As an example, Figure 8 shows the calculated cooling curves for a defined cooling sequence (1½ min. transfer time from the furnace to the water pool, 5 min. WC, 2 min. AC, 5 min. WC, 2 min. AC, ...) and a wall thickness of 200mm. The predicted values correspond quite well with measured curves from various depth of real components, see Figure 9.

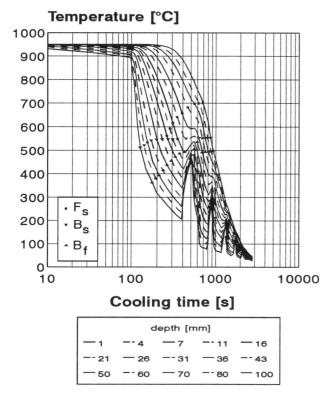

Fig. 8 - Computed cooling curves for different depth from the surfaces resulting from intermediate cooling water/air. The symbols indicate the start temperatures for ferrite and bainite transformation. Wall thickness = 200mm.

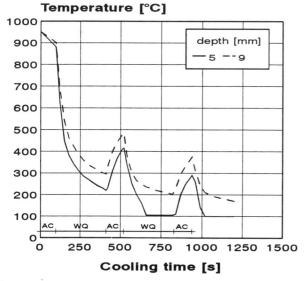

Fig. 9 - Measured cooling curves at various depth of a real turbine casing (compare with Figure 8).

The isothermal transformation (TTT) diagram has been calculated using the approach described above. In addition, an experimental TTT-diagram was measured to adjust the predicted diagram regarding the impurity drag effect (separation of the ferrite and bainite transformation field) for this kind of steel. This fundamental information on the transformation kinetics of this steel type was then used to predict the microstructural evolution during the intermediate cooling, as described in Figure 3. The calculated transformation start temperatures are indicated in Figure 8. The microstructure across the wall after cooling, as well as the calculated hardness are shown in 10. Again, there is an excellent agreement with the measured values. The fact, that the maximum ferrite amount is not in the center (as also observed on a hollow drill sample taken from a real casing) is caused by an additional isothermal transformation at a depth of about 60mm during the first reheating in air.

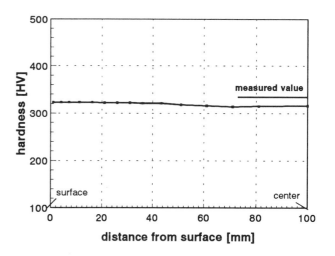

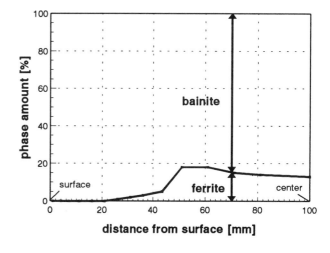

Fig. 10 - Predicted microstructure and hardness profile across the component wall with a thickness of 200mm.

The computer modelling was supported by an extensive experimental programme, wherein the transformation behaviour was measured by a dilatometer. Using a GLEE-BLE machine, cooling curves as measured on real components were applied on small test samples. The results were directly compared with those of hollow drill samples from original casings. Other topics of concern were the investigation of toughness and creep strength behaviour under varying conditions (austenitizing temperature, cooling rate, chemical composition and desoxidation practice). A detailed description is given in [16].

After some parameter studies, proper cooling sequences could be derived which assure cooling conditions to give mainly bainite. Furthermore, the cooling sequences were adjusted for different wall thicknesses, so that component with smaller wall thickness, like steam valves can also be "microstructural engineered" as described for turbine casings. In addition, turbine casings and other industrial components have usally geometries with different wall thicknesses, like small outlets and flanges, which are exposed to a significantly higher cooling rate at the surface. This circumstance has also to be considered.

A comparison of the resulting working range using the intermediate cooling technique with oil and water quenching is shown in Figure 11, taking into account the given constraint conditions concerning the microstructure.

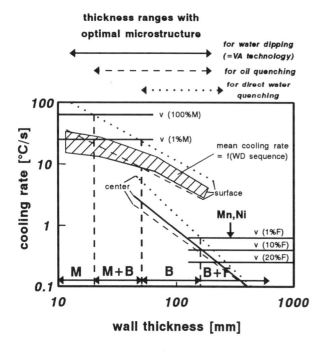

Fig.11 - Determination of optimal cooling conditions to achieve homogeneous bainitic structure for a range of wall thicknesses (comparison of water, oil and water/air quenching).

Prediction of HAZ-Microstructure in weldments

Microstructural prediction of the heat affected zone of weldments also provides an improved understanding of the weldment properties [17-19]. The main objective is to predict the weld thermal cycle, the austenite grain growth, the precipitation kinetics, the amount of microstructural constituents and the hardness as a function of the distance from the fusion line.

The thermal cycle can be described analytically using the well-known Rykalin equation, modified as described in the German standard SEW-088. This approach takes into account the influence of sheet thickness, welding parameters (heat input) and preheating temperature. The weld thermal cylces for a structural steel (StE 355 according to DIN) in the coarse grained, fine grained and intercritical subzone of the HAZ is shown in figure 12. The calculations were done using the "HAZ Calculator" [20].

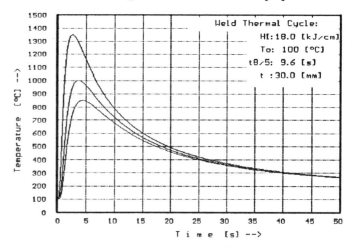

Fig. 12 - Weld thermal cycles for a structural steel (StE355) in the coarse grained, fine grained and intercritical region of the HAZ at a given heat input.

Again, using the fundamental approach as described above, the microstructure can be predicted for different zones [21]. Although the cooling rates are rather similar, the transformation is strongly influenced by the austenite grain size, which is a function of weld thermal cycle, i.e. the peak temperature and the time above A_{c3} or the distance from the fusion line. In other words, the reduced number of nucleation sites in the coarse grained zone is responsible for a larger amount of martensite there. For a given heat input, the results of the microstructural computation are shown in Figure 13.

A more comprehensive study can be applied when the finite element method is used to predict the weld thermal field, the microstructural development and lately the residual stress. Figure 14 shows a typical FE-mesh for weldment calculations, which has been generated using the pre- and postprocessing package IDEAS. The microstructural changes in each element are calculated using the model described above. Similar approaches have been reported in [22].

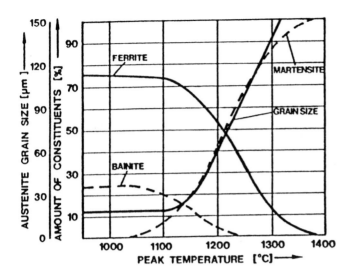

Fig. 13 - Prediction of HAZ microstructure as a function of the distance from the fusion line.

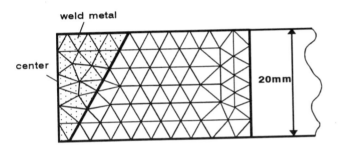

Fig.14: FE-mesch used to calculate the microstructure and residual stresses in a weldment.

CONCLUSIONS

Due to the modular structure and fundamental approaches applied, the model is capable of solving manifold industrial problems. Some components can be easily replaced, adjusted or extended to meet the requirements. Such a comprehensive model is a powerful tool because the various influencing parameters and synergistic effects are simulated on the basis of physical metallurgy. The model can be applied for material development as well as for the optimization of heat treatment technologies.

Future developments are devoted to extensions concerning high-alloyed steels as well as micro-alloyed steels, coupling with precipitation kinetics, and transformation kinetics of deformed austenite and acicular ferrite.

ACKNOWLEDGMENTS

The computer calculations for the optimization of the cooling sequence of heavy cast steel turbine casings were made within the framework of the Christian Doppler Laboratory for Computer Modelling of Metallurgical Processes and Processing Technologies of the Austrian Industries in Graz. We would like to acknowledge the contribution of Mr.Schmidt and Mr.Simschitz within the CDL activities. Furthermore we like to thank the foundry of VOEST ALPINE Stahl Linz GesmbH, which made the practical investigations and measurements of the cooling curves on water-dipped casings and which financed the experimental microstructural investigations. The permission of Dr.S.E. Feldman of Minitech Limited of Hamilton, Canada to include the Minitech Hardenability Predictor within the software is also gratefully acknowledged.

REFERENCES

1 - Doane, D.V., J.S.Kirkaldy (eds.), "Hardenability concepts with applications to steel", AMS-AIME, Warrendale, (1977).

2 - Macherauch, E., V.Hauk (eds.), "Residual Stresses in Science and Technology", DGM-Verlag, Oberursel, (1987,1991).

3 - Denis, S., S.Sjöström, A.Simon, Metall. Trans., 18A, (7), 1203-1212, (1987).

4 - Fletscher, A.J., "Thermal Stress and Strain Generation in Heat Treatment", Elsevier Applied Science, London, (1989).

5 - Kirkaldy, J.S., Metals Handbook, Vol.4, 10th ed., 20-32, (1991).

6 - Gergely, M., S.Somogyi, T.Reti, T.Konkoly, Metals Handbook, Vol.4, 10th ed., 638-656, (1991).

7 - Buchmayr, B., J.S.Kirkaldy, J. Heat Treatment 8, 127-136, (1990).

8 - Kirkaldy, J.S., Metall. Trans. 4, 2327-2333, (1973).

9 - Kirkaldy, J.S., E.A.Baganis, Metall. Trans. 9A, 495-501, (1978).

10 - Kirkaldy, J.S., D.Venugopalan, in "Phase transformation in ferrous alloys, A.R.Marder, J.I.Goldstein (eds.), AIME Philadelphia, 125-148, (1984).

11 - Kirkaldy, J.S., S.E.Feldman, J. Heat Treatment 7, 57-64, (1989).

12 - Hsu, T.-R., "The finite element method in thermo-mechanics", Allen&Unwin, Boston, (1986).

13 - Christianus, D., K.H.Keienburg, H.König, F.Staif, VGB-Werkstofftagung, Essen, (1983).

14 - Schuster, F.A., G.Ibinger, H.Rudelsdorfer, Österr. Gießereitagung, 25./26.4.1974, Leoben, Austria.

15 - Schuster, F.A., G.Köfler, Berg- u. Hüttenmänn. Monatshefte 127, (1), 1-6, (1982).

16 - Giselbrecht, W., B.Buchmayr, H.Cerjak, Austrian Research Project, FFF-3/6948, final report, (1992).

17 - Harrison, P.L., R.A.Farrar, Int. Mat. Reviews 34, 35-51, (1989).

18 - Buchmayr, B., Int.Conf. Computer Technology in Welding, TWI, Brighton, paper 32, (1990).

19 - Buchmayr, B., "Computer in der Werkstoff- und Schweißtechnik-Anwendung von mathematischen Modellen", Fachbuchreihe Schweißtechnik Bd.112, DVS-Verlag, Düsseldorf, (1991).

20 - Buchmayr, B., Proc. Conf. Trends in Welding Research, Gatlinburg, Tenn., ASM Int., 327-331, (1990).

21 - Buchmayr, B., H.Cerjak, IIW-Conf. Weld Quality - Role of Computers, Pergamon Press, 43-50, (1988).

22 - Watt, J.F., L.Coon, M.Bibby, J.Goldak, C.Henwood, Acta metall. 36 (11), 3029-3035 & 3037-3046, (1988).

Proceedings of the First International Conference on Quenching & Control of Distortion, Chicago, Illinois, USA, 22-25 September 1992

Prediction of Macro- and Micro-Residual Stress in Quenching Using Phase Transformation Kinetics

S. Das, G. Upadhya, and U. Chandra
Concurrent Technologies Corporation
Johnstown, Pennsylvania

Abstract

Residual stresses are induced in steel and other high strength alloys during quenching. These residual stresses arise from two different sources; (a) the large temperature gradients and the accompanying plastic deformations on a macroscopic scale, and (b) the solid state phase transformations and the attendant volumetric dilatation as well as the plasticity on a microscopic scale. The latter effect is also referred to as transformation plasticity.

Even though the traditional thermo-mechanical or macro-residual stresses constitute a major part of the total residual stresses, the contribution of the phase transformation effects is significant. Indeed, it has been demonstrated that neglecting these effects can result in grossly inaccurate predictions.

This paper presents a comprehensive methodology based on finite element analysis for the prediction of quenching related macro- and micro-residual stresses. An industry standard, general purpose finite element code is enhanced to account for the micro-residual stress effects. These enhancements include, the tracking of relative fractions of various phases using the theory of transformation kinetics and the computation of additional strains due to volumetric dilatation and transformation plasticity. The importance of considering the phase transformation effects during quenching and the effectiveness of the method is demonstrated through examples.

QUENCHING OF A PART made of steel or other high strength alloys often results in significant distortions and residual stresses. These effects are, in general, undesirable and must be controlled. This requires a thorough understanding of the causes and the capability to predict these effects. The residual stresses in quenched parts result primarily from two

sources; (1) inhomogeneous plastic deformation in the part at a macroscopic level, and (2) metallurgical changes or transformations at a microscopic level. The occurrence of quench related distortions and residual stresses at the macroscopic level can be easily explained. Moreover, with the availability of commercial general purpose finite element codes, these effects can be predicted routinely for even the most complex geometries [1]. The focus of this paper is on the prediction of residual stresses due to the metallurgical or phase transformations. This subject is much more complex, and its finite element treatment has only recently become possible. The phase transformation is accompanied by two effects, (1) a volumetric change, and (2) a pseudo-plasticity or the so-called transformation plasticity. Rigorous treatments of the subject, including the derivation of relevant mathematical expressions, are given by Greenwood and Johnson [2], and Magee and Davies [3].

Early attempts at numerical prediction of distortions and stresses due to quenching were confined mainly to modeling the generation of macroscopic plasticity in the workpiece [4,5]. Rammerstorfer et al. [6] recognized the significance of transformation plasticity, but treated the subject in an approximate manner. Specifically, these authors ignored the treatment of microstructural evolution and artificially modified the coefficient of thermal expansion and yield strength of the material to account for volumetric dilatation and transformation plasticity. Denis et al. [7] and Leblond et al. [8] have provided thorough mathematical treatments of microstructural evolution and transformation plasticity.

More recently, Watt et al. and Henwood et al. [9,10] have utilized an earlier work by Kirkaldy and Venugopalan [11] to study the decomposition of austenite in welds. Oddy et al. [12] have further generalized Greenwood and Johnson's work [2] to triaxial stress state and partial phase transformations. In an unrelated effort, Sawamiphakdi and Kropp [13] have used a commercial general purpose finite element code, ABAQUS [14], to model Greenwood and Johnson's [2] expression for transformation plasticity. The details of the phase transformation kinetics in Sawamiphakdi and Kropp paper are very inadequate; apparently these authors did not track the evolution of microstructure and, instead, read off the values of volume fraction of various phases from the TTT diagram.

The present work adopts the simplicity of Sawamiphakdi and Kropp's work [13], and uses ABAQUS [14] as the primary computational tool. However, instead of reading the values of various phase fractions from a TTT diagram, it mathematically tracks the evolution of microstructure. For the transformation plasticity part, this paper utilizes the generalized Greenwood-Johnson expression proposed by Oddy and co-workers [12]. These mathematical expressions are coded in a special user subroutine, UMAT, which is then coupled with ABAQUS [14]. This subroutine also includes the coding for volumetric dilatation. In this manner, a very simple yet complete treatment of phase transformation effects during quenching has been made possible. The validity of this methodology is demonstrated by application to some simple examples.

Analytical Approach

From the point of view of a heat transfer engineer quenching is a very simple heat transfer process. A metallic specimen is dipped into a cold fluid and is cooled very fast. Conduction is the primary mode of heat transfer within the metal. The heat loss from the surface of the metal can be written in terms of a film heat transfer coefficient. This heat transfer coefficient is not very easy to determine. Experimental measurements and inverse heat conduction techniques have been used in the past to obtain accurate estimates of the film heat transfer coefficient. More details about some of these inverse heat conduction techniques can be found in the works of Osman and Beck [15] and Das and Mitra [16]. The focus of this work, however, is not on determining the exact value of the heat transfer coefficient. It will be assumed that this value is already known, which makes the thermal analysis a simple exercise. Thus, instead of going into the details of the thermal analysis, the effects of the variation in temperature in the material will be discussed here in some detail.

Due to the change in the temperature, the microstructure and the stress state inside the specimen changes. All the major phenomena occurring during the quenching process and their interdependencies can be summarized as follows:

1) *Thermal stresses*: Due to the large temperature gradients present in the part during the quenching process, thermal stresses are produced in it. At any point, this stress value varies with time depending on the variation of the thermo-mechanical properties with temperature and the cooling rate. The influence of temperature on stresses is profound but the effect of stresses on temperature is negligible for all practical purposes. Thus, an uncoupled thermo-mechanical analysis is enough to track the stresses correctly.

2) *Microstructure evolution*: The primary microstructure present at high temperature is austenite. Depending on the cooling rate of the specimen at any particular location, this primary phase transforms to bainite, pearlite, ferrite and martensite. The TTT (time, temperature and transformation) diagram can be used to determine how this transformation

occurs. During the evolution of microstructures latent heat is exchanged. This affects the thermal cycle by introducing/removing latent heat.

3) *Microstructure induced stresses* : The evolution of microstructures influences the stress field substantially. Solid state phase transformation is accompanied by volume variation, and transformation plasticity. Even though, in most phase transformation stress calculations, volumetric strain and transformation plasticity are treated as two separate quantities, they are not unrelated effects. The term 'volumetric strain' is used to represent the change in volume due to transformation per unit original volume. Whereas, 'transformation plasticity' refers to the strains produced by the interaction between the stresses generated due to transformation of individual grains and the macroscopic stress field that already exists in the system. Thus, there could be situations when the volumetric strain is non-zero but the transformation plasticity is zero. These phenomena affects the stress field during the transformation from a face centered cubic to a body centered cubic structure.

The subject of thermal stress analysis in elastic and plastic regimes has been well documented (see [6]) and will not be repeated here. Instead, in the following sections we concentrate on aspects (2) and (3) mentioned above, i.e. the tracking of the evolution of microstructure and determination of microstructure induced stresses.

Calculation of Microstructure Evolution

Austenite decomposition has been modeled earlier by Kirkaldy et al.[11], based on a steady state approach using reaction kinetics for solid state transformations. Using this approach it is possible to predict TTT curves for low alloy steels with a reasonable degree of accuracy. Watt et al. [9] have used this approach in combination with the austenite grain growth models developed by Ashby & Easterling [17], to predict microstructures in the HAZ of weldments. In the current work, the approach developed by Watt et al. [9] will be used to calculate the relative phase fractions of the daughter phases obtained from austenite during quenching.

The critical temperatures for solid state transformations in steels are:

1) *The A3 temperature*: This is the temperature below which austenite starts to decompose to ferrite. Ferrite develops by nucleation at the austenite grain boundaries, and then growth into the austenite grains.

2) *The eutectoid temperature A1*: This is the temperature below which austenite starts to transform to pearlite. This reaction occurs in competition with formation of pro-eutectoid ferrite.

3) *The temperature of bainite start, B_S:* When the temperature falls below B_S, the ferrite and pearlite formation reactions stop and austenite transforms to a bainite structure.

4) *The temperature of martensite start, M_S* : When the temperature is below M_S, the remaining austenite which has not yet transformed to the earlier phases transforms to martensite in a diffusion less transformation accompanied by volume change.

The various transformation temperatures mentioned above can be quantitatively determined from the given composition of steel. The rate of formation of these phases at any given temperature can be estimated by using the Watt algorithm. Accordingly, the rate of evolution of any phase i is given by:

$$\frac{dX_i}{dt} = B(G,T)X_i^m(1-X_i)^p \qquad (1)$$

where T is the temperature, B the effective rate coefficient, G is the austenite grain size, and m and p are semi-empirical coefficients (known for the various phases in steel), and X_i is the fraction of phase i.

Thus equation (1) can be used to track the evolution of microstructure if one can determine the thermal history of the location of interest. The coefficients m and p are set to less than one to assure convergence in a form that is derived from a point nucleation and impingement growth model. The rate coefficient B includes the effect of grain size on the density of eligible nucleation sites. It also includes the amount of austenite supercooling and the effect of alloying elements and temperature on diffusion. The actual expressions for B, m and p for various phases are available in reference [9].

Calculation of Transformation Stresses

The total strain rate can be written as the sum of the individual components of the strain rates as [12] :

$$\dot{\varepsilon} = \dot{\varepsilon}^E + \dot{\varepsilon}^P + \dot{\varepsilon}^T + \dot{\varepsilon}^{\Delta V} + \dot{\varepsilon}^{TrP} \qquad (2)$$

The various components in above equation represent strain rates due to elastic, plastic and thermal loading, volumetric change and transformation plasticity, respectively. The incremental transformation plasticity can be written as [12] :

$$\Delta\varepsilon^{TrP} = \frac{5S_{ij}\Delta V}{4YV}(2 - 2X_n - \Delta X)\Delta X \qquad (3)$$

where X_n is the amount of austenite already transformed, ΔX is the fraction being transformed at a particular step, S_{ij} is the deviatoric stress tensor, Y is the yield stress of the weaker phase and $(\Delta V / V)$ is the volumetric strain occurring during phase transformation. Austenite is the weaker phase during these transformations. The strain increment due to the volumetric dilatation can be written as :

$$\Delta\varepsilon^{\Delta V} = \frac{\Delta V}{3V}\Delta X \qquad (4)$$

The last two terms in the strain rate equation (eq. 2) are non-zero only in the temperature range where the phase transformation occurs. The phase fractions are calculated from the phase transformation kinetics model outlined in the previous section. The fractions are then used in equations 3 and 4 to update the transformation strain increments and the volumetric dilatation strain increments. The net strain increment values are then used to calculate the induced stresses.

A flowchart showing various steps in the computational scheme employed in this work is given in figure 1. A transient thermal analysis is first performed within ABAQUS [14] to determine the temperature history at each point in the workpiece. This information is supplied to the user subroutine, UMAT, which computes the fraction of each daughter phase according to eq.1 and the strain terms related to transformation plasticity, volume change and the thermal strain. The total strain increment is then used to calculate the stresses.

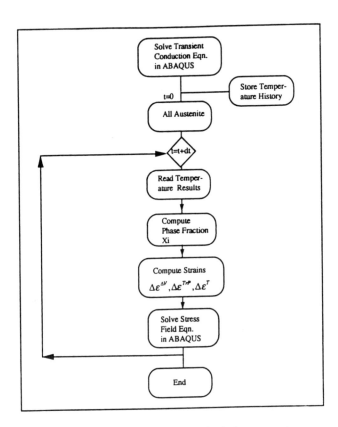

Figure 1: Flow chart showing the calculation procedure.

Examples

To verify the solution scheme described above, a problem with results available in published literature was first solved. This problem, taken from the work of Sawamiphakdi and Kropp [13], involved the quenching of an infinite cylinder. They used the equivalent material property method with negative values for the coefficient of thermal expansion in the transformation range. The trends observed in our calculations match very closely with the results reported in reference [13]. This example has not been included here to maintain brevity. In the remainder of this section two simple examples are discussed in detail.

Example 1. This example deals with the quenching of an infinitely long plate. The only finite dimension is its thickness. The stress analysis for this is performed with and without including the transformation effects. Figure 2 shows a slice taken from this plate. The thickness of the plate is 914.4 mm, and one half of it is modeled by ten 8-noded bi-quadratic plane strain elements. This problem has been solved in the ABAQUS [14] examples manual without considering the effects of transformation plasticity. The material properties of this steel plate and the surface heat transfer coefficient are listed in Table I.

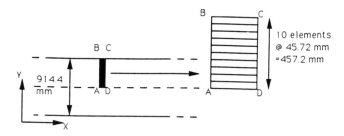

Figure 2 : Geometry and FEM model for example 1

Table I. Material Constants Used in Example 1

Property	Value	Units
Young's modulus	206.8	GPa
Poisson's ratio	0.3	
Yield Stress	248.2 for T <= 121 C; $$248.2\left(1-\frac{T-121}{1111.1}\right)$$ for T > 121 C	MPa
Density	7832	kg/m^3
Thermal Conductivity	58.8	W m^{-1}K^{-1}
Specific Heat	600	J kg^{-1}K^{-1}
Film Heat Transfer Coefficient	193.1	W/m^{2}K

In Figure 3 the evolution of these stresses in the longitudinal direction (shown by the X axis in figure 2) is shown for a point at the centerline of the plate. Figure 4 shows a similar variation at the outer surface of plate. These figures show that inclusion of the transformation effects makes significant difference in the stress history and the final residual stress in the material. Points where the simple thermal stress analysis predicts tensile residual stress ends up with compressive residual stress when the transformation effects are included. The opposite of this is observed at points where the thermal stress analysis predicts compressive stresses. This change in the final residual stress occurs since the phase transformation effects oppose the effects of the thermal strains.

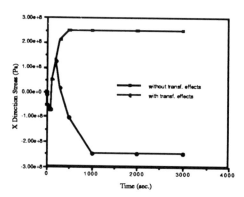

Figure 3: Comparison of stresses at A with and without transformation effects.

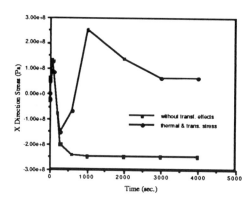

Figure 4: Comparison of stresses at B with and without transformation effects.

Example 2. In this example a 100mm long and 30mm diameter cylinder made of 4140 steel is analyzed. In this problem temperature dependent properties, obtained from reference [18] were used for all calculations. The symmetry in the geometry of the cylinder is utilized and only one quarter of it is modeled by 32 axi-symmetric finite elements. Figure 5 shows a schematic of the discretized quarter cylinder. Heat transfer occurs both in radial and axial directions. The heat transfer coefficient at the two outer boundaries is assumed to be 2000W/m^2 K. As in the previous example, temperature history data obtained from the thermal analysis is used to do mechanical analysis. First, the thermal stresses are calculated. Then, a similar calculation is done by including the transformation effects along with the thermal gradient effects. As shown in the Figure 5, two points are chosen to study the transient and residual axial (in the direction shown by the Y axis in the figure) stresses.

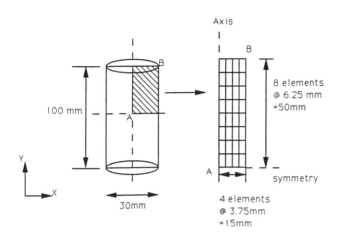

Figure 5: The geometry and the FEM model for example 2

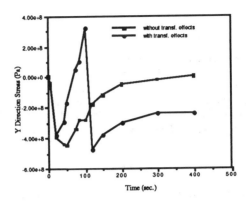

Figure 7: Comparison of stresses at A with and without Transformation effects.

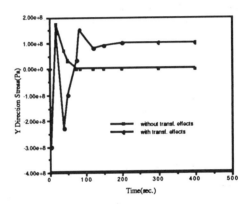

Figure 8: Comparison of stresses at B with and without Transformation effects.

The CCT curve obtained from the microstructure prediction routine for 4140 steel is plotted in Figure 6. As can be seen from this graph, the cooling rate at location B in example 2 is higher than the same at location A. But in both cases, the final product is not 100% martensite, but a mixture of ferrite, bainite and martensite. The location A undergoes more of ferrite and bainite transformation compared to location B, and hence there will be significantly less martensite at location A than at location B.

The computed stresses at these two points are shown in Figures 7 and 8. In Figure 7, the stress histories for point A are compared while in Figure 8, the same is done for point B. Beyond the elastic limit the thermal residual stress is controlled by the yield stress and the hardening parameters of the material. In this case the thermal analysis shows that the elastic limit is not exceeded in the material at any temperature. That is why the final thermal residual stress is calculated to be zero. However, if transformation effects are taken into consideration plastic stresses are introduced into the system which are independent of the macro-mechanical stresses. This is the final residual stress that is observed for both point A and B.

Summary

This paper presents the mathematical treatment of phase transformation effects during quenching of high strength steels, using the finite element method. A commercial, general purpose, three-dimensional finite element program, ABAQUS, is used as the primary computational tool. The code is further enhanced via its user subroutine, UMAT, to account for the decomposition of austenite to its daughter products, volumetric dilatation and transformation plasticity. A scheme based on phase transformation kinetics is used to track the decomposition of austenite. For the treatment of transformation plasticty affects, an analytical expression is used. The resulting methodology is then illustrated by its application to some simple examples. The example dealing with an infinite plate highlights the contribution of transformation plasticity in the presence of large macro stresses. The example of a cylinder of finite length, on the other hand, highlights the contribution of transformation plasticity in the absence of macro residual stresses.

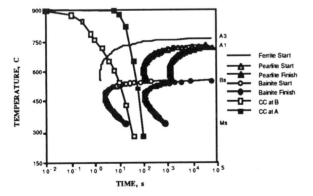

Figure 6: Calculated CCT curve for 4140 steel with prior austenite grain size (ASTM) =8

Acknowledgment

This work was conducted by the National Center for Excellence in Metalworking Technology, operated by Concurrent Technologies Corporation (formerly Metalworking Technology, Inc.), under contract to the U.S. Navy as part of the U. S. Navy Manufacturing Technology Program. The authors take this opportunity to acknowledge all the support and encouragement provided by Ms. Mary Jane Kleinosky.

References

1. Chandra, U., oral presentation, 118th TMS Annual Meeting, Las Vegas, Nevada (1989)

2. Greenwood, G.W. and R.H. Johnson, Proc. Roy. Soc., 283, 403-422 (1965)

3. Magee, C.L. and R.G. Davies, Acta Metall., 20, 1031-1043 (1972)

4. Ishigawa, H., J. Thermal Stresses, 1, 211-222 (1978)

5. Burnett, J.A. and J. Padovan, J. Thermal Stresses, 2, 251-263 (1979)

6. Rammerstorfer, F.G., D.F. Fischer, W. Mitter, K.J. Bathe and M.D. Snyder, Comp. & Struct., 13, 771-779 (1981)

7. Denis, S., E. Gautier, A. Simon and G. Beck, Materials Science and Technology, 1, 805-814 (1985)

8. Leblond, J.B., G. Mottel, J. Devaux and J.C. Devaux, Materials Science and Technology, 1, 815-822 (1985)

9. Watt, D.F., L. Coon, M. Bibby, J. Goldak and C. Henwood, Acta. Metall., 36, 3029-3035 (1988)

10. Henwood, C., M. Bibby, J. Goldak and D. Watt, Acta. Metall., 36, 3037-3046 (1988)

11. Kirkaldy, J.S. and D. Venugopalan, "Phase Transformation in Ferrous Alloys", p. 125, A.R. Marder and J.I. Goldenstein, eds., AIME Publications, (1984)

12. Oddy, A.S., J.A. Goldak and J.M.J. McDill, Proceedings 2nd International Conf. on Trends in Welding Research, Gatlinburg, TN (1989)

13. Sawamiphakdi, K. and P.K. Kropp, Proceedings of ABAQUS User's Conference, 421-438 (1988)

14. ABAQUS, Hibbitt, Karlsson and Sorensen Inc. (1989).

15. Osman, A.M. and J.V. Beck, J. of Heat Transfer, 112, 843-848 (1990)

16. Das, S. and A.K. Mitra, to appear in J. of Comp. Physics.

17. Ashby, M.F. and K.E. Easterling, Acta Metall., 30, 1969-1978 (1982)

18. "Metals Handbook . Vol. 1 Properties and Selection : Iron and Steel", B.P. Bardes ed., American Society for Metals, Metals Park, Ohio (1985)

Proceedings of the First International Conference on Quenching & Control of Distortion, Chicago, Illinois, USA, 22-25 September 1992

Quench Analysis of Aerospace Components Using FEM

R.I. Ramakrishnan
Wyman-Gordon Co.
North Grafton, Massachusetts

Abstract

Quenching (rapid cooling of parts from high temperatures) following a solution heat treatment is an important manufacturing process used to control the physical and mechanical properties of the part. In the case of nickel-base superalloys commonly used for the turbine disks in jet engines, the mechanical properties are affected by the cooling rates from the solution temperature. In general, higher cooling rates yield better mechanical properties. However, high cooling rates are associated with large surface heat fluxes which may result in high temperature gradients near the surface and large temperature differences between the thick and thin sections of the part. These conditions usually induce thermal stresses which may cause distortion beyond acceptable limits and in extreme case may result in cracking the part. It is very useful for the designers to have critical information regarding temperature evolution, cooling rates, residual stresses and distortion even before the first piece is heat treated.

Finite Element Method (FEM) has proved to be a powerful tool for the process modeling of quenching. This paper briefly describes the methodology developed at Wyman-Gordon using the inverse and direct techniques based on FEM to quantify the effects of quenching. The inverse technique estimates the rate of extraction of heat (surface heat fluxes) to characterize a particular quenchant using transient temperature measurements taken at a few appropriate interior locations in the quenched part. The direct technique then uses the estimated surface heat fluxes as boundary conditions for the heat transfer analysis to predict the temperature evolution and cooling rates in the entire cross-section of the part.

Details of the experimental setup for an oil quench and fan cooling (forced convection) of a cylindrical disk and the validation of the experimental results are presented.

SUPERALLOY DEVELOPMENT RESPONDS to the need for materials with excellent creep and fatigue resistance at high temperatures. Historically, these needs have been most acute in aircraft jet engines and other gas turbines, although applications exists in heat exchangers and other high performance heat engines. Nickel-base superalloys currently comprise over 50% of the weight of advanced aircraft engines. Superalloys for turbine disks in general are given a solution and aging heat treatment as follows: (1) A solution heat treatment to control grain size and fully or partially dissolve the γ' so that it can subsequently reprecipitate on a finer, more homogeneous scale to increase alloy strength, (2) An aging heat treatment to precipitate additional γ' and/or other phases such as carbides and borides at the grain boundaries.

Rapid cooling of components from solutioning temperature is often an important part of the heat treatment process and is used to control the γ' morphology and the physical and mechanical properties of the parts. The rate of heat extraction at the surface during the quenching process is controlled mainly by the quenching medium. Typical quench media include water, oil, salt baths, polymers, and forced and static air. The following properties/factors of the quench media affect the rate of heat extraction which influences the cooling rates in the parts:

   (a) Temperature.
   (b) Thermal properties such as specific heat, heat of vaporization and conductivity.
   (c) Viscosity.
   (d) Agitation (flow of the quenchant around the part.)
   (e) Air velocity in the case of forced convection.

The mechanical properties are affected by the cooling rates from the solution heat treatment temperature. In general, higher cooling rates yield better mechanical properties. However, high cooling rates are associated with large surface heat fluxes, and since most aerospace parts have complex geometries with thick and thin cross sections, rapid cooling from high temperatures results in high temperature gradients in thick sections and large temperature differences between the thick and thin sections. These temperature gradients induce stresses which may cause distortion beyond acceptable limits, and in extreme cases may result in cracking the part. The magnitude and distribution of the residual stresses may be critical when it comes to the machining of the part to its final shape and during the life of the part in service.

Heat treatment of nickel-base superalloys is, to a large degree, an art; although a very sophisticated one. A lot of work has been done by a number of researchers [1-11] to quantify the effects of quenching. A methodology has been developed at Wyman-Gordon Co. using inverse and direct techniques based on finite element method to simulate the quenching process for axisymmetric parts. This procedure has proved to be a cost effective tool to predict cooling rates, thermal stresses and distortion of the part.

Estimation of Surface Heat Fluxes During Quenching - Inverse Method

The first step in the analysis is to obtain the required boundary conditions; in this case the surface heat fluxes (i.e. rate of extraction of heat) that characterizes the quenchant and equipment under a specific operating condition. In a "conventional" heat transfer problem (also called the direct problem), the interior temperature distribution of the part is calculated when the boundary conditions are known. Conversely, in the inverse heat conduction problem, the unknown surface heat fluxes are estimated by using transient temperature measurements taken at a few appropriate interior points in the quenched part.

Experiments. (a) Oil quench, and (b) fan cooling (forced convection) experiments were conducted at the heat treatment facility of Wyman-Gordon Co. using actual production equipment. A hollow disk (figure 1) of the material (nickel-base superalloy) under investigation was instrumented with 13 Type K (Chromel-Alumel, 3.175 mm diameter, 304 Stainless Steel sheath) thermocouples.

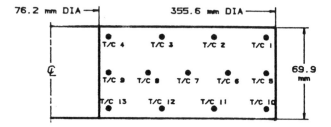

Figure 1: Cross section of the disk showing thermocouple locations

The thermocouples were held in place using compression fittings threaded onto the part surface, and were connected to a computer assisted data acquisition system. For the oil quench experiment, the disk was placed on a special heat treatment tray, loaded in a rotary furnace, and heated till the temperature in the disk stabilized at about 1149±5.6°C.

The disk was then removed from the furnace, placed on a vertical elevator and quickly immersed into an oil bath. The oil in the quench tank was agitated, and the bath temperature was maintained between 21°C and 24°C throughout the experiment. The disk was quenched in horizontal position and the whole process was automated. As the disk cooled, temperature from all 13 thermocouples was recorded every three seconds until the maximum temperature reading dropped to about 93°C.

For the fan cooling (forced convection) experiment, the above heat-up procedure was repeated. The disk was then removed from the furnace, placed in a fan cooling station and was cooled using air at ambient temperature and high velocity forced through two ducts onto the top and bottom surfaces of the disk. As in the previous experiment, temperature from all 13 thermocouples was recorded every three seconds until the maximum temperature reading dropped to about 260°C.

Inverse Heat Conduction Code. QUENCH2D [11], a code based on an inverse heat conduction method was used to process the experimental data. This code was developed at Michigan State University for the solution of general non-linear, two-dimensional inverse heat transfer problems. The objective of QUENCH2D is to provide estimates of the surface heat flux distribution as a function of surface temperature and space by using transient temperature measurements taken at a few appropriate points inside the quenched part. A piecewise-polynomial function, which can be a combination of same or different degree polynomials is used for the parametrization of the spatial distribution of the unknown surface heat flux. This greatly improves the flexibility of the specification of heat flux functional form. Two-dimensional planar or cylindrical axisymmetric geometries with irregular boundaries can be analyzed using this code. QUENCH2D uses the finite element code TOPAZ [12] modified for a direct problem solver subroutine for the calculation of the temperatures and sensitivity coefficients. The use of a powerful code such as TOPAZ gives greater power, generality and reliability for QUENCH2D.

Inverse Problem Analysis. After examining the experimental data, the entire surface of the disk was divided into four distinct sub-surfaces (figure 2) viz., inner diameter, top, outer diameter, and the bottom surface. A linear function was used for the parametrization of the spatial distribution of the unknown surface fluxes for each of the four sub-surfaces. Using the temperature information from all the 13 thermocouples from each experiment, the surface heat fluxes for the oil quench (figure 3) and the fan cooling (figure 4) were estimated as a function of surface temperature using QUENCH2D.

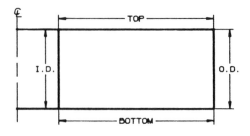

Figure 2 : Discretization of the disk surface

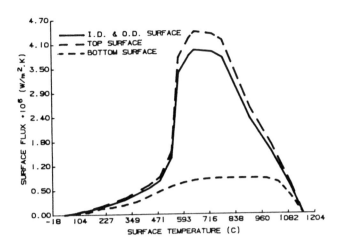

Figure 3 : Surface heat fluxes for oil quench

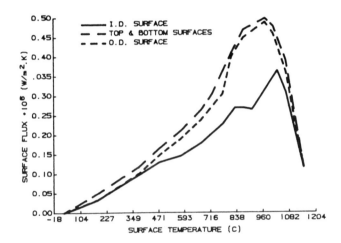

Figure 4 : Surface heat fluxes for fan cooling

Direct Problem Analysis [14]

The estimated surface heat fluxes were validated using finite element code ABAQUS [13] to check if the actual experimental results could be reproduced for both the cases. ABAQUS is a general purpose, production oriented, finite element code that can solve a wide range of non-linear problems. Thermal stress/displacement analysis was also carried out to predict stress evolution, residual stresses, and distortion of the part during the oil quench.

The direct problem was solved in two steps: (1) heat transfer analysis to predict the evolution of temperature distribution during the quench, and (2) thermal stress/displacement analysis to predict the stress evolution, residual stresses, and distortion. ABAQUS provides a wide range of element types for different analyses. Four node, linear interpolation, axisymmetric, and four node, bilinear interpolation, axisymmetric elements were used for the heat transfer and stress/displacement analyses respectively.

ABAQUS also offers a feature where users may specify certain input data through user defined subroutines (written in Fortran.) For the heat transfer analysis, SUBROUTINE DFLUX [13,14] was developed and used to input the non-uniform, space and temperature dependent distributed surface heat fluxes.

For the material under investigation, all thermo-physical properties such as the thermal conductivity, thermal expansion coefficient, specific heat, Young's modulus, Poisson's ratio and stress-strain curves were specified as functions of temperature. The stress-strain relationships were described using data from on-cooling tensile tests. An elastic-plastic material model using the distortion energy theory of von Mises was used as the yield criterion. The assumptions in the analyses were as follows: (1) Problem was axisymmetric, (2) No internal heat was generated, (3) Initial temperature was uniform, and (4) Microstructural changes during the quench process were not incorporated in the model. For the oil quench analysis, it was also assumed that the disk hits the oil instantaneously.

Results from Direct Problem

Heat Transfer Analysis. Temperature histories were tracked at six points (figure 5) that coincided with thermocouple locations in the experimental piece.

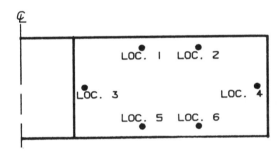

Figure 5 : Points for temperature tracking

Figures 6 and 7 shows the comparison between the experimental and analytical temperature histories for each location during the oil quench and the fan cooling experiments, respectively. It can be seen that there is a good agreement between experimental and analytical temperatures. It is possible to predict the temperature distribution for the entire cross section with transient temperature measurements taken at a few locations in the interior of the part.

Additional information such as instantaneous cooling rates (first derivative of cooling curves), and average cooling rates (between any two user defined temperature levels) may be obtained for different locations in the disk. Figure 8 shows contours of average cooling rates between 1121°C and 1066°C during the oil quench.

The top half of the disk has higher cooling rates because of the big difference in the rate of heat extraction between the top and the bottom surfaces. This is a characteristic of the oil quench tank and the heat treat tray/elevator arrangement.

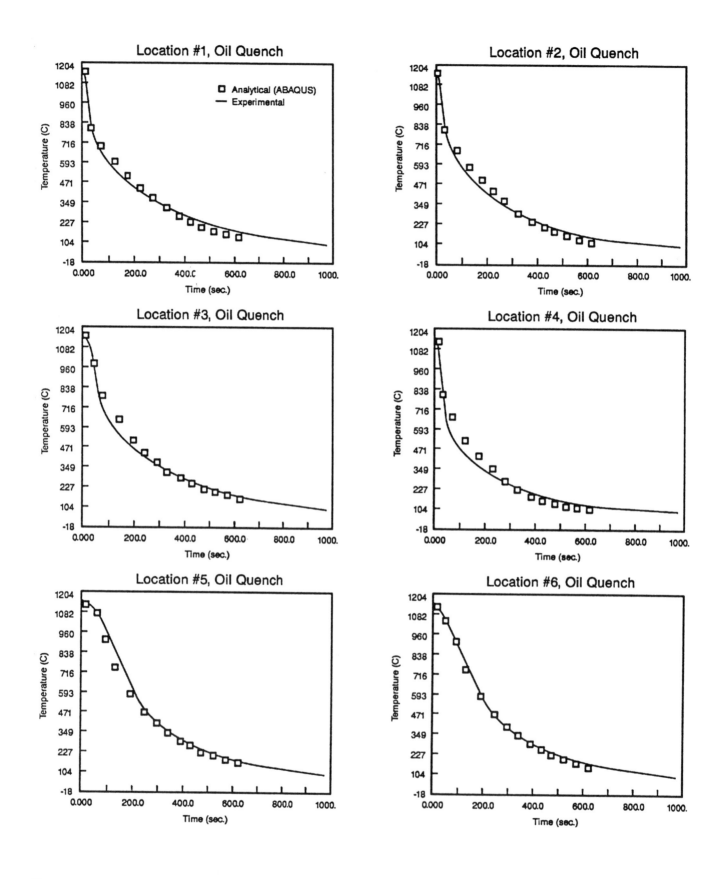

Figure 6 : Comparison of experimental and analytical temperature histories during oil quench

Figure 6 : Comparison of experimental and analytical temperature histories during oil quench (cont'd)

238

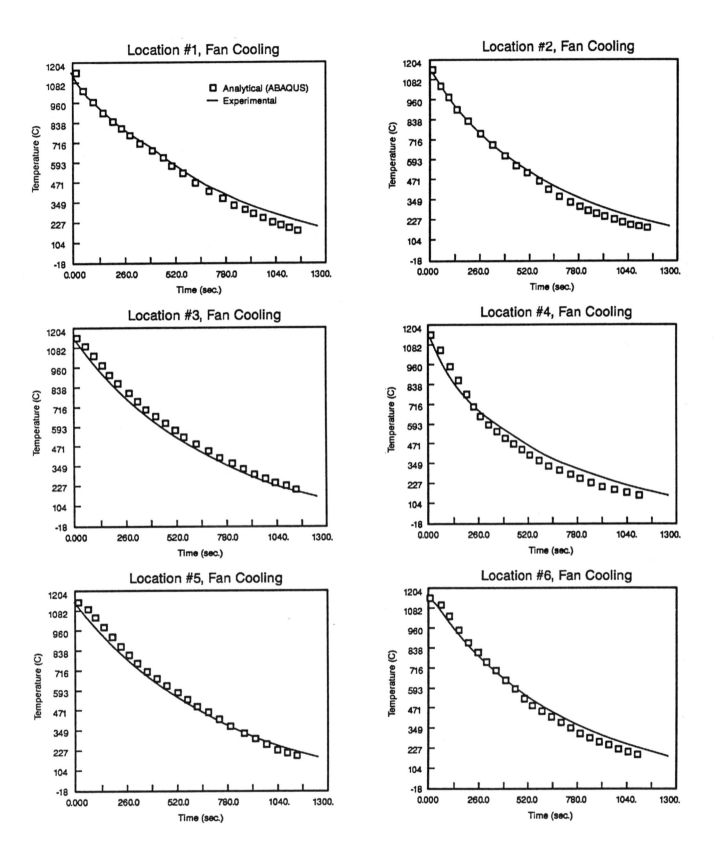

Figure 7 : Comparison of experimental and analytical temperature histories during fan cooling

Figure 7 : Comparison of experimental and analytical temperature histories during fan cooling (cont'd)

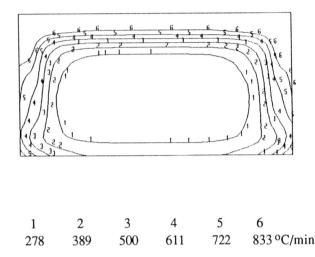

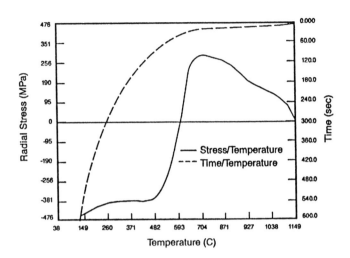

1	2	3	4	5	6
278	389	500	611	722	833 °C/min

Figure 8 : Contours of average cooling rates between 1121°C and 1066°C during oil quench

Thermal Stress Analysis (Oil Quench). High stress areas (figure 9) were identified from the thermal stress analysis for oil quench. Figure 10 shows the plot of the radial stress component/temperature and time/ temperature histories for the top surface. The radial stress component was also tracked for the bottom surface, whereas the hoop and axial stress components were tracked for the inner and outer surfaces respectively. The peak stress on each of the four surfaces is tabulated in table 1. $\sigma11$, $\sigma22$ and $\sigma33$ are the radial, axial, and hoop stress components respectively.

Surface	Peak Stress (MPa)	Time of Occurrence (s)
Top	325.7 ($\sigma11$)	20
Bottom	252.5 ($\sigma11$)	70
Inner Dia.	339.5 ($\sigma33$)	21
Outer Dia.	218.7 ($\sigma22$)	21

Table 1 : Peak stresses during oil quench

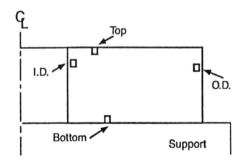

Figure 9 : High stress areas

Figure 10 : Radial ($\sigma11$) stress component/temperature and time/temperature histories for top surface

The stresses peak relatively early during the quench. This is because of the high temperature gradients set up near the surface during the initial stages of the quench. The influence of the magnitude and the rate of change of stresses are under investigation to predict quench cracking.

Figure 11 shows contours of the residual hoop stress component. Similar contours may also be obtained for the residual radial and axial stress components [15]. Figure 12a shows the original geometry, and figure 12b shows the final distorted geometry at the end of the oil quench magnified 5 times. Knowledge of residual stresses and distortion are very useful to study their impact on subsequent machining of the part to its final shape. During machining, when material is removed from parts having residual stresses, the internal stresses re-equilibrate which causes further distortion of the part.

Since the thermal stresses in a part are wholly a function of the alloy's thermo-physical properties and temperature gradients, minimizing thermal stresses requires reducing the temperature gradients and temperature differences in the parts during quenching since no control can be exercised on the material properties. This can usually be achieved by reducing the severity of the quench.

The application of the methodology with surface heat fluxes for different quench media such as oil, water, and forced and static air allows a judicious selection of the quench medium and in some cases a combination of quench media. It should be noted that quenchants with high rates of heat extraction result in rapid cooling which sets up high temperature gradients in thick sections, and large temperature differences between thick and thin section in the part. These conditions aggravate the residual stresses, distortion and cracking problems.

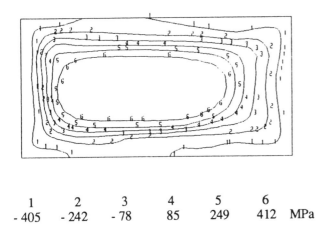

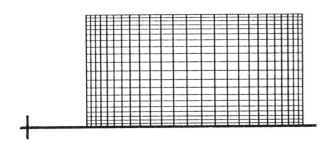

1	2	3	4	5	6	
- 405	- 242	- 78	85	249	412	MPa

Figure 11 : Contours of residual hoop ($\sigma 33$) stress component due to oil quench

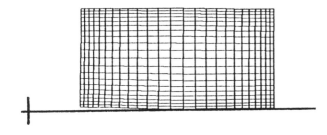

Figure 12a : Original geometry

Figure 12b : Distorted geometry (magnified 5 times) at the end of oil quench

Conclusions

Heat treatment is an important manufacturing process used to control the physical and mechanical properties of the parts, and the knowledge of cooling rates, propensity to crack, residual stresses and distortion due to quenching has always been of interest to designers. Heat treatment of superalloys has traditionally been based on experience and trial and error.

A more scientific approach using inverse and direct techniques based on FEM has been described to quantify the effects of quenching. Even though a simple hollow disk is considered here, the same approach can be extended to more complex shapes. The application of the methodology with a known database of the surface heat fluxes for different quench media such as oil, water, and forced and static air allows a judicious selection of the quench medium and in some cases combinations of quench media. This technique has been successfully used to evaluate optimum heat treatment cycles and geometries, and the up-front analyses eliminate the traditional costly and time consuming trial and error approach. This technique is also being used to preclude quench cracking, and evaluate effects of cooling rates on mechanical properties for certain alloys. In addition, simulation of the machining process has been carried out to evaluate the effect of the thermal residual stresses on the final distortion of the components.

Acknowledgements

I would like to thank Dr. James V. Beck and Dr. Arafa M. Osman for their assistance with QUENCH2D.

References

(1) Bates C.E., "Predicting Properties and Minimizing Residual Stress in Quenched Steel Parts," J. Heat Treating, Vol. 6, No. 1, 1988, pp 27-45.

(2) Price R.F. and Fletcher A.J., "Determination of surface heat-transfer coefficients during quenching quench of steel plates," Metals Technology, May 1980, pp 203-211.

(3) Wallis R.A., Bhathena N.M., Bhowal P.R. and Raymond E.L., "The application of Process Modeling to Heat treatment of Superalloys," 65th Panel Meeting, AGARD -Structures and Materials Panel, Turkey, Fall 1987.

(4) Totten G.E., Dakins M.E., Ananthapadmanabhan K.P. and Heins R.W., "Cooling rate curve area: a new measure of quenchant performance," Heat Treating, December 1987, pp 18-20.

(5) Trujillo D.M. and Wallis R.A., "Determination of Heat Transfer from Components During Quenching," Industrial Heating, July 1989, pp 22-24.

(6) Chevrier J.C., (1973), Thèse Nancy.

(7) Moreaux F., (1973), Thèse Nancy.

(8) Chevrier J.C., Moreaux F. and Beck G., 1973, Int. J. Heat Mass Transfer, 15, pp 1631-1645.

(9) Park J.E. and Ludtka G.M., "Calculation of Surface Heat Transfer Parameters During Quenching from Surface Thermocouple Signals," ASME- WAM, San Francisco, CA, December 10-15, 1989.

(10) Persampieri D., Roman A.S. and Hilton. P.D., "Process Modeling for Improved Heat Treating," Advanced Materials & Processes,3/91, pp 19-23.

(11) Osman A. and Beck J.V., 1989, "QUENCH2D - A General Computer Program for Two-Dimensional Inverse Heat Transfer Problems,"Michigan State University.

(12) Shapiro A.B., 1984, "TOPAZ - A Finite Element Heat Conduction Code for Analyzing 2-D Solids," Lawrence Livermore National Laboratory, Livermore, CA.

(13) ABAQUS Users' Manual (Ver.4.9, 1991), Hibbitt, Karlsson & Sorenson, Inc., Pawtucket, Rhode Island.

(14) Ramakrishnan R.I. and Howson T.E., "Quench Analysis Of Superalloys Using ABAQUS," ABAQUS Users' Conference, Newport, Rhode Island, May 27-29, 1992.

(15) Ramakrishnan R.I. and Howson T.E., "Modeling the Heat Treatment of Superalloys," JOM, June 1992.

Proceedings of the First International Conference on Quenching & Control of Distortion, Chicago, Illinois, USA, 22-25 September 1992

Crack Susceptibility of Nickel-Copper Alloy K-500 Bars During Forging and Quenching

M.L. Tims, J.D. Ryan, and W.L. Otto
Concurrent Technologies Corporation
Johnstown, Pennsylvania

M.E. Natishan
Naval Surface Warfare Center
Annapolis, Maryland

Abstract

Brittle fracture is sometimes observed after forging and heat treating large nickel-copper alloy K-500 (K-Monel) shafts. Inspection of failed fracture surfaces revealed that the cracks were intergranular and occurred along a carbon film that had apparently developed during a slow cooling cycle (such as during casting of the ingot stock). Cracks were observed in both longitudinal and transverse directions and arrested just short of the outer bar diameter. To minimize the development of fracture, conditions which promote carbon segregation (such as slow cooling through the 700 to 600 °C range) should be avoided. In addition, the bars should be forged at the highest metallurgically allowed temperature and at the slowest allowed deformation rate. A few large reductions are better than many small reductions. Heating should be done at a slower rate to avoid thermal shock while water quenching above 980 °C should be avoided for bars having a diameter greater than 5.0 cm.

NICKEL-COPPER ALLOY K-500 is a solid solution of copper in a nickel matrix with a face-centered cubic (FCC) crystal structure which is strengthened by complex $Ni_3(Al/Ti/Fe)$ intermetallic precipitates to a room temperature yield strength of approximately 620 MPa. Its high strength, ductility and general corrosion resistance make it a frequent material of choice for fasteners, couplings, connecting rods, etc. in marine and chemical handling applications. (See Table 1 for a typical chemical composition of K-500.) In general, K-500 components have had a successful service history. Until recently, only infrequent intergranular failures had occurred and those were attributed to hydrogen embrittlement due to close proximity to anodes in seawater.

Center line cracking has recently been observed in forged alloy K-500 bars [1]. These cracks were found during the final processing procedures, either during finish machining operations or by ultrasonic examination after final machining. The cracks appeared as extensive longitudinal or transverse, flat cracks which initiated at the center of the forging and arrested just short of the outer diameter of the forging. This fracture behavior suggests that a compressive residual stress existed at the surface of the failed bars.

Table 1: Chemical Composition of Nickel-Copper Alloy K-500 [5]

Nickel (plus Cobalt)	63.0 min
Carbon	0.25 max
Manganese	1.5 max
Iron	2.0 max
Sulfur	0.01 max
Silicon	0.5 max
Copper	27.0 - 33.0
Aluminum	2.30 - 3.15
Titanium	0.35 - 0.85

Figure 1 shows a flow chart of K-Monel bar processing during forging and heat treatment. Several different routes are possible depending upon the condition of the bar after forging and the practices of the forge shop doing the work. The bars are initially heated, followed by heavy forging reductions. Reheating is done as necessary to maintain the desired bar temperature. Light reductions then follow and are used to reshape the finished bar to a more round cross-section. Water quenching is supposed to follow forging [2]; however, air cooling is sometimes inadvertently substituted. The heat treatment operation consists of solution annealing, water quenching, aging, and then air cooling. In some cases, solution annealing is skipped if, for example, the bar temperature is hot enough after forging.

This paper concentrates on the water quench and solution annealing procedures and their effect on the stress state in the bars. Metallurgical findings have been previously reported [1] and are summarized here. Detailed descriptions of the development of flow stress and workability curves and a ductile fracture analysis of the forging process have also been reported previously [3, 4]. A summary of the ductile fracture study is presented here.

Metallurgical Investigations

Failure analyses showed the failure mode to be brittle, intergranular fracture with the exposed grain surfaces covered with a carbon film. This structure was confirmed to be unalloyed carbon (graphite) using scanning Auger spectroscopy techniques [1]. Further experimentation confirmed that these grain boundary carbon films were the cause of the premature, brittle, intergranular failure of the forged bars. Tensile testing was conducted in an effort to reproduce this fracture mode. Results showed that intergranular fracture due to grain boundary graphite was a rate-dependent fracture mode. Research is continuing to determine whether the rate-dependence is due to a creep mechanism or to a graphite enhanced hydrogen embrittlement mechanism. Results to date indicate that creep is the more likely candidate.

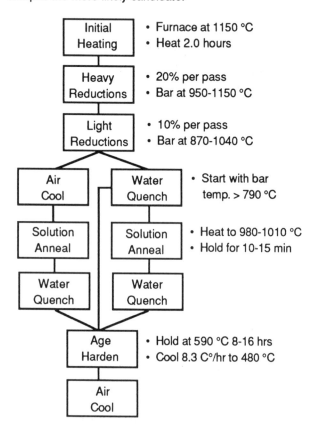

Figure 1: Process Flow Chart for Forging Nickel-Copper Alloy K-500 Shafts

There are several producers of alloy K-500 in the United States which supply stock to a large number of forging shops. The majority of failures identified have been traced back to one producer but no conclusions have been made regarding the relationship between production techniques and the mechanism of failure. The difference in material for the various producers occurs in the melt and casting processes. Material from only two producers was chosen for this study, for comparative purposes, since material from one had a history of cracking problems while the material from the other did not.

The first producer uses an argon oxygen degassing furnace after melting in an electric open air furnace and then bottom pours simultaneously into a cluster of six molds. The melt is then allowed to cool slowly, resulting in directional solidification and segregation of elements, with the lower melting temperature elements segregating to the grain boundaries and to the central portion of the ingot. The second producer whose material was used in this study, uses an electroslag remelt (ESR) process in which an ingot is remelted by immersion in a molten-flux bath which is resistance heated by a current passing between the base plate and the electrode. Drops of molten alloy pass through

the molten flux where further refining takes place before the drops fall into a pool of molten alloy which is progressively solidified in a controlled manner. This process increases the cooling rate such that segregation is minimized, resulting in a more homogeneous microstructure and chemistry.

The microstructures resulting from each of these processes differs in grain size and "cleanliness" of the material. The ESR processed material studied in this program had a very small average grain size (15 microns) and a very uniform precipitate and inclusion distribution. The air melted and cast material studied had a larger average grain size (195 microns) and a bimodal grain size distribution. The precipitate distribution was much less uniform than the ESR material with evidence of grain boundary segregation of carbides and a "free" carbon phase.

Additional studies showed that carbon segregation occurs most rapidly during cooling from 700 to 600 °C. In this same temperature range, K-500 experiences a dramatic dip in ductility as measured by percent elongation [5].

Ductile Fracture Study

A quantifiable method of determining when ductile fracture occurs for a wide variety of different processes has been the subject of much interest [6]. No single criterion has yet been found which is universally applicable to all processes and types of materials; however, work is being directed to help understand many of these uncertainties. A ductile fracture analysis was completed for the forging of the K-Monel bars, because it was speculated that ductile fractures could serve as crack initiation sites for the brittle fractures observed in the failed bars.

Table 2: Temperature-Dependent Thermophysical Properties of Nickel-Copper Alloy K-500

Temperature (°C)	Thermal Conductivity (W/m/K)	Specific Heat (J/kg.K)
21	17	419
93	20	448
204	23	477
316	26	490
427	29	502
538	32	523
649	35	553
760	38	590
871	41	657
982*	44	779
1093*	46	900
1204*	49	1022

*Data extrapolated for these temperatures

CONSTANTS:

density = 8442 kg/m^3 Poisson's ratio = 0.32
emissivity = 0.50
coefficient of thermal expansion = 1.7×10^{-5} / C°

All forging studies were made assuming plane strain and isothermal conditions. Since stress and strain fields are independent of bar radius for a given percent reduction, the results reported here apply to bars of any size. Effects of bar temperature and deformation rate were studied along with the reduction per pass and the total number of passes. The stress and strain histories resulting from forging were required for ductile fracture analyses. All forging studies were completed using NIKE2D - a public domain, non-linear finite element analysis [FEA] code from Lawrence Livermore National Laboratory [7]. Modifications to NIKE2D were made to include various ductile fracture criteria [8].

The Oyane et al. ductile fracture criteria [6] was used here. This criteria attempts to predict ductile fracture based upon a porous plasticity model. It therefore attempts to model void coalescence and growth. Stress-strain behavior and workability data were developed using a servo-hydraulically controlled universal testing machine. Conditions representative of forging operations were studied [3].

Brittle Fracture Study

Those processing steps most likely to lead to brittle fracture include the quenching and heat treatment steps. Therefore, a thermal stress analysis during these processing steps was completed. Thermal analysis was completed using version 4.4 of ANSYS - a commercial FEA code [9].

Heat Treatment. The temperature distributions within the bars during water quenching, air cooling, and heat treatment were computed for two-dimensional (2D) models. The resulting thermal response was used to determine the distribution of thermal stresses. In the case of air cooling, the thermal stresses were expected to be minimal and therefore were not computed. However, the time required to cool through the 700 to 600 °C range was of interest since carbon segregation is known to occur in this temperature range [1]. Of particular interest were those bars having diameters of 20 cm, since this is a common pre-forged bar size. Smaller bar diameters (2.5, 5.0, and 10.0 cm) were also studied.

For all heat transfer studies, the temperature- dependent thermal properties used are shown in Table 2. It was assumed that these values were applicable to both ESR and ingot-cast materials. Bilinear curve fits were used to describe the thermal stress states (see Table 3). The appropriate thermal boundary conditions were computed for the lateral face only. These conditions were also applied to the free ends of the bar. Radiation effects were included by adjusting the film coefficient accordingly. Heat lost to tooling and the heat generated during plastic working of the bar were ignored in the analyses. The bars were assumed to start with a uniform temperature as indicated by the specific case studied. Axisymmetry was assumed in all cases.

Table 3: Bilinear Curve Fit to Stress-Strain Behavior of Nickel-Copper Alloy K-500

Temp. (°C)	Yield Strength (MPa)	Elastic Modulus (MPa)	Tangent Modulus (MPa)
-20	352	179954	724
650	303	124106	3627
820	159	47229	43
930	70	32199	87
1040	40	55158	49
1150	24	51022	28

Pure, non-agitated water was assumed for the quenching medium. The temperature of the water was assumed to remain constant at 27 °C. Using standard boiling heat transfer analyses [10, 11, 12], the temperature dependent film coefficient was computed as listed in Table 4. The fluid temperature (T_∞) assumed in boiling heat transfer analyses is the boiling temperature of the fluid medium.) However, as the mode of boiling changes to one where liquid water

contacts the bar, the appropriate fluid temperature changes to that of the liquid water. So that only one fluid temperature would be required in the FEA models, the first set of values listed in Table 4 were modified. These modified values (also shown in Table 4) were determined by equating the heat flux for both sets of film coefficients (and fluid temperatures) for a given surface temperature. In other words, film coefficients associated with boiling heat transfer were adjusted by a multiplying factor equal to the ratio of temperature difference with respect to boiling and the temperature difference with respect to liquid water:

$$h' = h\,(T_S - 100\ °C)\,/\,(T_S - 27\ °C). \qquad (1)$$

Equation 1 applies during boiling heat transfer conditions where T_S is the local surface temperature in °C. The resulting modified film coefficients are also shown in Table 4. An additional 10 percent increase was applied to the modified film coefficients. This was done to account for some agitation which is anticipated within the quenching medium.

Table 4: Temperature-Dependent Film Coefficients During Water Quenching

Surface Temperature - T_S (°C)	Standard Film Coefficient - h (W/m^2/K)	Modified Film Coefficient - h' (W/m^2/K)
27*	539	591
66*	846	931
100*	1119	1232
101	1590	28
105	8517	636
109	25381	3089
113	51216	8506
117	77221	16188
132	10448	3503
147	1891	812
161	573	284
176	211	119
260	186	142
343	176	148
427	172	153
510	171	159
593	175	165
677	180	176
760	190	187
843	207	204
927	238	238
1010	294	301
1093	321	329
1177	350	358

NOTE: When using h for $T_S < 101$ °C, the appropriate fluid temperature (T_∞) is 27 °C. When using h for $T_S \geq 101$ °C, T_∞ should be 100 °C. On the other hand, $T_\infty = 27$ °C for all T_S when h' is used.

*Values listed for these temperatures are valid for a bar of 5.0 cm diameter. Film coefficients for other bar diameters (D) should be scaled by: $(5/D)^{0.25}$, where D is in cm.

To simplify the introduction of residual stresses developed during quenching into the solution annealing cycle, the solution annealing cycle was simulated immediately following that of the quenching cycle. The furnace used for solution annealing was assumed to operate at 1025 °C while having a constant film coefficient of 149 W/m^2/K. A heating time of 2.5 hours was assumed for all bar sizes. This was done for ease of FEA programming and in no way suggests an appropriate solution annealing cycle for small bars.

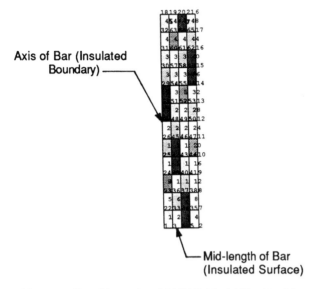

Axis of Bar (Insulated Boundary)

Mid-length of Bar (Insulated Surface)

Figure 2: Two-Dimensional ANSYS Model Used for Thermal Stress Analyses Showing Node and Element Numbers

Some studies were aimed at investigating the effects of air cooling. In all cases, the air temperature was assumed to be constant at 27 °C. Natural convection effects were assumed and then increased by 50 percent to account for inadvertent movement of air in the vicinity of the bar. The convective film coefficient (h_c) varied according to the bar diameter (D) as follows:

$$h_c = 13.1 * (20/D)^{0.25}. \qquad (2)$$

Appropriate units for film coefficient are W/m^2/K while the units for diameter are cm. Radiation effects were added by adjusting the convection film coefficient. Because of the highly non-linear nature of radiation, the modifications to the film temperature were dependent upon the local surface temperature (T_s) as follows:

$$h_{tot} = h_r + h_c \qquad (3)$$

where,

$$h_r = \varepsilon\sigma \, [T_s^2 + (300 \text{ K})^2] \, (T_s + 300 \text{ K}) \qquad (4)$$

and

h_{tot} = total equivalent film coefficient,

h_r = equivalent radiation film coefficient,

ε = surface emissivity, and

σ = Stefan-Boltzmann constant.

Figure 2 shows a sketch of the FEA mesh used for the heat transfer studies. Here 2D isoparametric elements were used. Because of symmetry in the length direction, only half of the bar (free end to mid-length) is represented. Because of axisymmetry, only half of a diametral cut is shown. The full solid bar is therefore generated by rotating this slice about its left edge through 360 degrees and then mirroring the resulting solid about its bottom edge. The appropriate thermal boundary conditions are insulated along the

symmetry boundaries (the bottom and left edges in Figure 2) with convection along the exposed boundaries (the top and right edges). Of particular interest with this mesh will be the effects at nodes 1, 4, and 2 along the mid-length and nodes 18, 20, and 6 along the free end. Temperature histories will be studied in detail here. Stresses acting throughout elements 1, 4, 45, and 48 are of interest. Another region of interest will be element 37 since the maximum stresses are frequently located in this area. (Elements are small regions within the mesh, in this case rectangles, and are indicated by numbers located at the centroid of the individual region. Nodes refer to points within the mesh and are indicated by the number just above and to the right of the corners of the elements.) Conditions demonstrating the greatest tendency for high stresses were naturally of most interest here. These conditions included large diameter bars and high pre-quench temperatures.

Thermophysical properties for K-Monel were found for temperatures up to 870 °C [5, 13]. These data were then extrapolated to obtain estimates at higher temperatures. Although a potential danger exists when extrapolation is used, no high temperature data were available. Since no solid phase transformations occur beyond 870 °C, some degree of confidence exists with these extrapolated values. Temperature-dependent thermal properties used for the present analyses are given in Table 2.

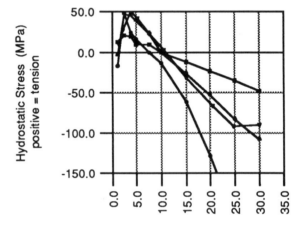

Percent Reduction

→ 930 °C; 0.1 / sec   → 1150 °C; 0.1 / sec
→ 930 °C; 10.0 / sec   → 1150 °C; 10.0 / sec

Figure 3: Hydrostatic Stress at the Bar Center During Forging

Results and Discussion

Ductile Fracture. Figure 3 shows the in-process hydrostatic stress history at the center of a bar during forging with two rigid, parallel, flat dies. (Hydrostatic stress is the mean of any three mutually perpendicular normal stresses at a given point. It is invariant with direction.) Various temperatures and strain rate conditions are represented with each showing similar trends. At very low reductions (1.25 percent), the center of the bar is still within the elastic range and, therefore, does not experience large stresses. At 2.5 percent reduction, the outward movement of the free sides of the bar produces a high tensile hydrostatic stress at the center of the bar. As reduction continues, the compressive stresses acting in the vertical direction begin to grow. This acts to reduce the hydrostatic stress at the center. When the bar has been reduced by 11 percent, the hydrostatic stress is compressive in all cases. Beyond this reduction, the hydrostatic stress at the center of the bar continues to grow in a compressive manner.

Tensile hydrostatic stresses have been related to internal damage and ductile fracture [6, 8], while large compressive hydrostatic stresses are thought to reduce damage and heal porosity. Figure 3 therefore suggests that reductions of more than 11 percent may help to heal small fractures or porosity defects near the center of a bar which is being forged with flat, parallel dies. Higher strain rates and lower temperatures result in higher hydrostatic stresses at the center of the bar during forging. Figure 4 shows the residual hydrostatic stress at the center of the bar for a given reduction. A similar trend can be seen between the in-process and residual hydrostatic stresses: large reductions provide a more favorable stress condition (i.e., a lower tensile stress) at the center of the bar. Moderately low reductions (2 to 5 percent) leave an especially unfavorable residual stress along the center of the bar. For larger reductions, an improved residual stress state results at the center of the bar. Reductions of 20 percent are practical using current production equipment. Reductions of this amount would also provide a favorable residual stress state in the forged bars. Although the forging residual stresses were not used as the initial stress conditions in the thermal stress analyses, the resulting stress pattern would contribute to containing cracks to the central portion of the bar.

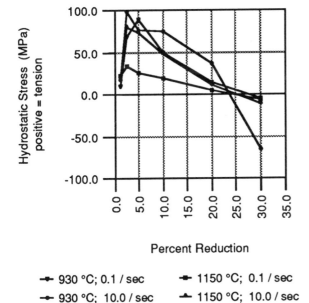

Figure 4: Residual Hydrostatic Stresses at the Bar Center After Forging

Separate upset test specimens were used to determine conditions under which ductile fracture occurred in K-500 specimens of ESR and air cast material [3]. Clearly, the ESR material showed better workability at all temperatures and strain rates tested. During deformation, coalescence and joining of voids accumulates until a ductile fracture occurs. Due to randomness of the grain shape and the distribution of voids, some variation is expected in the exact conditions

which lead to ductile fracture under specific processing conditions. The accumulated damage value using the Oyane et al. ductile fracture criteria [6] which lead to failure in the upset test specimens for air melted material at 930 °C and 10.0 sec^{-1} was 0.623 (no units). Air melted bars were seen to reach this value for a small number of light reduction passes for low forging temperatures and high strain rates [4]. Figure 5 shows a map of ductile fracture accumulation after one forging pass for air-melted material at 930 °C and 10.0 sec^{-1}. Notice that the maximum damage is located at the center of the bar. After this pass, the bar would be rotated and further reductions would be taken. Damage would accumulate during these subsequent passes. Ductile fracture was predicted to occur at the bar center from the FEA results after six reducing passes under the specified conditions. Consequently, under these conditions which are within the specification for forging K-Monel bars [2], ductile fracture in the bars is very likely. The resulting ductile fractures can then serve as crack initiation sites for a brittle fracture which may accompany the thermal processing that follows forging.

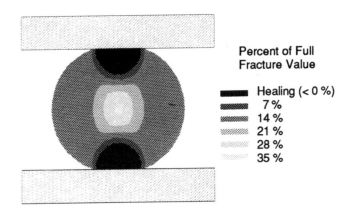

Figure 5: Ductile Fracture Map of a Nickel-Copper Alloy K-500 Bar at 10 % Reduction, 930 °C, and 10.0 sec^{-1} Using the Oyane et al. Ductile Fracture Criteria

Brittle Fracture. Figure 6 shows the temperature history for the array of nodes mentioned above (see Figure 2) for a 20-cm diameter, 122-cm long bar starting at 1150 °C. These nodes respond as expected to a sudden change in surrounding fluid temperature. Node 6 reacts the quickest followed in order by nodes 20, 18, 2, 4, and 1. The first 6000 seconds of the simulation represents quenching while the last 9000 seconds (time 6000 to 15000 seconds) represent the subsequent solution anneal. The knee in the cooling curve occurs when the mode of boiling heat transfer changes. Down to a local temperature of 176 °C, a stable film of vapor exists on the surface of the bar. This vapor film inhibits the transfer of heat and therefore results in a relatively low film coefficient (see Table 4). Between 176 °C and 117 °C, this vapor film begins to break down and liquid water intermittently comes into direct contact with the surface of the bar. Since liquid water is far superior in transporting heat, the bar begins to cool more quickly permitting even more liquid water to contact the bar's surface. Soon the bar is almost completely covered with liquid water and very rapid cooling takes place; hence, the sudden drop in temperature forming a knee in the cooling curves. This transition in boiling heat transfer is complete at 117 °C. Between 117 and 101 °C, stable nucleate boiling occurs. Below 101 °C, boiling is overshadowed by natural convection effects. Since a constant fluid temperature and film coefficient are used during heating, the curves take on the appearance of a decaying exponential as predicted by Newton's Law of Cooling.

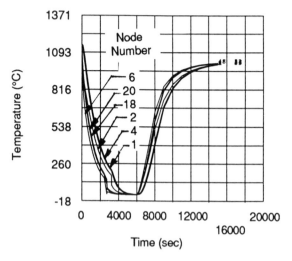

Figure 6: Temperature History for Selected Locations Within a Bar During Quenching and Subsequent Solution Annealing

Geometry, material properties, and temperature gradient are the principal ingredients in determining thermal stresses. Here, geometry and material properties are fixed. Causes of temperature gradients are therefore of great interest.

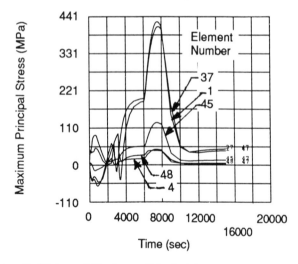

Figure 7: History of Maximum Principal Stress for Selected Locations Within a Bar During Quenching and Subsequent Solution Annealing

The maximum principal stresses at the centroids of elements 1, 4, 37, 45, and 48 are plotted in Figure 7. In the early stages of cooling (0 to 500 seconds), the stresses are low since the bar is hot and therefore has a low yield point. Plastic yielding occurs at this stage of cooling as evidenced by a non-zero residual stress (i.e., non-zero stress values at 6000 sec). During the first 2000 seconds, the center of the bar is in compression while the outside is in tension. This occurs since the outside cools faster and, therefore, tries to

contract more than the inside. To maintain material continuity, the inside is forced into compression while the outside is forced into tension. A stress reversal begins at about 2000 seconds which is before the knee in the cooling curve of Figure 6. When this knee is encountered at 2700 to 3200 seconds, the stress reversal is momentarily halted. The curves in Figure 7, therefore, oscillate somewhat at this point. After natural convective cooling has begun, the stress state within the bar heads towards its residual distribution.

At 6000 seconds, the residual stress values are reached. These values represent the stresses after complete cooling to room temperature and prior to solution annealing. The residual maximum principal stress pattern throughout the bar is depicted in Figure 8. Notice that the maximum is located about one third of the distance from the free end towards the mid-length of the bar (close to element 37). This maximum value is about 241 MPa. Values in the high 140 to low 210 MPa range can also be seen along most of the center line of the bar. These values are well above the 8.6 MPa stress needed to fracture a graphite rod. However, due to the complexities of the atomic structure at the grain boundaries of Monel K-500 where the carbon film resides, it is uncertain if a sustained stress of this magnitude will lead to a brittle fracture. Figure 8 also shows that the residual maximum principal stress at the edge of the bar is compressive. This compressive stress may help to confine the cracks to the interior of the bar.

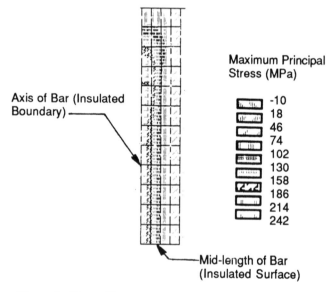

Figure 8: Fringe Plot of Residual Maximum Principal Stress After Water Quenching

The maximum principal stress along the center line of the bar tends to act in an axial direction at all times. This suggests that any cracks that would develop during water quenching would appear as transverse cracks. The values of the stress components tending to cause longitudinal cracks are only about half of those acting in the axial direction. Therefore, longitudinal cracks are much less likely during water quenching unless the material is significantly more susceptible to brittle failure along a longitudinal direction (perhaps due to the orientation of the carbon film).

During the early stage of the subsequent reheating for solution annealing, the outside layers of the bar are heated the most rapidly and, therefore, thermally expand more than the inside portion of the bar. These thermal stresses act in the same direction as the residual stresses from quenching.

Consequently, the tensile stress at the center and the compressive stress at the outside edge both grow during the early stages of solution annealing. A peak is reached at about 1500 seconds into heating when the center of the bar is 260 to 370 °C. The outside temperature of the bar is 320 to 480 °C at that time. Beyond 1500 seconds of heating, the temperature gradient begins to diminish causing a drop in the thermal stresses. Further drops in the stress state can be attributed to the lower elastic modulus at higher temperatures (see Table 2). In addition, real bars are thought to stress relieve during solution annealing [1]. Again, the maximum principal stress is an important factor in brittle fracture. The peak value during heating is about 490 MPa which, again, is located near element 37. This value is much higher than that experienced during quenching alone. It is also high enough to be a likely cause of brittle fracture along the carbon film which often forms along the grain boundary. Again, the complexities of the atomic structure along the grain boundary having a carbon film are not well understood. Possibly, small cracks induced during forging are large enough to serve as initiation sites for the catastrophic brittle failure of the bar during quenching and/or heat treating. Another possible stage of failure is during heating prior to forging where the furnaces are typically operated at 1150 °C.

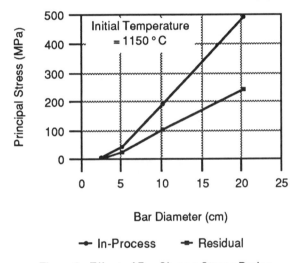

Figure 9: Effect of Bar Size on Stress During Water Quenching

Another area of interest is the effect of bar diameter on thermal stresses during quenching. Bar diameters of 20, 10, 5.0, and 2.5 cm were studied. Thermal stresses which developed are shown in Figure 9. It is clear that the bar diameter has a large effect on both the in-process and residual stresses at the center of the bar. Since the largest bars experience the largest thermal stresses, and since only large bars have been noted to fail [1], these results suggest that the thermal shocks induced during quenching and subsequent heating may be the cause of failure.

For all cases shown in Figure 9, the bar started with a uniform temperature of 1150 °C. At this temperature, Monel K-500 is soft enough to yield plastically under a small stress (less than 28 MPa for strain rates associated with thermal stressing). Plastic deformation has obviously occurred for those bars which were greater than 2.5 cm in diameter as evidenced by a non-zero residual stress. Values shown in Figure 9 represent the largest value of principal stress that the bar experiences during both water quenching and the

subsequent reheating for solution annealing. In all cases, this maximum occurred during the early stages of reheating.

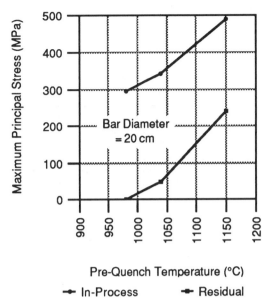

Figure 10: Effect of Pre-Quench Temperature on Stress During Water Quenching

The effect of pre-quench temperature on thermal stresses for 20-cm diameter bars is shown in Figure 10. As expected, higher temperatures result in higher stresses. Virtually no residual stresses exist in bars that are quenched at 980 °C, while large residual stresses exist for bars which start at 1150 °C. In all cases, the additional thermal stress added by reheating is about 280 MPa. The consistency of this difference makes sense since the material is within the elastic range when the maximum stresses are induced. This value is more than that developed under quenching alone for all temperatures studied. Therefore, in addition to starting water quenching with bars that are at a low temperature, milder heating cycles will help to reduce the probability of cracking.

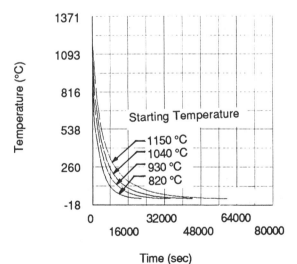

Figure 11: History of Bar Temperature During Air Cooling for Various Initial Temperatures

249

The thermal history of an 20-cm diameter bar during air cooling is shown in Figure 11. Although a 2D analysis was completed, the range in temperature was minimal and these curves accurately represent any point in the bar for a given pre-quench temperature. Since a small temperature gradient was calculated, no significant thermal stresses are anticipated. An important point to gather from these plots is the amount of time that the bars are within the temperature range 700 to 600 °C. This is the temperature range where rapid carbon segregation occurs. This segregation can form the deleterious carbon film which has been related to previous failures [1]. This points to the need for eliminating any form of air cooling through the 700 to 600 °C range including during facility breakdowns or other delays. Regardless of the starting temperature, the bars are within this range for about 1000 seconds (17 minutes). This may be sufficient time to develop a deleterious carbon film [1].

Conclusions

Based upon the results given above, the following conclusions are presented:

1. The most likely sources of the stresses needed for fracture of Monel K-500 during the forging process steps have been identified as those resulting from forging residual stresses and those associated with thermal transients. In both cases, a tensile stress develops in the central portion of the bar which can lead to fracture. Although Monel K-500 is a ductile material, the carbon layer which can build up at the grain boundaries of individual Monel K-500 grains is brittle and serves as a site for brittle fracture growth. This problem is particularly acute for large diameter bars which are water quenched from a high temperature. Step heating is recommended, not only for solution anneal reheating, but also for the initial heating from room temperature to forging temperature.

2. Ductile fracture during the forging of ingot-cast Monel K-500 is possible but only under the worst case of temperature and strain rate. Higher temperatures provide improved workability. Hydraulic presses are better than forging hammers when working with ingot-cast Monel K-500.

3. Reductions of 20 percent are much more favorable than reductions of 10 percent. For flat, parallel dies, reductions of 2.5 to 5.0 percent create an especially unfavorable stress pattern near the center of the bar, regardless of temperature or strain rate. This means that small reductions, which are used only for cosmetic purposes, can be more damaging than those reductions which are used for major reductions in diameter. Passes which are used for improving the roundness of the bar should be eliminated on the forging presses, if possible. Rod rolling would greatly improve the stress state throughout the bar if roundness is required.

4. Significant differences were noted in the workability of ingot-cast versus ESR material: the ESR material was much more workable.

5. Air cooling does not induce significant temperature gradients. Therefore, only small thermal stresses are present during air cooling. Significant development and growth of a carbon film along grain boundaries is possible during air cooling, however.

6. The cracks on the failed K-Monel shafts arrested just short of the surface due to the residual stresses left in the bar from both forging and heat treating. Both processing steps leave a highly compressive stress field near the surface of the bar. This stress inhibits the growth of brittle cracks and therefore cracks are normally confined to a central portion of the bar.

Acknowledgments

This work was performed for David Taylor Research Center (currently, Naval Surface Warfare Center - Carderock Division) under contract number N61533-89-C-0063.

References

1. Natishan, M. E., Sparks, E. R., and Tims, M. L., "Intergranular Cracking in Nickel-Copper Alloy K-500 Due to Grain Boundary Carbon," Proceedings of the International Symposium for Testing and Failure Analysis, Los Angeles, California, 29 October - 2 November 1990.
2. "Procedure for Forging and Heat Treating Nickel-Copper-Aluminum (K-Monel)," NAVSEA Specifications 134-5515-33.
3. Tims, M. L., Otto, W. L., and Ryan, J. D., "Development of Flow Stress and Workability Data for Monel Alloy K-500," Metalworking Technology, Inc. report, April 24, 1991.
4. Tims, M. L., Otto, W. L., and Ryan, J. D., "A Study of Ductile Fracture Tendencies During Open Die Forging of Monel K-500 Bars," Metalworking Technology, Inc. report, April 24, 1991.
5. Monel: Nickel-Copper Alloys, product brochure from Inco Alloys International, Inc., Huntington, West Virginia, 1984, pp. 31-50.
6. Clift, S. E., Hartley, P., Sturgess, C. E. N., and Rowe, G. W., "Fracture Prediction in Plastic Deformation Processes," International Journal of Mechanical Science, Volume 32, Number 1, pp. 1-17.
7. John O. Hallquist, NIKE2D -- A Vectorized Implicit, Finite Deformation Finite Element Code for Analyzing the Static and Dynamic Response of 2-D Solids with Interactive Rezoning and Graphics, User's Manual, Lawrence Livermore National Laboratory, UCID-19677, Rev. 1, December 1986.
8. Bandstra, J. P., "Metalworking Simulation and Formability Using the Finite Element Method," Autofact '90, Detroit, Michigan, 12-15 November 1990.
9. DeSalvo, Gabriel J. and Gorman, Robert W., ANSYS Engineering Analysis System User's Manual, Swanson Analysis Systems, Inc. Houston, Pennsylvania, May 1, 1989.
10. Incropera, Frank P. and DeWitt, David P., Fundamentals of Heat Transfer, John Wiley & Sons, New York, New York, 1981, pp. 467-482.
11. Eckert, E. R. G. and Drake, Robert M. Jr., Analysis of Heat and Mass Transfer, McGraw-Hill Book Company, New York, New York, 1972, pp. 549-566.
12. Karlekar, Bhalchandra V. and Desmond, Robert M., Heat Transfer, West Publishing Company, St. Paul, Minnesota, 1982, pp. 674-685.
13. Handbook of Thermophysical Properties of Solid Materials, Revised Edition, Volume II: Alloys, The MacMillian Company, New York, 1961, pp. 459-472.

Proceedings of the First International Conference on Quenching & Control of Distortion, Chicago, Illinois, USA, 22-25 September 1992

Factors Affecting Distortion in Hardened Steel Components

M.A.H. Howes
INFAC/IIT Research Institute
Chicago, Illinois

Abstract

Heat treatment is a vital part of metal processing, but it is also one of the chief causes of rejected components and parts that must be reworked. The problem arises from the nature of the process. Surface hardening and heat treating of steel often require that it be heated to high temperatures, held for long periods, and then rapidly cooled by quenching. These drastic measures generate high mechanical properties, but they can also cause the part to change shape unless conditions are closely controlled.

This paper explores the factors that must be considered in understanding how distortion occurs in steel parts that must be hardened and quenched. Distortion is of great interest to the manufacturers of precision parts that must be heat treated to develop optimum properties sometimes at the expense of quality.

AT LEAST SEVEN MAJOR FACTORS lead to dimensional problems after quenching:

1) variation in structure and composition present in the steel blank
2) movement due to the relief of residual stress
3) creep of the part at temperature under its own weight
4) gross differences in section causing differential heating and then during cooling by quenching
5) volume changes due to phase transformations
6) differences in heat abstraction within the part during quenching due to variable agitation and part geometry
7) coefficient of expansion during heating and cooling.

These factors must be controlled during processing so that the final shape is predictable by the manufacturer.

The first variable that must be considered is the material source, starting with the steel supplier. Compositional variations in the cast ingot can cause different responses during heat treatment. The processing of the steel into the form required by the manufacturing process (forgings, bar stock, plate, tubes, etc.) can cause further variations and may leave high levels of residual stress, which may be removed partially or completely by normalizing or another stress-relieving process. Since these heat treatments are usually done in large batches, the product is often variable and produces further variable results from part to part, which causes different responses in subsequent processing.

Components are almost always made by processes that leave residual stresses in the material. Any forming process or machining process leaves stress systems that will be relieved by a dimensional change during a heat treatment. Thus, if the part is heavily stressed prior to heat treatment, the shape will change due to this factor alone. Processing should be arranged so that virtually stress-free parts are heat treated. Variations in heat treatment parameters such as case carbon level and processing temperatures will also cause final shape and size differences.

Parts harden because of phase transformations that occur usually with accompanying volume changes. These are predictable, however, if all parts undergo the same transformation sequence. Quenching is often carried out as a batch process, leading to variations in response since the quench severity will often not be the same for all parts of the batch.

Many attempts are being made to predict distortion in simple shapes by first considering coolant characteristics, heat transfer coefficients, flow rates, agitation, surface geometry, material conductivity, latent heats, etc. This information is then combined with phase transformation data and deformation and stress calculations using Poisson's ratio, yield strength, work hardening modulus, and density changes. To be of practical use these predictions should recognize the variabilities that occur in parts during processing. Once manufacturers recognize that variability in materials and processes does have important effects on precision parts at later stages in manufacturing, better process controls and the tracking of quality will be accepted as being beneficial and cost-effective.

Material

Steel supplied to the manufacturers of precision parts is typically either forgings or rolled products, which are made from ingots or continuously cast products. In rolling and forging, the steel is heated into the 1050°-1200°C (1900°-2200°F) range and then worked by hammering, pressing, or rolling to break down the cast structure and produce a homogeneous cross section in both composition and structure. However, these effects are never totally eradicated and careful examination shows that they persist into finished components. Figure 1 shows typical solidification patterns that occur when steel solidifies in an ingot. Solidification proceeds from the walls and finishes in the center of the ingot. Note the geometric effect of processing a square ingot to finish with a cylindrical-shaped component. This causes the possibility of different structures occurring at 90-degree positions. It is also common to observe differences from the top to the bottom of the ingots and also from the first ingots to be cast from a heat of steel to the last ingots to be poured. As the ingot solidifies, a compositional gradient also arises for each element in the alloy. Manganese, for example, in a nominal 1.0% manganese steel could vary from 0.8% at the surface to 1.2% in the center. These variations cause variable responses in hardenability, microstructure after heat treatment, residual stress levels, and consequently distortion.

If the source of steel supply is consistent and the steel is processed the same way every time, these effects cause consistent, predictable behavior that is acceptable. However, if the steel is coming from different melt shops, rolling mills, and forgers each with different processing schedules, heat treatment responses can vary—often without apparent explanation. Most steel is hot rolled, and after rolling it is allowed to cool in air on a "hot bed." This causes a difference in cooling due to conduction of heat from the bottom of the bar and convection cooling from the top. If the bar

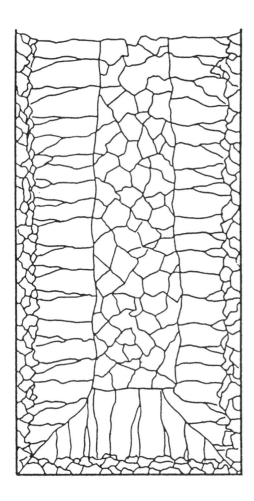

Fig. 1 - Ingot solidification pattern.

is allowed to cool completely in this position, the top of the bar will have residual tensile stresses that will tend to bend the bar and make straightening necessary. Straightening can produce very high levels of residual stresses, and a stress-relieving treatment must be carried out. For critical components a complete understanding of the processing that the steel has undergone must be obtained by discussion between suppliers and users.

Production of quality components can be considered as a value-added chain, and everyone in the chain from the material maker onwards should have a good working knowledge of how the chain links together to produce the final components. Too often each link operates in almost total ignorance of even the adjacent links, much less the entire chain. For example, does the steelmaker know that **how** he makes his product can cause problems in controlling part dimensions later? This awareness is vital to ensuring final quality.

Processing Before Heat Treatment

Steel supplied by the steel processor to the machining operation is often required to be in a stress-free condition. When a steel is hot-worked, the material cannot work harden and the steel can be processed easily until the temperature of the steel falls into the cold-working range where work hardening stresses start to build up and increase the hardness. After metalworking, the forgings or rolled products are often given an annealing or normalizing heat-treatment in order to reduce hardness so that the steel may be in the best condition for machining. These processes also reduce the residual stresses in the steel.

Annealing and normalizing are terms used somewhat interchangeably, but they do have specific meaning. Both terms imply heating the steel above the transformation range. The difference lies in the cooling method: Annealing requires a slow cooling rate, the thought being that equilibrium conditions are approached; normalized parts are cooled faster in still, room-temperature air. Annealing can be a lengthy process but produces relatively consistent results, whereas normalizing is much faster (and therefore favored from a cost point of view) but can lead to variable results depending on the position of the part in the batch and the variation of section thickness in the part being stress-relieved.

Subcritical treatment can also reduce stresses by overtempering previously hardened parts. This type of treatment is also favored because lower temperatures are used and the results can be more uniform, but grain-refined structures are not produced.

Good and Bad Residual Stresses. As discussed above, residual stresses result not only from heat treatment but also from cold-working steel by deformation caused by metalworking, machining processes, etc. Within any steel part there is a balanced stress system consisting of tensile and compressive residual stresses. If the finished part has the compressive stresses at the surface, under normal tensile loading these stresses increase the strength of the part and are thus beneficial. Processes like shot peening also are used to increase surface compressive stresses to improve performance and compensate for structural defects. This type of residual stress is intentional and is part of the design.

The problem arises when a metal part has a residual stress system prior to heat treatment. Then a shape change will occur which is predictable only if the magnitude and distribution of the stress system is known.

The parts presented for heat treatment should not only have the correct dimensions but should also have a consistent residual stress pattern. Ideally, the part should be absolutely stress-free so that movement due to this cause can be disregarded, but in practice some final machining passes are necessary before heat treatment. The best compromise is to completely stress-relieve the part before the final machining cuts are made. With care, the stresses caused by the initial machining stages will not cause the part to go out of tolerance before final machining and the final operations can be done relatively gently and thus cause a minimum of stresses to be retained in the part. Several stress-relieving treatments may be necessary during initial machining to prevent dimensions from going out of control. If a part with a preexisting stress system is machined and has thus had some of the stresses removed, the system will constantly rebalance itself by changing geometry. However a part is made, if it is made consistently the results are predictable.

Effect of Heat Treatment on Distortion

Surface Hardening. The major heat treatment used for high-quality parts is a case hardening process designed to form a hard surface layer on the gear surface. This layer gives the part a hard, wear-resistant finish but also sets up a compressive stress system at the surface which helps to resist fatigue failures. In components having gear teeth and bearing races the type of fatigue encountered is usually pitting fatigue in areas of high loading and slipping under concentrated contact.

The conventional process is carburizing, although nitriding is used for parts particularly susceptible to distortion. Carburizing involves the diffusion of carbon from a gas atmosphere while the part is heated to about 925°C (1700°F) in an atmosphere on vacuum furnace. Carbon is introduced to a level of 0.70 to 1.00% at the surface. After carburizing, the part is quenched, usually in oil to produce a hard martensitic layer on the surface. The diffusion times used are usually in the range of 4 to 20 h depending on the temperature of treatment and the case depth required. The case depth required by the designer is related to the size of the part and is often deeper with increased size of the part to produce the correct residual stress pattern. Nitriding involves the diffusion of nitrogen from a gas atmosphere in the temperature range of 495°-565°C (925°-1050°F). This may be done in an atmosphere furnace or in vacuum ion-nitriding equipment. After nitriding, the parts are hard without quenching, and they have some degree of compressive stresses due to compound formation in the surface layers. Nitriding takes anywhere from one day to one week due to the slow diffusion rates. Since nitriding is performed at relatively low temperatures and quenching is unnecessary, distortion is a minor problem; accordingly, nitriding will not be considered any further in this discussion. Another diffusion process sometimes used is carbonitriding, the simultaneous diffusion of carbon and nitrogen, generally for lower cost parts.

Selective Direct Hardening. As an alternative to carburizing and nitriding, a hard case may be produced by selective direct hardening. Instead of increasing the carbon content of the steel surface by diffusing extra carbon into the case, a steel composition that already contains 0.4-0.6% carbon is selected. This steel could then be through-hardened and tempered in a furnace, but a hard case and tough core can also be produced by selectively heating the surface where the case is required to a controlled depth and leaving the core in the original hardened and tempered condition. Both methods produce a case, one by controlling depth of carbon or nitrogen diffusion and the other by the depth of heating.

Effect of Phase Changes. Whichever method of surface hardening is selected (except for nitriding), the mechanism of hardening remains the same. For simplicity, only three phases will be considered in the following discussion:

- Pearlite, a lamellar eutectoid mixture of body-centered cubic ferrite and orthorhombic cementite (Fe_3C)
- Austenite, a face-centered cubic structure, relatively soft and stable
- Martensite, a body-centered tetragonal, hard metastable phase.

When pearlite, which probably exists within the microstructure of a part prior to heat treatment, is heated to carburizing temperatures, the dense austenite phase is formed so the percentage volume will decrease by 4.64 - 2.21 × %C. A carburizing steel contains about 0.10 to 0.25%C in the core, so if on heating, the steel contains 0.10%C, the volume will decrease by (4.64 -2.21 × 0.1)% or 4.2%. Since the heating for carburizing is relatively slow and both phases are relatively soft, the part will accommodate this change. The linear changes are approximately one-third of the volume change or, in this example, 1.4%. For comparison, if a 25 mm (1 in.) long part is being machined to a tolerance of 0.13 mm (0.005 in.) (a wide tolerance), this represents an allowable deviation of only 0.5% on the length of the part. In general, therefore, linear changes due to phase transformations cannot be ignored.

There are further complications. An implicit assumption in the above discussion is that the carbon content of the steel is uniform across the cross section. This may not be true because segregation occurs and also some heat treatments call for two case depths on a part. Thus, when the part is reheated to diffuse in the second case, the carbon content can vary from 0.1% to 1.0%, leading to a range of linear change of 0.8%-1.4% in different areas of the part.

Carburizing is usually carried out using the "Boost-Diffuse" method: The furnace atmosphere is first controlled at a high carbon potential, and later in the cycle the atmosphere carbon potential is reduced to the desired level. Depending on how the Boost-Diffuse cycle is controlled, different carbon profiles are produced, the result being different transformation temperature gradients with different phase expansion characteristics. Changing the type of carburizing cycle can change the distortion behavior.

Creep During Heat Treatment. Any heat treatment process that requires heating the part to above the transformation temperature causes the part to lose most of the strength that it possesses at room temperature. A part subjected to extended times at elevated temperatures (as in carburizing) could creep under its own weight unless it is properly fixtured and supported. Long slender parts are best suspended vertically. If this is not practical, the support should have the same contour as the component resting on it.

Nonuniform Heating and Quenching. When a part is designed, most enlightened designers recognize the need to keep the section sizes as uniform as possible in order to minimize temperature gradients and the tendency to produce high stresses due to differential expansion and contraction during heating and quenching. If a part is made with features such as gear teeth, however, it is unavoidable that these areas will have higher surface-to-volume ratios than the rest of the parts and that gear teeth will often tend to heat and cool faster than the rest of the section. Due to the coefficient of expansion the base of the tooth will be restrained by the rim, and this area will tend to go into compression during heating and tension during cooling. Similar effects will take place else-where in the part wherever there is a change of section.

Quenching Carburized Parts. The effects on dimensions during hardening can be even more severe. After carburizing, components can be direct quenched after cooling from the carburizing to the hardening temperature or they can be slow cooled and reheated and then quenched to cause hardening to occur. Whichever method is used, the hardening results from the transformation of austenite to hard martensite which is accompanied by a volume expansion of 4.64 - 0.53 × %C percent. This translates to a linear range of 1.37%-1.53% expansion due to differences in carbon content.

If a component is quenched under conditions of uniform heat abstraction, the situation may be represented by Figure 2 (1). This figure represents a cylindrical shape that has been carburized and heated to a temperature above the austenitizing temperature and then quenched. The isochronal lines show how the surface cools faster than the center of the section because heat is abstracted from the surface by the quenching media. This trend continues right through the quenching process. Figure 2 also shows lines representing the start of martensitic transformation (T_S) and the finish of transformation (T_F). It will be seen that these temperatures are depressed as the case carbon increases. The net result is that

transformation of austenite to martensite starts at the case/core interface with an expansion, as martensite is formed. The case is the last material to transform and the expansion to martensite causes compressive stresses because the core is already transformed and restrains the tendency for the case to expand.

If the T_F temperature is below room temperature, austenite will be retained. This lowers the surface hardness and will cause dimensional instability if very low temperatures are encountered in service since more austenite will transform, resulting in a subsequent volume increase. Usually a subzero treatment is performed after carburizing and quenching if the carburized case is likely to retain excessive austenite. In practice, a small amount of retained austenite (about 15%) is often thought to be desirable and can improve the performance of the part. It is normal to temper the freshly formed martensite.

Although Figure 2 assumes "conditions of uniform heat abstraction" at the surface of the part being quenched, this is a most unlikely occurrence in practice. One of the major difficulties in distortion calculations is knowing what the heat abstraction is at all parts of the surface of the quenched body. The work of Pilling and Lynch (2) showed that the mechanism of heat transfer during quenching may be considered in three stages. Quenchants typically go through the three stages of cooling as the part cools. During the first or "vapor" stage, cooling is slow due to an insulating blanket of gas. This gradually breaks down and quenching enters the second or "agitated boiling" stage, where heat abstraction is the highest. The final stage is one of "convection cooling" in liquid, which is entered below the boiling point of the quenching fluid.

What actually happens during quenching is that two or even three of these stages can be occurring in different areas of the part surface at the same time. Phase transformations then take place over a wide time interval with the resulting volume expansions and the rigid martensite phase occurring at different times.

Consider a part being quenched. The following factors will affect heat removal:

• The importance of mechanical agitation during quenching has been stressed by many studies, including the earliest comprehensive account of quenching by French (3). Some natural movement of the quenching media always occurs when hot parts are quenched by virtue of the convection currents set up in the liquid. This liquid movement past the hot part will slow down as the temperature difference between the liquid adjacent to the metal surface and the bulk of the media decreases. It is therefore desirable that the quench bath be agitated, but the quenchant cannot be directed at all areas of the surface at the same time. The part has to be in a flowing stream of quenchant, which has the effect of having maximum cooling on the surface first impacted but has minimum effect on the downstream surfaces.

• Geometry of the part will affect cooling according to direction of the quenchant stream relative to major areas of flat surface. Russell (4) showed that during the cooling of a sphere, without agitation of the quenchant, the bottom cools at a slower rate than the top since the convection currents in the liquid are concentrated around the top of the sphere (Figure 3a). Russell's work extended to cylinders shows that a long cylinder (Figure 3c) has a greater proportion of effective vertical cooling surface than a short cylinder (Figure 3b), and if the volume-to-surface-area ratios are equal the long cylinder will cool faster.

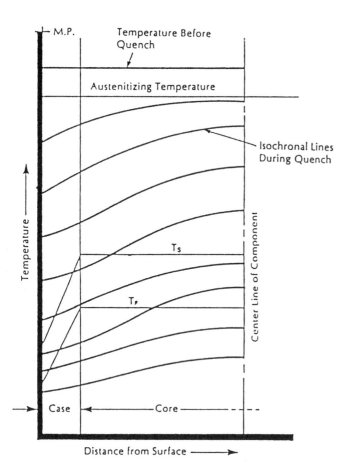

Fig. 2 - Temperature distribution in steel parts during quench after carburizing.

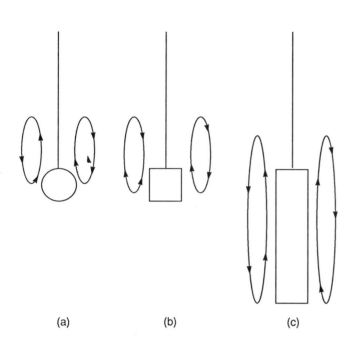

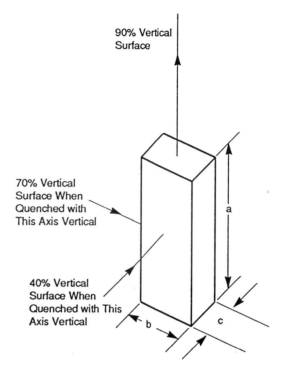

Fig. 3 - Liquid convection currents existing around (a) a sphere, (b) a short cylinder, (c) a long cylinder.

Howes (5) constructed a set of three rectangular specimens as shown in Figure 4. According to the direction of quenching, 40%, 70%, or 90% vertical surface could be exposed to the quenchant. After quenching, center cooling with 40% vertical surface exposed took about one third longer than with 90% vertical surface exposed.

Deterioration of the quenchant may also affect heat abstraction rates. It may occur after long service and must be taken into account if distortion is to remain under control. Quenchants should be tested at intervals dictated by experience and should be replaced or reconditioned as necessary.

Press Quenching. Press quenching involves quenching one part at a time by restraining it in dies while controlling the quenchant flow to different parts of the surface until the part is fully cooled to a predetermined temperature. Since a batch of parts is not involved and quenchant flows are controllable within a wide range, this quenching method should offer the best possibilities for control. It is an expensive quenching method, and considerable experimental work is often needed to optimize quenching conditions. It is widely used for precision parts that must have exact dimensions after quenching.

Although the quenched parts are under closely controlled conditions while in the press, very often the part is transferred from the reheat furnace to the quench press manually. This leads to variability due to differences in operators. The operator may have power assistance for handling the part, but the only way to ensure documented consistency is a well-designed and controlled robotized transfer system with the reheat furnace under computer control.

Specimen No.	Dimensions, mm (in.)			
	a	b	c	V/A
1	51 (2.0)	17 (0.667)	8 (0.333)	0.1
2	102 (4.0)	34 (1.333)	17 (0.667)	0.2
3	152 (6.0)	51 (2.0)	25 (1.0)	0.3

Fig. 4 - Dimensions of the three rectangular test pieces that were quenced in three directions.

Components Made from Direct Hardening Steels

Many components are not carburized and hardened but are made from a medium carbon steel (0.4-0.6% C). These steels can be through-heated to above their austenitizing temperature, and then quenched and tempered to produce a hard tempered martensitic structure. This treatment causes transformation to proceed from the surface to the center of the section and often results in unfavorable tensile residual stresses at the surface and increased tendency to crack during quenching or later during grinding.

An alternative method of treating medium carbon steels is selective hardening. Energy is transmitted quickly into the surface by applying induction, laser, or electron beam heating on the specific areas where surface hardening is required. The surface layer is thereby heated to above the austenitizing temperature and will later become the hardened case. When the energy is turned off, rapid cooling progresses (due to self-quenching) and, as in carburizing, the case is the last to transform; the restraint induces residual compressive stresses as the surface expands during transformation from austenite to martensite. This situation is illustrated in Figure 5, showing how the final results are similar to those obtained with a carburized case.

The advantages of selective hardening using high energy sources are that the process avoids the lengthy carburizing period and can thus be completed in seconds or minutes. Also, since only a small portion of the part is heated at any time, most of the structure is cold and rigid enough to preclude distortion. From a practical point of view, these processes can often be in the manufacturing line—avoiding the batching used for carburizing and other diffusion processes, eliminating part transportation, and making inspection easier. Part-to-part consistency is also likely to be better. Consistency of part structure will only be achieved, however, if processing conditions are kept constant. The energy applied to the work surface must be carefully controlled as to power level and time since too little energy results in a shallow or no case whereas too much can cause surface melting. Both conditions can affect final dimensions.

The nonuniform heating can cause high stresses due to the coefficient of expansion of the steel. This may be enough to upset thin sections and cause distortion after the self-quench or applied liquid quench. Another difficulty is that sometimes parts are "scanned" by induction, flame, or laser and where the scan ends, an area with tensile residual stresses can exist. The hardening process should be carefully designed to prevent or minimize this effect.

Machining After Heat Treatment

A finishing process such as grinding or hard-finishing is often required to correct dimensional changes caused by heat-treatment. The tendency is to try to use parts as heat-treated without touching the surface again (except perhaps for lapping) because in this condition the part can have a much higher fatigue strength (6). This is particularly true for parts loaded under concentrated contact such as bearings or gears.

For parts with close tolerances, however, the component size must be brought under control in the finish grinding or hard turning stage. This leads to a dilemma—i.e., if excess material is left on the part prior to heat treatment, there will be enough stock to enable the size to be brought under control. However, if too much is taken off, the most effective regions of the carburized (or nitrided) case are removed. Figure 6 shows a gear with excessive material being removed from a tooth after heat treatment.

In the example shown in Figure 7, the tooth has distorted to the right, and to correct the profile, excess stock has to be ground from the right side of the tooth. This has several serious consequences. First, the lack of uniformity in case depth leads to uneven residual stress distribution. Second and worse is that the gear appears satisfactory in a nondestructive inspection even though the performance of the gear will be less than optimum. Third, a considerable thickness of material has to be removed during grinding, increasing the probability of grinding burns. Some problems that are blamed on grinding can doubtless be traced back to badly controlled heat treatment and the prior machining processes.

Thus, the effects of the heat treatment process have to be considered before and after the process in both the soft machining and hard finishing stages if the final dimensions and structure are to be successfully controlled.

Prediction of Distortion

The many factors described demonstrate that distortion prediction is no easy task, but some first attempts can be made. A suggested approach would be as follows:

1) Definition of Shape. The dimensions of the part can be defined preferably as three-dimensional CAD presentation with the ability to present the part as a solid model.

2) Construction of a Finite Element Network. The solid model can be translated into a finite element network, the complexity of which will be determined by the size, shape, and form of the part. Experience will show the optimum number of mesh elements to be used. This will also be related to the available computer power and the desired accuracy.

3) Quenching Calculations. The cooling of every network cell during the quench can be calculated based on the known characteristics of the quenching media and the part. These will include the heat transfer coefficient of the quenchant under the conditions

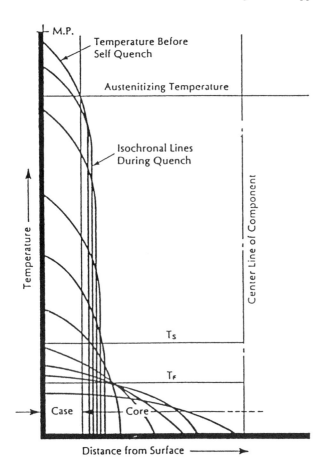

Fig. 5 - Temperature distribution in steel parts after laser heat treatment.

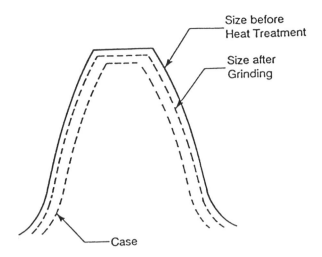

Fig. 6 - Schematic of material ground from gear tooth.

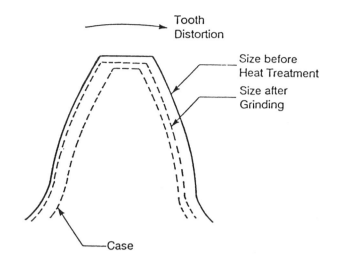

Fig. 7 - Schematic of material ground from a distorted gear tooth.

specified by the designer and those naturally present in the quenching system, as well as the items below:

- initial temperature of the part
- temperature of the quenchant
- cooling characteristics of the quenchant
- agitation of the quench media
- orientation of quenchant flow to surface being cooled
- complexity of surface features
- conductivity of the part
- latent heat data.

4) Calculation of Phase Transformations. Phase transformations can be calculated using time-temperature-transformation data available on the source material for the part. For carburized and nitrided steels the differences in composition and thus hardenability and phase boundary temperatures must be allowed for. Continuous cooling data will be the preferred source.

5) Calculation of Deformation and Stress. These calculations can be made with available data on Poisson's ratio, yield strength, work hardening modulus, and density changes due to phase changes. All these data are obviously temperature dependent. The output of these calculations will be a predicted final shape and the level of residual stresses.

As part of the predictive methodology, it is necessary to be able to calculate how every element of the part being quenched will respond in the chosen quenchant. The vast literature dealing with quenching shows that many attempts have been made to develop a method of predicting cooling curves (7-13). These methods are often unsatisfactory because of the difficulty in obtaining the data on which some of the mathematical analyses are based. The following assumptions are often made:

- That Newton's law of cooling is true under all quenching conditions. This law is usually expressed in the form:

$$dQ/dt = C \, (T_s - T_m)$$

where C is assumed constant, T_s = surface temperature of the part, and T_m = temperature of the quenchant. Heindlhofer (12) has shown that this law is not strictly true unless $(T_s - T_m)$ is small, which is not a valid assumption under normal quenching conditions.

- That physical properties of the material are constant over the entire temperature range considered, which is another invalid assumption.

- That the heat transfer coefficient of the quenchant is constant. However, this factor has been shown to vary during quenching (3). It will also be dependent on quench bath temperature and degree of agitation.

The complexity of issues and assumptions makes it necessary to have complete data available on the geometry of the part being quenched, on the material used for the part, and on the quenchant. Most methods of calculation have only been applied to simple symmetrical shapes, which are useful as exercises but are rarely reproduced in engineering practice.

Conclusions

Although many factors must be considered in predicting distortion, a level of understanding has been reached to enable development of the data for building the relationships between the variables from which to construct predictive equations. A major difficulty may be in calculating heat abstraction during the quench because of the many variables involved. Validation of proposed distortion prediction methods will be a complex process since the number and range of variables will be extensive. The benefits of a successful prediction method would be most useful to the manufacturing and heat treatment industries, and even a more comprehensive understanding of the interaction between manufacturing and processing factors would be a big step forward.

References

1. Dawes, C., and R. J. Cooksey, *Iron Steel Inst. (London) Spec. Rep.* **95**, p. 77 (1966).

2. Pilling, N. B., and T. D. Lynch, *Trans. Am. Inst. Min. Metall. Eng.*, **62**, 669 (1920).

3. French, H. J., *Trans. Am. Soc. Steel Treat.*, **17** (May), 694 (1930).

4. Russell, T. F., *Iron Steel Inst. (London) Spec. Rep*, **24**, p. 283 (1939).

5. Howes, M.A.H., Ph.D. dissertation, London University, p. 38 (1959).

6. Howes, M.A.H., and J. P. Sheehan, *SAE (Soc. Automot. Eng.) Tech. Paper* 740222, 1974.

7. Sinnot, M. J., and J. C. Shyne, *Trans. Am. Soc. Met.*, **44**, 758 (1952).

8. French, H. J., and O. Z. Klopsch, *Trans. Am. Soc. Steel Treat.*, **9**, 857 (1926).

9. Russell, T. F., *Iron Steel Inst. (London) Spec. Rep*, **14**, p. 149 (1936).

10. Post, C. B., and W. H. Fenstermacher, *Trans. Am. Soc. Met.*, **33**, 19 (1944).

11. Jackson, R. and R. J. Sarjant, J. B. Wagstaff, N. R. Eyres, D. R. Hartree, and J. Ingham, *J. Iron Steel Inst. (London)*, **150**, 211 (1944).

12. Heindlhofer, K., *Phys. Rev. [Ser. 2]*, **20**(3), 221 (1922).

13. Paschkis, V., *Trans. Am. Soc. Met.*, **37**, 216 (1946).

Effect of Carburizing Process Parameters on Distortion of AISI 4118 Gears

S. Saboury
Bradley University
Peoria, Illinois

Abstract

The inherent nature of dimensional variations due to carburizing/quenching process is well documented (1). New governmental emission and noise regulations for diesel engines, combined with the increasing demand for efficiency, require better control over the dimensional variations resulting from manufacturing processes. This work is the result of an investigation to determine the effect of carburizing process variables on the distortion variations of two gear geometries during carburizing. Two groups of AISI 4118 helical gears were carburized and oil quenched. The first group, the smaller pitch-thicker face gears were free-quenched while the second group, were press-quenched. Prior to cutting, some blanks from each group were normalized to investigate the effect of initial microstructure on the distortion after carburizing. In the carburizing process, the effects of carburizing temperature and the quench temperature were investigated. Lead error measurement were utilized as the scale of distortion variation. These measurements were compared between before and after carburizing. The results were statistically analyzed and lead error standard deviations, F-values, and Cpk ratios were compared to evaluate the effect of carburizing process variables on the distortion variations. Amongst the factors studied, increasing the quench temperature showed the most significant effect on reducing the lead error variations and maintaining the quality for both gears. Lowering the carburizing temperature had a more pronounced effect on the free-quenched gears while starting with a normalized structure was more effective on controlling distortion of the press-quenched gears. The carburized surface hardness, case depth and microstructural features were all maintained within specifications, after changes made in the quench temperature or initial normalizing of structure. Lowering the carburizing temperature however, resulted in a slightly shallower case depth than specified.

New global standards for reduced diesel engine emissions and noise levels has created a formidable challenge for manufacturers in the 1990's. This along with the ever present need to reduce cost and increase reliability, coupled with refined design techniques, results in an urgent need for components which can be manufactured with a high degree of consistency in mechanical properties and dimensional accuracy. Gears are products that are often carburized and hardened to achieve the mechanical properties needed for high wear applications. Due to the complex geometries and the volume changes involved in the hardening process, gears are particularly difficult to manufacture with the required degree of consistency. One way to address this need for improved consistency in gears is to reduce the variations caused by the heat treatment process. This will yield a gear with more consistent hardness and microstructural characteristics which are the key factors in determining part performance. Dimensional control is also directly related to variations caused by the heat treatment process and for gears, improved dimensional control translates into quieter operation and longer gear life, often without the need for additional costly finishing processes such as grinding after hardening (2). This paper is intended to present and explain experimental results from a study focused on optimizing specific heat treatment process control variables. The quantitative effects of modifying these variables on dimensional variations will be presented and discussed for two AISI 4118 gears with different geometries and quench techniques.

All heat treated components undergo distortion. During carburizing, carbon is diffused into the part surface by heating the component to the austenitizing temperature in a controlled carbon atmosphere. The addition of carbon causes the surface austenite grains to grow faster than the core grains. After diffusing carbon into the surface to the desired depth and concentration, the temperature is lowered to approximately 28°C (50°F) above the material's upper critical temperature and then rapidly quenched in an oil medium. During this quenching process, due to thermal and transformational volume changes distortion is expected and observed. As martensitic transformation proceeds, the internal stresses created by the volume change inhibit the completion of transformation and a portion of austenite is "retained" after quenching. This unstable retained austenite would decompose at a later stage through a heating process, like tempering, or a mechanical stress, like loading at operation, or a combination of both. This delayed transformation would in turn lead to further distortion. The amount of retained austenite then, would be another important factor to control distortion (3). The major factors affecting the amount and morphology of austenite decomposition products, as well as that of retained austenite are austenite morphology and composition, and temperature-time variations during cooling of austenite.

In the carburizing process, starting with machined (green) gears, the major initial factors affecting distortion would consist of methods of stacking and fixturing of parts, residual stresses due to prior thermomechanical treatments (e.g. hot forging), machining processes, composition, microstructure and their variations through the component. During gas carburizing, these initial factors would continuously interact with factors like temperature at different carburizing and diffusion zones, the amount and depth of diffused carbon, the carbon potential of the carburizing gas and the geometry of the component, to control kinetics of austenite growth and compositional changes. Finally, during quenching, each element of austenite grain, depending on its composition, morphology and restriction to expand, responds to the cooling cycle imposed to transform or to remain as "retained austenite". It should be noted however that in spite of the problems created by excessive retained austenite, small amounts (<15%) are often desirable in the case to allow mating surfaces of gears in mesh to conform, thereby avoiding the creation of areas of stress concentration caused by a non-uniform load distribution, (1).

Existing Process

Gears. Two gears with significantly different geometries were selected for this study, (Table I). Both are high volume helical diesel engine timing gears with a helix angle of 18 degrees, made from AISI 4118 hot-forged blanks manufactured to AGMA Class 5 dimensional control.

The hot-forged gear blanks are turned, hobbed, shaved, inspected and cleaned prior to carburizing. Both gears are carburized identically. The quench oil for both, the dunk tank and press quench is supplied from the same reservoir. The grid for Gear 1W9302 is loaded with 17 pieces per grid, laying flat, with a gross weight of 138.8 Kg(306.1 lb) and free quenched. The grid for Gear 2W7339 has 23 pieces, loaded flat with spacers between the gears. Each gear is

removed individually from the grid and immediately press quenched.

Furnace System. The furnace used is a "push" type continuous gas furnace with six microprocessor controlled zones for accurate control of temperature and carbon potential. Zone 1 allows the gears to preheat to austenitic temperature. Zones 2 to 4 are carburizing zones, zone 5 is the equalizing and zone 6 is the quenching zone. The carbon atmosphere consists of an endothermic "RX" gas.

Quench System. An interconnected system, consisting of a 2400 gallon dunk quench tank and a 8000 gallon compensating tank is utilized for quenching gears after gas carburizing. The Houghton "Quench K" oil used for quenching is analyzed quarterly to determine suitability for continued use. The compensating tank provides circulating oil for the dunk tank as well as supplying oil to the two Gleason press quench machines used for gear 2W9302. A single propeller agitator circulates oil around baffles, through the load and between the two tanks at 9000 gpm. Dunk quench oil temperature is maintained at a range of 49-74°C (120-165°F) and the press quench oil is held between 43-54°C (110-130°F). Quenching time is about 30 seconds and for both gears, a residual temperature of 127°C (261°F) is the basis for the time in quench.

Experimental Factors Studied

Higher Quench Temperature. An increase in the quench oil temperature would result in a slower cooling rate and a smaller temperature gradient in the component during quenching. After carburizing and because of this process, the component has a higher carbon content on and near the surface. This leads to a lower Ms and Mf temperature closer to the surface. It is desired that the transformation of core austenite takes place prior to that of the surface, since otherwise, the core volume expansion would be resisted by the hard surface, causing tensile residual stresses on the surface (4), distortion, decreased fatigue resistance and possibly quench cracking. With a cooler quench medium and higher cooling rate, the steeper temperature gradient between the case and the core can prevent the core austenite from transforming first (5-7). On the other hand, a very high oil temperature may also cause problems. If the base material does not have a high enough hardenability, a lower cooling rate may result in the formation of "undesirable" products from austenite, and unacceptable core structure and properties. This does not seem to be a problem with steels commonly used for gear manufacturing such as AISI 4118 or AISI 8620 due to high hardenability. Also if the Ms temperature of the case decreases substantially due to excessive carburizing, there is a chance of producing unacceptable amounts of retained austenite near the surface.

Table I. Gear Characteristics

PART NO.	PITCH DIA. mm(in)	OUTSIDE DIA. mm(in)	FACE WIDTH mm(in)	HUB I.D. mm(in)	NO. OF TEETH	GEAR WEIGHT Kg (lb)	GEARS PER GRID	QUENCH TYPE
1W9302	213.66(8.41)	217.35(8.56)	46.4(1.83)	152.4(6)	80	5.71(12.6)	17	Free
2W7339	280.43(11.04)	284.15(11.19)	17.8(0.7)	38.1(1.5)	105	4.26(9.4)	23	Press

For the AISI 4118 steel, the Ms temperatures estimated (see Table II) indicated that raising the quench temperature to 93°C (200°F) would not result in excessive amounts of retained austenite or other undesirable products. This temperature was adopted for determining the effect of raising quench temperature on the lead error variations.

Table II. Ms Temperatures

Location	Ms Temp °C (°F)
Core (0.2%C)	388(730)
Case (0.6%C)	254(490)

Lower Carburizing Temperature. Due to the increased carbon content during carburizing, the surface austenite grows at a higher rate than the core austenite (2). This excessive growth can lead to two potential problems. First, during quenching a larger austenite grain can result in a less uniform expansion, creating residual stresses and distortion (1). Secondly, a larger austenite grain size can result in an increased amount of retained austenite after quenching (1,3). The retained austenite can transform, at least partially, at a later stage due to either a temperature change such as tempering or the application of stress under functional service, or a combination of the two factors. The nature and the extent of this delayed transformation, is difficult to predict and further distortion would be expected even after the gears are put to service. At lower carburizing temperatures austenite grows at a lower rate and the problems explained would be potentially reduced (3). The only disadvantage of decreasing the carburizing temperature is the need to compensate for the lower temperature by increasing the carburizing time. A compromise between the potential improvement in quality and the required additional time using Harris relationship (8) led to a decision to decrease the carburizing temperature from 927°C (1700°F) to 900°C (1650°F) and to increase the time of carburizing by 40 percent.

Normalizing the forged blanks. The hot-forged blanks used for cutting the gears usually have an elongated directional morphology with relatively large prior austenite grains. During carburizing, this initial morphology can result in the formation of elongated large austenite grains in preferred orientations, causing distortion of the geometry at this stage (1). Furthermore, upon quenching this structure, an anisotropic volume expansion can lead to the formation of residual stresses and further distortion (9). A normalizing treatment would produce an equiaxed structure with a finer grain size leading to a more uniform volume change during phase transformations (5). The normalizing heat treatment also relieves the residual stresses caused by hot-forging, thus making dimensional control easier during machining (9). The process selected for normalizing gear blanks consisted of slow heating to 927°C (1700°F) under controlled atmosphere, holding at this temperature for one hour and air cooling. The total furnace time was about 3 hours.

Design of Experiment

Dimensional variations are commonly evaluated using "out-of-round" diameters, non-parallel faces and distortion of the lead angle (lead error). Since these variations are related (2), lead error of the helical gear teeth was selected to represent the distortion for this study. Due to the limited number of variables, experiments were designed to evaluate the effect of any single variable or any combination of two or three variables, on dimensional control. Eight groups from each gear geometry were identified by grid numbers to be carburized and quenched accordingly, as shown in Table III.

Table III. Gear Test Groups Identification

Grid No.	Normalized Forgings	Raise Quench Temp	Lower Carb Temp	Code Name
CtGr	NO	NO	NO	CG
1	YES	NO	YES	N*C
2	NO	NO	YES	C
3	YES	YES	YES	N*C*Q
4	NO	YES	YES	A*C
5	YES	YES	NO	N*Q
6	NO	YES	NO	Q
7	YES	NO	NO	N

The lead error was measured on both sides of three teeth located 120 degrees apart, for each gear, before and after heat treatment. Depending on group sizes, this resulted in between 60 and 90 lead error measurements for each of the 16 grids. The measurements were used for statistical analysis of lead error variations.

Results Analysis and Discussion

The results of statistical analysis for the two gears studied are given in Tables IV and V. The letter "N" under the "Code Name" identifies groups normalized prior to machining, the letter "C" shows the groups carburized at the lower temperature of 900°C (1650°F), and the letter "Q" represents groups quenched in the higher 93°C (200°F) temperature oil. The letters "CG" represent the control group with no alteration to the existing process variables. Standard deviations were calculated using all lead error values for each grid. These values were used to find the variance ratios (F) by comparing "before" and "after" carburizing lead error variances for each test grid. The lead error standard deviation and F variance ratios are shown in Figures 1 and 2 for gears 1W9302 and 2W7339 respectively.

Table IV. Gear 1W9302 Lead Error Variations

Grid No.	Code Name	Std Dev Before	Std Dev After	"F" Ratio	Cpk Ratio Aft./Bef.
CtGr	CG	2.637	6.413	5.915	1.030
1	N*C	3.421	6.021	3.097	1.563
2	C	5.385	4.260	1.598	2.862
3	N*C*Q	2.692	6.813	6.407	1.109
4	Q*C	4.641	6.113	1.735	1.592
5	N*Q	2.756	6.030	4.786	1.252
6	Q	5.306	4.208	1.590	2.996
7	N	3.011	5.747	3.642	1.309

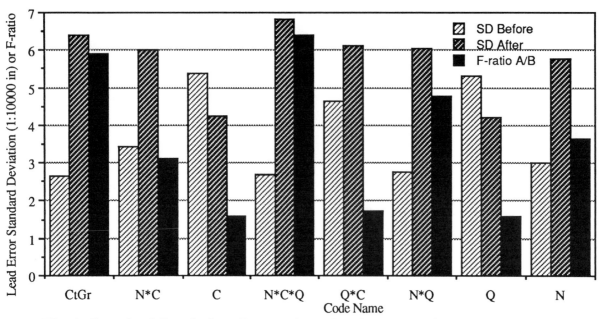

Fig 1. Standard Deviation Comparison for Gear 1W9302

Table V. Gear 2W7339 Lead Error Variations

Grid No.	Code. Name	Std Dev Before	Std Dev After	"F" Ratio
CtGr	CG	2.755	10.316	14.024
1	N*C	6.023	10.944	3.302
2	C	4.280	9.412	4.835
3	N*C*Q	5.661	9.570	2.857
4	Q*C	4.343	9.025	4.318
5	N*Q	6.008	11.796	3.855
6	Q	3.661	4.813	1.729
7	N	5.933	11.648	3.855

Table VI. Specified Values
for the Upper and Lower Size Limits

Gear	USL, μm(10^{-4}in) Before	After	LSL, μm(10^{-4} in) Before	After
1W9302	33(13)	61(24)	-33(-13)	-61(-24)
2W7339	25.4(10)	45.7(18)	-25.4(-10)	-45.7(-18)

Cpk is an accepted index of quality that measures how narrow the process spread is compared to the specification spread, modified by how well the process centers around the midpoint of the specifications (10). Hence, Cpk can be increased by two factors. First by moving the center of the process towards the midpoint of the specifications (in this case by reducing the absolute mean value of the lead error). Secondly by decreasing the process standard deviation (in this case, the standard deviation of the lead error values). If Cpk is less than 1, the process is spreading beyond at least one of the specification limits, resulting in defective products. If it is equal to or greater than 1, the closest specification limit is far enough from the process center so that virtually no product is being made beyond the specification limits.

In the present study, the Cpk values were calculated for the lead error before and after carburizing using the standard method (1) of choosing the smaller of the two values, Cpu and Cpl:

$$Cpu = (USL - Mean)/3\sigma$$

$$Cpl = (Mean - LSL)/3\sigma$$

where USL and LSL are the lower and upper size limit, respectively. The manufacturers specifications for these values of lead error are given in Table VI. In order to compare the effect of the process variables studied, the ratio of the Cpk after carburizing to Cpk before carburizing for gear 1W9302 was calculated (see Cpk Aft./Bef. column, Table IV). The graphical representation of Cpk ratios is shown in Figure 3.

Lead error variance changes in carburizing. For gear 1W9302 the standard deviation of lead error values for all cases tested are given in Table IV and shown in Figure 1 for both before and after carburizing. The F-value (variance ratio) is calculated by dividing the greater variance by the smaller variance of lead errors. A lower F-ratio is an indication of a smaller change in the variance (or standard deviation) of lead error, in the carburizing process. Increasing the quench temperature (Grid 6), shows the most significant improvement in decreasing the F-ratio from 5.91 (for the control group) to 1.59. Decreasing carburizing temperature (Grid 2) also shows an equally significant effect on maintaining a low F-ratio. Prior normalizing (Grid 7), although noticeable in lowering the F-ratio, seems less effective than the other two variables studied. The combined changes also show significant improvement in decreasing variability, however the improvements do not seem to be accumulative.

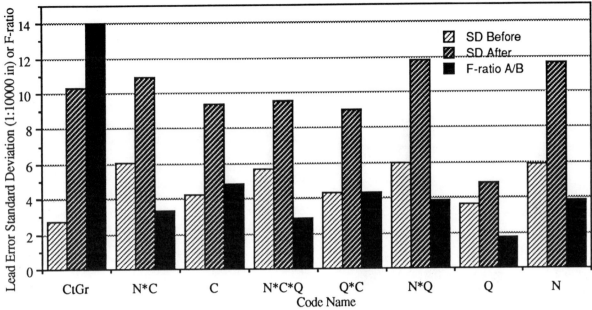

Fig 2. Standard Deviation Comparison for Gear 2W7339

The combination of increasing quench temperature and decreasing the carburizing temperature (Grid 4), results in a slightly higher F-ratio than any of the two changes alone. The combination of normalizing and lowering the carburizing temperature (Grid 1), results in an intermediate value for F, between the values for each individual effect, while the combination of normalizing and increasing the quench temperature (Grid 5), results in an F-ratio greater than each individual effect but still lower than the value obtained for the control group. The possible interactions between all three process changes, leading to a large F-ratio (Grid 3), is difficult to explain and may require a closer study of this combination.

For gear 2W3779, Table V shows the standard deviation and variance ratios calculated based on the lead error measurements. Figure 2 compares the standard deviation values and F-ratios between different grids graphically. The trends observed in this comparison indicate that any of the individual process changes studied, reduced the lead error variations significantly. As a single change, raising the quench temperature (Grid 6), was the most effective, while prior normalizing (Grid 7), and lowering carburizing temperature (Grid 2), were ranked second and third respectively. Combining any two changes also reduced variance ratios. The combination of normalizing and lower carburizing temperature (Grid 1), resulted in a smaller variation in standard deviation than any one of the single process changes (Grids 2 and 7). Combining the higher quench temperature with any one of the other two variables resulted in an intermediate effect. This may be due to a higher sensitivity of non-normalized gears to quenching temperature. The normalized structure would undergo a more uniform volume change during quenching and the effect of quench temperature although noticeable, would be less significant than that for non-normalized forgings. The gears carburized at a higher temperature are also believed to be more sensitive to quenching temperature. Although

lowering the carburizing temperature was beneficial through reducing the austenite grain growth, when combined with a higher quench temperature, it did not produce as low a variance ratio as observed in raising the quench temperature alone. The combination of all three variables (Grid 3), provided the lowest F-ratio except for the higher quench temperature alone (Grid 6).

It is important to note that any of the process changes studied or any combination of changes, reduced the variance ratio by a factor of between three to eight times.

Cpk-index Comparisons

The Cpk (demonstrated excellence) values for each condition (Grid) were calculated both before and after carburizing. The ratio shown in Table IV (Cpk Aft./Bef.) is the Cpk-index obtained for after carburizing divided by the value prior to this process. For the control group, a ratio close to one is indicative of similar Cpk values for before and after carburizing conditions. This does not, however, imply similar values of mean error or standard deviation from before to after carburizing, since the specified upper and lower size limits for these conditions are different, see Table VI. The ratios, nevertheless, can be used to compare the effect of process changes, on maintaining the quality of gears for different process combinations. Greater ratios are indicative of a better control over variations and mean values of lead error during the carburizing process. Comparing the Cpk ratios (Figure 3), a similar trend is observed as was discussed in comparing F-ratios, (Figure 1). Raising the quench temperature (Grid 6) or decreasing carburizing temperature (Grid 2), each seem to improve the Cpk ratio by almost three times, however, the effect of combination of these changes (Grid 4) is less significant. Normalizing gears, either as an isolated change in process (Grid 7) or in combination with higher quench temperature (Grid 5) or with decreasing carburizing temperature (Grid 1), improved the Cpk ratio, but showed a less pronounced effect than raising the quench temperature or decreasing the carburizing temperature. An overall comparison between all Cpk ratios indicates that any of the process changes studied alone, or any combination of changes, is capable of improving or at least maintaining the quality better than the existing procedure.

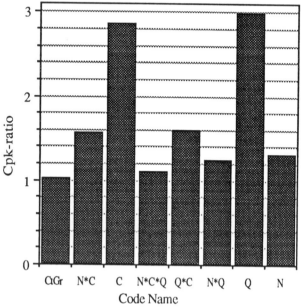

Fig 3. Cpk-index Ratio for Gear 1W9302

Conclusions

The effect of raising the quench temperature, decreasing the carburizing temperature and normalizing hot-forged gear blanks was studied on the lead error variations of two geometrically different helical gears, using different quenching techniques. As individual variables, raising the quench temperature had the most significant effect on decreasing the lead error variations, caused by carburizing, for both gears. The higher quench temperature also contributed significantly in maintaining quality, indicated by a three fold increase in the Cpk ratio. Normalizing the gear blanks and decreasing the carburizing temperature, each as individual variables, had also pronounced effects on decreasing lead error variations and improving Cpk ratios. These variables however, were not quite as significant as was the raise in quench temperature. The combination of two changes in process variables also demonstrated reduced variation in lead error, resulting from the process; however the effects were not accumulative. The combination of all three changes did not produced a clear process advantage for the smaller diameter gear either in controlling error variations or improving the Cpk ratio. None of the process changes studied, nor their combinations, adversely affected the mechanical properties or the microstructure to unacceptable levels. Lowering the carburizing temperature appeared to need a longer compensation time than calculated to maintain the required case depth.

References

1. Heat Treating, "ASM Handbook", Vol. 4, 10th ed., 1991
2. Rakhit, A.K., *Design Engineering Technology*, ASME Conf., Cambridge, MA, 1984, No. 84-DET-224
3. Mocarski, S., *Induction Heating*, v. 41, no. 6, Jan 74
4. K.E. Thelning, "Steel and its Heat Treatment", Butterworth, 1985
5. Parrish, J.A., *Metal Progress*, Sept. 1985, p 49-53
6. Child, H.C., *Heat Treat Met.*, No 4, 1981, p 89-94
7. Fujio, H. and T. Aida, *Bull. JSME*, Vol 22, No 169, July 79, p 1009-1016
8. Harris, F.E., *Metal Progress*, Vol 44, 1943, p 265-272
9. B.R. Wilding, "Heat Treatment of Engineering Components", Iron and Steel Institute, 1970, p 20-25
10. M.K. Hart and R.F. Hart, "Quantitative Methods for Quality and Productivity Improvement", ASQC Quality Press, Milwaukee, Wisconsin, 1989, p 186-214

Acknowledgments

The present work started as a senior design project at Bradley University, Peoria, IL. Chester McHenry, Steven Baker and Mary Hunt are acknowledged for their participation in the project. Bruce Whelchel of Bradley University is also acknowledged for the initial arrangements.

Dimensional Changes During Hardening and Tempering of Through-Hardened Bearing Steels

H. Walton
Torrington Co.
Shiloh, North Carolina

ABSTRACT

When one considers the complex interaction of stresses induced by thermal gradients and phase transformation it is understandable why bearing rings are prone to distortion and size change during heat treatment. The heat treaters job is made even more difficult if substantial stress is already present in the as machined components. If distortion occurs during subsequent operations the heat treater is the usual 'scape goat' even though the fault invariable lies elsewhere. This paper provides a review of the many factors contributing to size change-distortion and suggests some often overlooked remedies.

CAUSES OF DIMENSIONAL CHANGES

The many causes of both changes in part size and the occurrence of distortion are shown in the Cause/Effect diagram in Figure 1. From this it can be seen that although metallurgical factors are important there are many more "assignable causes of variability" that must be considered. For example the use of ceramic fiber blanket under heavy parts during heating has been found to be beneficial in reducing drag, and allowing the part to expand more freely thus reducing distortion. Shot blast cleaning of parts is another less obvious potential contributor to distortion. (Peen straightening is a widely used method for straightening hardened shafts for example.)

If the heat treat department is presented with a ring that has been machined to the correct size with the correct grind stock allowance and is free from any internal stress, it is possible to present in turn, a round part with balanced stock distribution to the grind shop. For example, a local machine shop required 50 pieces of a thin walled cylinder to be heat treated with minimum distortion. The part was 4.37 inches long 2.35 inches diameter with a wall section of 0.057 inches made from 52100 steel. An internal chamfer at one end determined that the part had to be free quenched. It was agreed that a 'rough machining/interstage anneal/final machining' sequence would be followed. The hardening cycle was:

A) Preheat to 450°C transfer to furnace
B) Ramp heat to 810°C and soak
C) Quench into oil at 150°C with agitators off
D) Air cool, wash and temper

Results Are:
Hardness 59 Rockwell; 0.003 inches maximum out of round (OOR); Concentricity 0.0008 inches maximum; Taper 0.002 inches end to end; in other words it is possible to heat treat even the most difficult part without

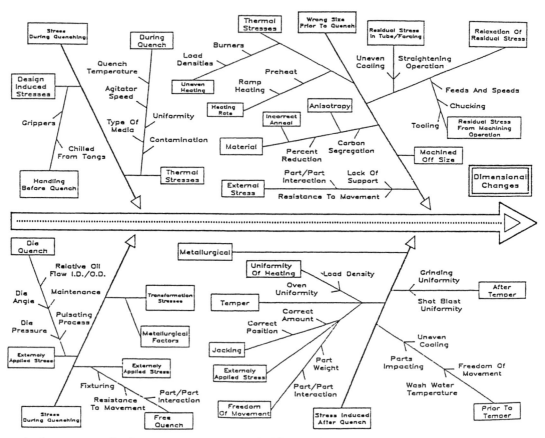

Fig. 1-Summary of the many causes of dimensional changes in a quenched and tempered steel component.

resorting to press quenching techniques, provided suitable precautions are taken to ensure that the effects of thermal and transformation stresses are minimized.

Both the preheat and ramp heating were used to ensure uniform heating. Preheating also allows controlled relaxation of any residual stress without 'macrodeformation'. For this particular cross section the slow quench with the agitators turned off would still be more than adequate to give full hardness. However, this practice should be used with extreme care. With a better understanding of actual quench rates of given parts in designed loads (number of parts per basket etc.) and knowing the hardenability of the steel it should be possible by adjusting the agitator speed to match the quenchability of the oil more closely to that desired to give through hardening without distortion.

The temperature of the oil, although not high enough for true marquenching was adequate to minimize transformation stresses (austenite to martensite). The majority of the transformation would have occurred during the relatively slow air cool.

Referring again to Figure 1, although many of the items are self explanatory some may require further explanation. Anisotropy for example, is characterized by the different property characteristics in the rolling direction compared with other directions. Although there seems to be some uncertainty as to whether the thermal expansion or contraction or both are controlled by anisotropy, it has been shown that many steels with carbides that are aligned in the direction of rolling will expand far more in that direction during heat treatment with a corresponding reduction in the width direction. For a ring manufactured from tube or forging, the rolling direction can be considered as being circumferential. If there is an increase in the expansion rate in this direction this will result in a corresponding increase in the diameter.

The degree of anisotropy could well be determined by the percentage reduction that the steel has undergone to arrive at the ring section and so this may be another factor that has to be taken into account when trying to understand the size variability between different components.

Whether the ring is machined from a tube or a forging will directly influence the final size of the ring. A ring machined from a forging could have a significantly greater increase in diameter during the quench than the same ring made from tube particularly if the tube had been cold drawn. This is because the residual stresses inherent in the tube stock after manufacturing relax during austenitization and lead to a reduction in the ring diameter. If the residual stresses are unbalanced then the ring may also be out of round after heating but before quench.

INTERNALLY CREATED STRESS

Thermal stresses and transformation stresses need to be clearly defined. Thermal stresses are introduced at any time the part is exposed to the application or the extraction of heat. Where a temperature differential occurs within the part there is a potential for distortion particularly if the temperature differential is nonsymmetrical. For example, the indexing sequence of the rotary hearth furnace should be designed to ensure that cold parts are never charged directly along side hot parts, thus avoiding the risk of the chilling action of one on the other and consequently an out of round condition.

Referring to Figure 2 it can be seen that when a given section is quenched, large temperature differentials occur throughout the section. During the initial stages of the quench the outside surface is cooling faster and contracting more than the core. Because the core material resists the contraction the surface region, which at this stage is still relatively plastic, will stretch. With continued cooling eventually a position is reached where the core is now cooling faster than the surface. The relatively cold and less plastic case now inhibits the core from shrinking. The

result is a core region under tensile and a surface region under compressive stresses (desirable for serviceability).

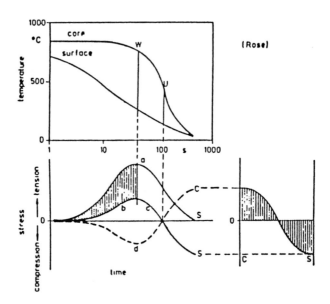

Fig. 2-Generation of thermal stresses.

However, the above description would only apply to the quenching of austenitic stainless steels or other materials that do not undergo phase transformations during cooling. In the case of carbon alloy steels, the sequence of events is complicated by the existence of transformation stresses. Austenite to martensite transformations which are necessary to produce the desired hardening characteristics regrettably also results in an appreciable volume increase of the steel. When the volume increases interact with the normal thermal contraction the resulting stress distribution can be difficult to interpret (Figure 3)[2].

Although Figure 3 is included to emphasize the potential complexities of the resulting stresses in a carbon alloy steel cooled at various rates, the interpretation of the diagram is left to the reader. Suffice to say that where there is a wide temperature differential when transformation to martensite starts-as in a) the resulting surface stress is in tensile (cracking!). This is to be compared with b) where although there is still 100% transformation to martensite through the cross section the

267

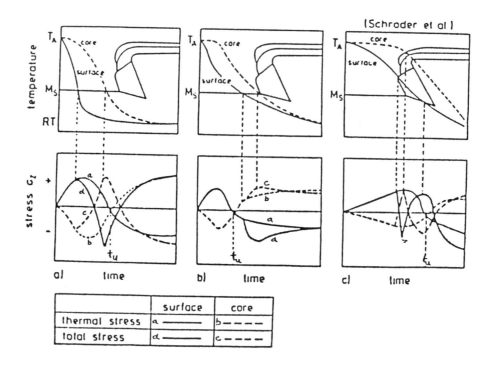

Fig. 3–Schematic CTT-diagrams with cooling curves for surface and core and calculated behavior of the corresponding thermal and total stresses.

temperature differential is much less at the start of transformation and the resulting surface stresses are now in the more desirable compressive state. Keep in mind that when a stress is introduced, there is a potential for movement and when the stresses are nonsymmetrical then distortion is expected. For example dissimilar cross sections in a given part will cool at different rates and from the above it can be seen that it is possible after quench that the surface of the thin section is under tension while the thick section surface is under compression. The development of transformation stresses and their corresponding effects are seen not only during quenching but also during subsequent operations.

Frehser and Lowitzer[3] found that when quenching various sizes of mild steel square slabs, (transformation stresses at a minimum) that the length of the slab decreased and the thickness increased. A greater magnitude of size change was obtained by a) increasing the temperature drop during the quenching (increasing austenitization temperature), b) by decreasing the high temperature strength

of the steel (the lower carbon steel the lower the high temperature strength and the more distortion experienced), and c) increasing the quench rate. In two separate papers[4,5] Tuguchi Analysis on the many factors influencing distortion during free quenching of rings concluded that the faster the agitation speed (the faster the quench rate) the less the growth and distortion of the rings. This nontraditional practice may be explained by the need to ensure all surfaces of components particularly those within a large batch load, are quenched at the same rate thus avoiding unbalanced stresses and associated distortion.

Some work at F.A.G. presented in a paper by Mayr[6] sheds more light on this topic. Figure 4 is a summary of the distortion results (OOR) obtained on 50 piece samples of rings made from 100Cr6 steel (52100 equivalent) after quenching by different methods. Medium roundness (U) and maximum deviation from roundness (U + 3S) are shown. Because of the more symmetrical cooling, axial quenching provides a rounder part than obtained from quenching vertically orientated parts. Salt bath quenching is superior to oil for two reasons:

1) The lack of a vapor blanket in salt; although resulting in a much faster heat extraction rate does ensure more uniform cooling. In oil, the vapor blanket may break down on different areas of the surface at different times thus creating unbalanced stresses.

2) By separation of the thermally induced stresses from the transformation induced stresses and minimizing the latter by allowing slow cooling through the martensite range (in air).

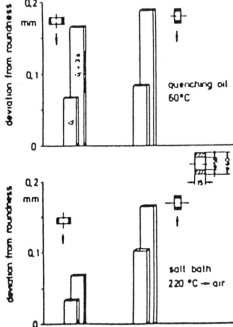

Fig. 4-Roundness deviation of rings of steel 100Cr6 as a function of quenching medium and quenching direction.

An interesting observation is that the vertically oil quenched sample shows less distortion than the vertical salt quenched sample. This can be attributed to a much greater temperature differential between the submerged and non submerged parts of the ring being present in the

vertical salt quench (non-vapor blanket) as compared with that obtained by oil quenching.

The depth of hardening in thick cross section components can have a considerable influence on the dimensional changes, distortion and resulting surface stresses. For a shallow hardening steel, where the depth of hardening will be strongly influenced by the effectiveness of the quenching operation, a wide range of size change for individual components from the same bath could be expected. With increasing hardenability and therefore the greater likelihood of through hardening even given the expected differences in quench rate, the more consistent size change would be expected.

Increasing alloy content and/or temperature may increase the proportion of retained austenite. Although the as quenched structure may show lower size change as a result of the higher level of retained austenite, there could be a greater size change -expansion- during tempering (austenite to martensite). The interrelationship between section size, and austenitizing temperature for two steels is shown in Fig. 5[1]. 145CrV6 and 90MnV8 are both high carbon steels with Jominy depths to Rc 50 of approximately 10mm (6/16") and 25mm (16/16") respectively.

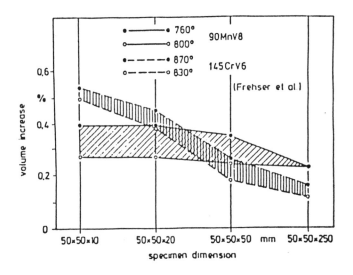

Fig. 5-Volume increase of the steels 90MnV8 and 145CrV6 as a function of austenitizing temperature and specimen dimension.

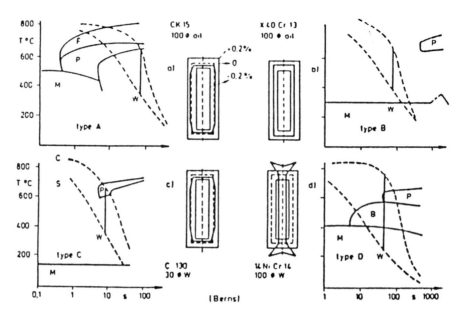

Fig. 6-Typical dimensional alterations caused by thermal and transformation stresses.

The complex interaction of thermal and transformation stresses is further illustrated in Figure 6[7] where cooling curves for the surface and core of cylindrical parts are superimposed on the continuous cooling transformation diagrams for four different steels. In each case, W indicates the time interval at which the greatest temperature differential between the core and the surface occurs. The relative position of time W with respect to these transformations curves has a direct bearing on the magnitude of size change. If W occurs after transformation (as in type A) or before transformation (as in type B) then large transformation stresses can be avoided and dimensional changes will follow the Ameens Rule[8] (dimensional changes resulting from temperature induced stresses will cause the shape of a component to become more spherical). Type B would be most prevalent in normal section 52100 components. Under type C where the core is transforming to non-martensitic products before the surface, thermal and transformation stresses are unidirectional so that geometrical changes similar to type A and type B are produced. However, in type D where at W the surface has transformed while the core is still austenitic, the thermal and transformation stresses are in opposition leading to an increase in the length direction and a decrease in the diameter of the cylinder (cylindrical solids form spools). The latter would be typical of the response expected after a relatively slow quench (slack quench) of a through hardened steel.

So how does the above help in determining what the uniform and nonuniform dimensional changes are going to be during the quenching of a particular component? In other words can we predict the growth or shrinkage expected and how far out of round the part is going to be? Some attempts have been made to answer these questions. However, because of the complex interaction of the many variables, particularly the steel chemistry and hardenability, there has been little success. And of course the externally applied temperature and stress differentials that can occur on different areas of a single component (parts touching, chilling action of tongs, variable oil velocity, etc.) is difficult to take into account. Before any meaningful model for how a given steel responds dimensionally to a known set of quench conditions can be considered, it is first necessary to deal with these external factors (assignable causes of variability).

One successful approach for controlling grind stock is to log inspection data before and after heat treatment on printed histogram reports and analysis of the information over many years. It is then possible to develop reasonably good predictions as to how each individual part would respond during a given heat treatment cycle (furnace, die set up, etc.). Because of the many differences in furnaces and quench conditions (heating rate, atmosphere type, oil type, etc.) for accurate predictability it is necessary to repeat the exercise at each plant location. With present day S.P.C. techniques this should be relatively easy.

EXTERNALLY APPLIED STRESSES

Before considering the subject of externally applied stresses, it is worth considering a practice which is gaining wider acceptance in manufacturing. Having seen that changes in cross section can have an influence on the relative dimensional changes, it should now be more understandable as to why some parts with complex cross sections behave the way they do. Little can be done in the quench operation other than by using extreme external forces in order to combat the unbalanced dimensional changes. Given all external factors are recognized and controlled within certain limits, a given component with a complex cross section will respond the same every time it is quenched. So why not accept the distortion and machine the part accordingly? That is, machine a race for example to a different angle so that the correct angle is obtained after heat treatment.

Press quenching, although the most costly procedure, is frequently used to ensure roundness and until we are able to closely control all the many factors listed in Figure 1 the method will continue to be used particularly on the large diameter thin section rings (ratio of diameter to cross section greater than 15). Several die methods are presently being used:

    1) Cone die quenching (or combination of cones and plates)

    2) Expanded mandrel die quenching
    3) Plug quenching
    4) Marsize fixtures

Marsizing is particularly useful for the high hardenability (T2) marquench parts with thin wall section. After removal from the low temperature salt bath, the part is allowed to air cool inside a solid ring. Progressive transformation of austenite to martensite during the cool down causes the part to grow. Contact with the solid ring rounds the part as transformation progresses. After transformation is complete, normal thermal contraction occurs and the parts may be easily removed from the die. For this method to be applied to the lower hardenability 52100 grades, careful control of steel chemistry and its corresponding influence on the Ms temperature would need to be exercised.

Cone die quenching because of its ease in setting up, its adaptability to many components within a given size range, and the high productivity when using multiple heads is the lowest cost die method most applicable to through hardening steel rings. The multihead presses are amenable to robot handling for extra productivity and more consistent processing. Because of the line contact in the ID, cone dies are not limited by the ID and OD shape and only with low width parts is it necessary to be particularly careful in selecting the die geometry; if the diameter of the low width part is too large for the particular dies (also not forgetting to allow for growth of the hot part) there is the possibility that the center flat of the dies may touch resulting in a loss of the rounding action and a slack quench from the lack of oil in the bore of the part. The many cone die quench variables includes die angle; die pressure; constant or pulsating pressure; speed of application; oil flows ID and OD; oil temperature; delay between applying die pressure and oil flow; time under pressure; cross section of part; transformation characteristics of steel; roundness of part prior to quench; etc. Of the many variables, by far the most important is the last item; roundness of part prior to quench. Cone die quenching is best considered as a clamp quench technique where the part is held to its original shape when placed on the die and little if any rounding action is going to occur. Obviously there will be some rounding on thin section parts the degree of which will be determined by the die angle or the die pressure. Dies with an angle of 30° are preferred because of the greater span of compatible part sizes possible. This is at the expense of a lower resolved radial rounding pressure. If the die pressure is increased then there is the danger particularly on taller parts of creating a barrel effect.

A major limitation of the plug or segmented die method is the low productivity. However, this has been overcome to a certain extent by quenching more than one part on the die at a given time. A second disadvantage is that more die sets are required to handle a given range of parts. If a three segment expanding mandrel is too small for a given part, then three point lobeing is possible; if too large 6 point lobeing is likely to occur. In order to realize the full benefit of Gleason quenching the operators need to be experienced in setting up dies, shimming, and dealing with the many nuances of the machine. Where Gleason machines are in use or are to be introduced, consideration should be given to the use of expert systems to ensure continuity in methods.

POST QUENCH TREATMENTS (LOCKED IN STRESS)

Either manual or hydraulic jacks are used extensively for rounding of large rings. As to how far to jack the part in order to accomplish a certain degree of movement depends largely on the experience of the operator. Degree of temper, steel type, cross section, etc. also play a significant role.

For rounding to occur in through hardened rings it is usually necessary to apply heat (tempering). Figure 7 illustrates the size change occurring during the tempering of ball bearing rings after shrinking onto various diameter shafts. For example, an over size value of zero means that the shaft diameter is the same size as the I.D. of the ring and

therefore no preload during temper would be expected. Any dimensional change is purely related to the mechanisms occurring during the tempering of the martensite (precipitation of the carbide) and transformation of retained austenite. Where the retained austenite level is zero the part will normally decrease in diameter. With retained austenite present (as in most 52100 steel rings) the opposite is the case.

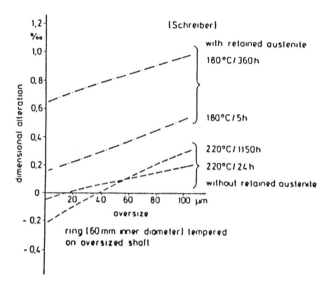

Fig. 7-Diameter changes for rings of the steel 100Cr5 tempered on shafts with different oversizes.

When a preload is present, there is a progressive increase in the size of the ring with a degree of over size of shaft. The longer the times at temperature, in cases where retained austenite is initially present, result in the greater size change, (probably because of transformation of stable retained austenite to bainite). In either case, if the preload can be sustained during the tempering operation, then the growth (rounding action when using jacks) continues with time.

Returning to the shrinkage expected during temper of non-prestressed components containing little or no retained austenite, although this is directly related to the relaxation of martensite accompanying the precipitation of carbides, the degree of relaxation at a given tempering temperature is also dependant on how much carbon is initially in solution in the martensite. This in turn is determined by the austenitizing temperature, time, and the corresponding solution of carbides.

Where rings have been rejected after grinding because of undersized O.D. or oversized bore, then reclamation may be possible by applying the above theories. For example, with the higher hardenability steels it has been found that by tempering for prolonged periods of time at normal tempering temperatures, an appreciable growth has been obtained (transformation of retained austenite to martensite or bainite). In components where retained austenite is already at a minimum such growth is unlikely. Some contraction of the latter components can sometimes be achieved by increasing the tempering temperature to 500° F or above provided that appreciable loss in hardness is not experienced. What works for a given part will not necessarily work with a different part out of the same steel. For these reclamation practices to be of value the individual heat treat department using them needs to keep a log on what treatment was used for what part and the results obtained. From this systematic tabulation general guidelines will become evident.

CONCLUSIONS

The aim of the foregoing paper has been:

A) To provide an insight into the almost impossible task faced by the heat treat department for those non-heat treaters who find it difficult to understand why after heat treat a part should have changed size.

B) To provide a better understanding of the theories involved so that the heat treater be more able to improve or introduce new size control techniques.

C) To emphasize that cost reduction programs aimed at reducing distortion of heat treated parts must take into account the overall interactive nature of the many variables involved and

in particular the condition of green rings prior to heat treatment.

REFERENCES

1- Rose, A., Harterci Tech Mill, 21(1966) 1-6

2- Shroder, R., Harterci Tech Mill, 39(1984) Heft 6 280-292

3- Frehser, J., Lowitzer, D., Stahl und Eisen 77 (1957) 1221-1233

4- Fuller, G.A., Atmosphere Carburizing at Less than 925°C (1700°F) for Distortion Control Carburizing Processing and Performance, Proceedings of an International Conference 12-14th July 89

5- Baldwin, F.E., Case Studies of Distortion Control Optimization Using Taguchi Analysis, Proceedings of an International Conference 12-14th July 89

6- Mayr, P., Dimensional Alterations of Parts Due to Heat Treatment, Arbeitsgemeinschaft fur Warmebehandlung und Werkstofftechnih, Lesumer HeerstraBe 32, 2820 Breman 77, ERG.

7- Berns, H., Verzug von Stahlen infolge Warmebehandlung, Zeitschrift fur Werkstofftechnik 8(1977) 149-157

8- Ameen, E., Dimension Changes of Tool Steel During Quenching and Tempering, Trans. ASM 28(1940) 472-512

9- Schreiber, E., Unerwunschte Nebenwirkunger der Warmebehandlung In: Grosch, J(Hrsg): Grundlager der technischen Warmebehandlung von Stahl

Proceedings of the First International Conference on Quenching & Control of Distortion, Chicago, Illinois, USA, 22-25 September 1992

The Effects of Quenchant Media Selection and Control on the Distortion of Engineered Steel Parts

R.T. von Bergen
Houghton Vaughob plc.
Birmingham, England

SYNOPSIS

The paper commences by defining the types of distortion that can occur during heat treatment, and is followed by a comprehensive review of the factors which influence the distortion of engineered steel parts during quenching. The paper provides an in depth examination of the effects of quenchant type (including mineral oils and polymer quenchants), quenching characteristics, operating temperature, method and degree of agitation and quench system design upon the nature of distortion. Methods of controlling distortion through effective quenchant media selection and control are discussed with numerous actual case histories.

INTRODUCTION

Distortion is perhaps one of the biggest problems in heat treatment.[1] Comparatively little information has been published on the subject and this paper is intended to provide a useful guide for the practical heat treater to help identify reasons for distortion and suggest possible corrective action which can be taken.

Distortion can be defined as a permanent and generally unpredictable change in the shape or size of a component during processing. It invariably costs money in terms of rectification or straightening processes, or in the provision of increased grinding allowances and may also result in difficulties during assembly or the scrapping of costly machined components. All too often the heat treatment department gets the blame for distortion, mainly because it is only during subsequent grinding or assembly operations that the distortion becomes apparent. However before considering distortion during heat treatment it is perhaps worthwhile briefly mentioning some possible causes for distortion before heat treatment.

DISTORTION BEFORE HEAT TREATMENT

Dimensional changes can occur in the soft condition for a number of reasons.

a) **Relief of internal stresses.** Relief of internal stresses during machining operations can lead to springing out of shape, particularly if components are made from cold worked or drawn steel.

b) **Excessive collet pressures.** This can lead to distortion due to work being deformed elastically during machining and springing back upon release from the collet. The machining of thin section bearing rings from cold reduced tube can result in this type of distortion.

c) **Machine slackness and failure to comply with drawing limits.** This may result in components being out of tolerance before they enter the heat treatment furnace.

d) **Fabrication techniques.** Processes such as welding or those involving cold plastic deformation such as stamping or extrusion can introduce stresses into the component which may cause distortion.

DISTORTION DURING HEAT TREATMENT

This can be classified into two types and it is important to distinguish between them.

1) Size distortion or movement.

2) Shape distortion or warpage.

Size distortion

It is well known that the volume of a component changes during heat treatment due to the structural changes occurring in the steel. These are influenced by steel composition, austenitising temperature, soaking time, quenching rate and tempering treatment.

Changes in heat treatment conditions which influence the microstructure will affect the finished size of the part. For example, if the quenching speed is slower than the critical cooling rate non-martensitic products will be formed. Alternatively, during hot oil quenching of bearing rings, changes in oil temperature may affect the proportion of retained austenite. Both will influence the overall size changes occurring in the component.

However for a particular steel and heat treatment conditions, these size changes are generally predictable and can therefore be allowed for in machining and grinding allowances.

Shape distortion

This is the most troublesome form of distortion since it is often unpredictable. There are many different forms of shape distortion.

Out of round	Taper
Out of flat	Dishing
Bending	Bowing
Buckle	Closing in of bores

Shape distortion can occur either during austenitisation or during quenching.

Shape distortion during austenitisation

a) **Relief of internal stresses.** Stresses introduced during prior metal forming or machining operations will be relieved during austenitisation. Annealing and straightening operations may be needed before heat treatment.

b) **Sagging or creep.** This can be due to inadequate support of the work in the furnace. Careful attention must be paid to the condition of jigs and fixtures, furnace hearths and work trays and to the way in which components are placed in the furnace.

> Long shafts should be suspended vertically, and rings should be laid flat. Overhangs on large complex components such as dies should be supported.

c) **Mechanical damage in furnace hot zone.** Thermal expansion of components can cause jamming across furnace hearths. Automatic vibratory feeding systems on continuous furnaces can force work into the hot zone, and hence distort components that are already at temperature.

d) **Temperature gradients in furnace.** Temperature variations due to blocked burners, faulty heating elements or poor atmosphere circulation can result in uneven heating rates and hence cause internal stresses in components. This is particularly important with large distortion prone components where it is important to ensure that these are in the central part of the hot zone where temperature uniformity is best.

e) **Component size variation.** Variations in component section size will lead to uneven heating rates, resulting in the possible formation of internal stresses. Wherever possible, low heating rates should be employed on critical components having large changes in section thickness in order to avoid uneven heating.

Shape distortion during quenching

Shape distortion during quenching generally occurs as a result of an imbalance of internal residual stresses, which, taken to its extreme limit can lead to cracking, either micro-cracking or bulk failure of the component.

A number of factors, either individually or in combination, can influence the nature and extent of shape distortion during quenching.

> Steel composition and hardenability
> Component geometry
> Mechanical handling
> Type of quenching fluid
> Temperature of quenchant
> Condition of quenchant
> Circulation of quenchant

1. **Steel composition and hardenability**

This dictates the critical cooling rate required to achieve specific structure and mechanical properties and governs the volume changes (and hence residual stresses) formed during quenching. (Fig. 1)

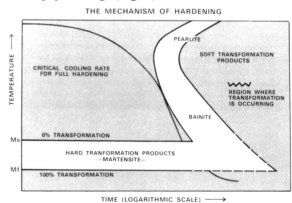

Fig. 1. Critical cooling rate for complete martensitic transformation.

276

In order to minimise distortion a quenchant should be selected which just exceeds the critical cooling rate of the steel and which provides a low cooling rate in the Ms to Mf transformation temperature range. However, it is often necessary to select a compromise in cooling rate to enable the processing of a wide range of steels of varying hardenability.

2. Component geometry

The section thickness of a component will influence the critical cooling rate required for full hardening and hence the selection of quenchant. Variations in section thickness on a particular component will lead to differential cooling rates and may result in variations in the level of residual stress with consequent distortion. Consideration should be given to quenching at the slowest possible speed dictated by the largest section or the possible use of hot oil quenching techniques.

3. Mechanical handling

In the austenitic condition steel is only about one-tenth as strong as it is at room temperature, as shown in Fig. 2 for a high carbon chromium ball bearing steel.[2]

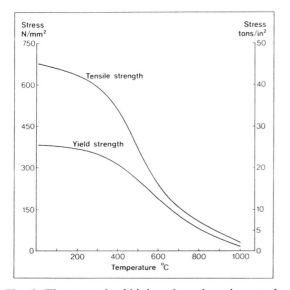

Fig. 2. The strength of high carbon chromium steel as a function of temperature.

Components are therefore very susceptible to mechanical handling damage when at elevated temperature. It is therefore important to handle components carefully, and to avoid dropping critical components into the bottom of the quench tank. Particular attention should be paid to quench chute design on continuous furnaces. Even inclined chutes can cause damage if thin section components strike pick-up slats on conveyor systems.

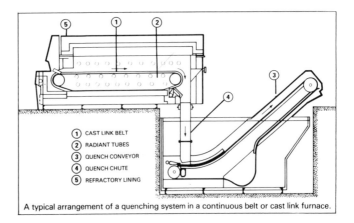

① CAST LINK BELT
② RADIANT TUBES
③ QUENCH CONVEYOR
④ QUENCH CHUTE
⑤ REFRACTORY LINING

A typical arrangement of a quenching system in a continuous belt or cast link furnace.

Fig. 3. Schematic continuous heat treatment installation.

An interesting example of this kind of distortion occurred at a manufacturer of automotive water pump spindles. Stepped shafts were quenched from a continuous furnace and dropped vertically approximately 3m down a quench chute onto the conveyor. Many of the shafts were bent and this was initially attributed to residual stresses due to the change in section size. However, closer investigation revealed the presence of flats on the end of certain shafts diametrically opposite the direction of bend, confirming that the distortion had been caused by mechanical impact. A re-design of the quench chute to prevent this impact damage eliminated the problem.

4. Type of quenching fluid

The three stages of cooling which occur during quenching i.e. vapour phase, boiling phase and liquid cooling phase are well documented and understood. However, the relative duration of the three stages and the cooling rates during each stage can vary widely for different quenchants and can consequently have a significant effect on distortion.

The ideal situation is for each stage of the quenching process to occur uniformly on the component to provide uniform transformation with minimum residual stress. Unfortunately this is not usually the case due to such factors as
- changes in component section thickness
- surface finish - the vapour phase tends to be more stable on smooth machined or polished surfaces, and break-up more readily with the onset of boiling at sharp corners, holes or surface irregularities.

Vapour Phase Characteristics

Although the quench rate in the convection phase is generally considered to be the most critical parameter because this is within the Ms - Mf temperature range where transformation occurs, the characteristics of the vapour stage can also be very important. This is perhaps best illustrated with an example of gear hardening.

Precision automotive transmission gears are frequently assembled with no further finishing operations to the gear teeth. Requirements for weight saving have led to progressive reductions in gear mass and tooth section with consequent potential for increased distortion. During oil quenching, vapour retention in tooth roots, combined with the onset of boiling on the flanks can cause undesirable "unwinding" of thin section gears. (Fig. 4) This can be minimised by the use of accelerated oils where special additives are used to reduce the stability of the vapour phase and promote boiling - thereby giving greater uniformity of cooling.

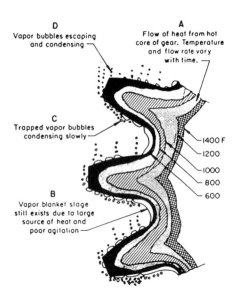

Fig. 4. Vapour retention in gear tooth roots during oil quenching.[3]

Convection Phase Characteristics

The quenching characteristics in the convection phase are best illustrated by the differential cooling rate curve. The cooling rate at 300°C has been fairly universally accepted as being critical for distortion since this is within the Ms - Mf temperature range of a wide range of engineering steels.

Typical values for some commonly used types of quenchant are shown below.

Quenchant		Cooling rate at 300°C (°C/sec)
Normal speed oil		5 - 15
Accelerated oil		10 - 15
Polymers	PAG	30 - 80
	ACR	10 - 25
	PVP	10 - 25
	PEO	10 - 30

It can be seen that PAG quenchants have higher cooling rates in the convection phase than quenching oils. Components are therefore more susceptible to the possibility of distortion and careful consideration has to be given to steel hardenability, component section size and surface finish before adopting a PAG quenchant. In general, PAG quenchants are suitable, and widely used for plain carbon, low alloy or carburised steels or higher alloy steel components of large section size.

More recently developed polymer quenchants such as ACR, PVP or PEO have much lower cooling rates at 300°C, similar to those of quenching oils. This has enabled the extension of the use of water-based technology to critical alloy steel components which would not otherwise be suitable for PAGs.

5. Temperature of quenchant

With modern quenching oil formulations, the temperature of operation does not have a significant effect on quenching speed.(Fig. 5)

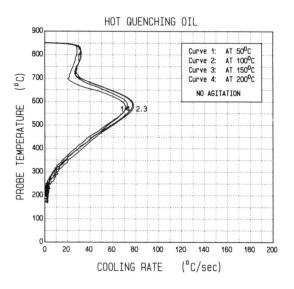

Fig. 5. The influence of oil temperature on quenching speed.

However, as is well known, the temperature of operation can have a dramatic influence on distortion and oils have been developed specially for use at elevated temperatures for mar-quenching or mar-tempering applications.

This process is generally applied to high precision engineered components, such as thin section bearing rings or automotive transmission gears or shafts where close dimensional control during quenching is essential. The process is designed to minimise the residual stresses generated during quenching by promoting uniform transformation.

During conventional quenching the surface, or the thinner sections of a component cool more rapidly than the centre or thicker sections. This can mean that some areas in the component have cooled to Ms and are hence beginning to transform whereas other areas are still in the soft, austenitic condition.(Fig. 6) When these subsequently transform volume changes may be restricted by the previously formed hard, brittle martensite, creating surface tension or imbalance of stress which can lead to distortion or quench cracking.

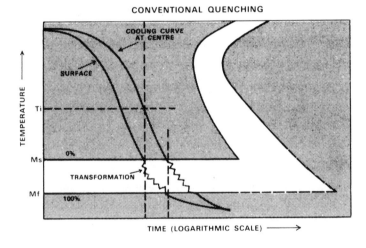

Fig. 6. The formation of stresses and distortion during conventional quenching.

During hot oil quenching, components are quenched into specially formulated oils generally at temperatures within the range 120°C - 200°C depending upon component complexity and distortion tendency and held at this temperature for sufficient time to allow for equalisation of temperature gradients across the section. The components are then withdrawn from the oil and during subsequent slow cooling in the furnace atmosphere transformation occurs uniformly throughout the section. (Fig. 7) This minimises the generation of internal stresses and reduces distortion.

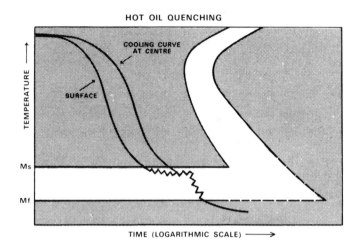

Fig. 7. Hot oil quenching techniques to reduce distortion.

Whilst the Ms temperature of typical engineering steels lies within the range 250°C to 350°C experience has shown that mar-quenching into hot oil at 150°C to 200°C can lead to significant reduction in distortion on complex precision engineered parts.

6. Condition of quenchant

The quenching characteristics of fluids can, and do, change during use for a variety of reasons. Most lead to an increase in quenching speed, particularly in the convection phase, and consequently to an increased risk of distortion and cracking.

Possible causes for changes in quenching speed include:

- contamination of quenching oil with water. As little as 0.05% of water in quenching oil has a very dramatic effect on quenching properties, by increasing maximum cooling rate and cooling rate in the critical convection phase.[4] (Fig. 8)

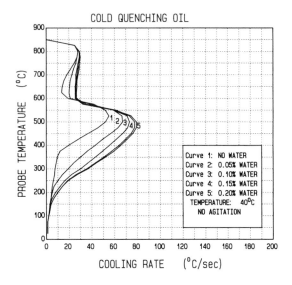

Fig. 8. The effect of water contamination in quenching oil.

- oxidation of quenching oil. Oxidation of mineral oils reduces the stability of the vapour phase and increases maximum cooling rate.[4] (Fig. 9)

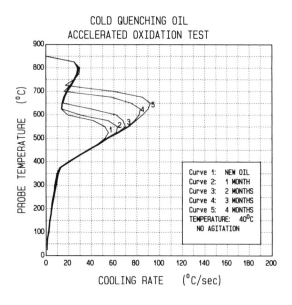

Fig. 9. The effect of oil oxidation on quenching characteristics

- contamination or degradation of polymer quenchant. Thermal degradation of polymer quenchants generally leads to an increase in cooling rate in the convection phase which can, to some extent, be compensated for by an increase in concentration. (Fig. 10)

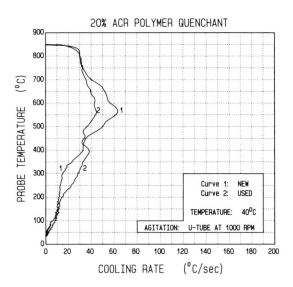

Fig. 10. The effect of polymer degradation on quenching characteristics.

It is therefore important, particularly in critical applications where close control of distortion is essential, to implement regular monitoring of the quenching fluid. Physical tests for acidity and water content in quenching oils, and the concentration of polymer quenchant solutions should be considered the minimum, and in practice, periodic evaluation of quenching characteristics is highly desirable.

7. **Circulation of quenchant**

Circulation of quenchant, either by pump or propeller is very important, to maintain uniform bath temperature and assist in the breakdown of the vapour phase of the quenching process.

The degree of agitation has a significant influence on the cooling rate of both quenching oils and polymer quenchants, but a different influence on distortion.

Quenching oil

As agitation increases there is a decrease in the duration of the vapour phase, an increase in maximum cooling rate and perhaps most important of all, an increase in the cooling rate in the convection phase.(Fig. 11) This latter effect may increase the risk of cracking and therefore excessive agitation should not be used.

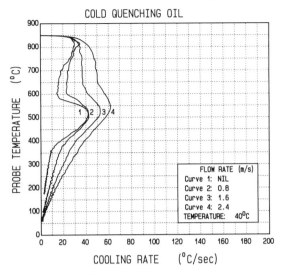

Fig. 11. The effect of agitation on quenching oil.

Polymer quenchant

With polymer quenchants, while there is a pronounced effect on the vapour phase and maximum cooling rate, increasing agitation has very little effect on the cooling rate in the convection phase.(Fig. 12) Vigorous agitation is normally recommended to ensure uniform quenching characteristics and this can be applied without increasing the risk of cracking.

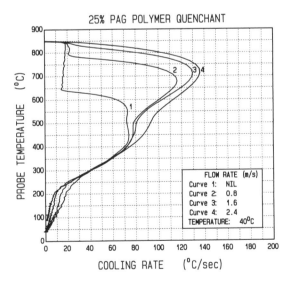

Fig. 12. The effect of agitation on polymer quenchants.

The direction of flow of the quenchant over the workpiece surface can also influence distortion.

This can perhaps best be illustrated by a recent case history in the bearing industry which demonstrates that uni-directional flow of quenchant over a surface can cause problems.

During the oil quenching of cylindrical bearing outer rings in an integral quench furnace, a predominant upward flow of quenchant through the charge resulted in the collapse of the bore towards the top of the ring and failure to clean up the OD during subsequent grinding.

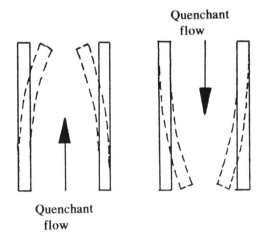

Fig. 13. The effect of quenchant flow on distortion.

Reversing the flow of quenchant produced the opposite effect. The answer to the problem was found by reducing the quenchant flow to the minimum required for circulation through the coolers and providing agitation by alternate up and down movement on the quench lowerator. Result -uniform contraction of bores as would be expected from theory.

SUMMARY

To summarise, distortion during quenching is a complex subject which can be influenced by many factors including steel composition and hardenability, component geometry and surface finish, mechanical handling and quench system design, type of quenching fluid, temperature of quenchant, condition of quenchant and amount of circulation.

The significance of distortion depends very much upon the type of component and the need for subsequent processing operations. For example, a few mm bend on a large bar or forging may be of no consequence or capable of simple rectification, whereas movement at micron level on precision gears or bearings receiving minimal subsequent processing may render the components completely unfit for service.

If distortion should be a problem much can be done to minimise it by careful and systematic examination of possible causes. I hope that my paper today will be of help to the practical heat-treater in identifying some of the causes and suggesting possible solutions.

REFERENCES

1. Quenching Principles and Practice, Houghton Vaughan plc, UK.

2. G.E. Hollox and R. T. von Bergen, Heat Treatment of Metals, 1978.2.

3. MEI Course 6, Heat Processing Technology, ASM, 1977.

4. R. T. von Bergen, Heat Treatment of Metals, 1991.2.

Proceedings of the First International Conference on Quenching & Control of Distortion, Chicago, Illinois, USA, 22-25 September 1992

Distortion Prediction Using the Finite Element Method

G.J. Petrus, T.M. Krauss, and B.L. Ferguson
Deformation Control Technology, Inc.
Cleveland, Ohio

Abstract

A PC-based, nonlinear finite element code was used to simulate quenching of a 4140 steel forging. Quenching conditions were varied to examine the influence of these conditions, specifically the heat transfer coefficient, on distortion and residual stress. The quenching conditions were selected on the basis of variations reported in the literature for oil and water quenching. The modeling results showed that the final microstructural distribution and amounts of various phases were dependent on the rate of cooling, as expected. Directly tied to the microstructural response were the levels of residual stress and the magnitudes of shape distortion.

Introduction

The heat treat industry has to cope with a business climate of increased competition and more customer demands in terms of lower cost, more stringent part specifications, and tighter control of geometry. As the heat treater strives to fulfill these requirements, process control and a greater understanding of the effects of process variation on heat treat response become more critical.

One of the leading problems facing the heat treater today is the problem of distortion during quenching. This is especially true as parts are being produced to net or near-net dimensions prior to final heat treatment. Indeed, in microalloyed steels, heat treatment is often combined with the cooling step from forging, thus eliminating the traditional heat treat operation, while placing great emphasis on control of shape and microstructure. Distortion during rapid cooling can be seen in a very wide range of applications and materials. The problem may be slight and require a minimal straightening operation or may be severe enough to scrap the part entirely. Cracking during quenching represents an extreme case of distortion which also results in large amounts of scrap each year.

The key to addressing distortion lies in understanding and predicting distortion problems as quickly, easily and accurately as possible. This paper reports the use of finite element analysis as a tool to reliably predict and analyze distortion problems. Finite element software that runs on an IBM compatible PC was used to simulate quenching of a 4140 steel forging. Process variables were altered to investigate the influence of these variations on the quenched microstructure, residual stress pattern and distortion.

The finite element method (FEM) is a powerful analytical tool for modeling the effects of externally applied loads on the internal response of a component. For example, nonlinear thermo-mechanical FEM codes capable of addressing both material and geometric nonlinearities are especially well suited for simulating distortion of components subjected to heat treatment. Nonuniform heating and cooling is standard fare in commercial heat treatment cycles. Temperature gradients within a part promote internal stress which can result in plastic deformation of the component, fracture being an extreme condition, and in residual stresses. FEM is an appropriate tool for simulating both the heating and cooling cycles, and for predicting the level and distribution of residual stresses and part distortion.

A word of caution is required concerning finite element analysis. There are three major sources of error to recognize in utilizing FEM. The first two sources are set by the software applied by the user. These include the mathematical formulations behind the constitutive equations and the solver, and their software implementation. The user minimizes these by software selection. The other, and typically more significant source of error, is completely under the control of the user -- model definition. The user is responsible for model definition, which includes a discretization of real world geometry, definition of boundary conditions, and selection of material property values.

Heat Transfer

The first key to the success of distortion prediction is accurate calculation of the heat transfer during the process. Property data include specific heat and thermal conductivity as functions of temperature for each phase present, latent heats given off during phase changes, and densities of the material phases. Handbook values for 4140 steel were used, including the phase transformation curve for microstructural characterization [1-7]. The quenchant itself is not explicitly modeled, so no direct properties of the quenchant are defined. However, the surface heat transfer coefficient controls the heat flux from the workpiece into the quenchant and is an input property of the quenchant. This film coefficient is a function of the quenchant temperature, the surface temperature of the material being quenched, the surface condition of the workpiece, and quenchant agitation. It is easy to see that this one

variable plays a major role in defining the quenching process. The film coefficient values used for this study are listed in Table I as a function of surface temperature.

Table I.
Surface Heat Transfer Coefficient

Surface Temperature, C	h, kW/(m^2*hr*C)
10	0.500
205	2.561
315	5.122
400	5.122
510	4.592
620	2.561
815	1.024
870	0.500

The forged shape that was selected for modeling is shown in Figure 1. The mesh used to examine the temperature distributions during quenching is also shown in Figure 1. The workpiece was assumed to be at a uniform austenitizing temperature of 843° C (1550° F), and the quenchant was at 60° C (140° F). Two cooling conditions were modeled. The baseline condition considered that heat extraction was uniform from all surfaces of the part. The other cooling condition changed the film coefficient for the bottom of the workpiece. Considering the results presented in [3], the variations induced were quite conservative. Figure 2 shows a comparison of the two cooling conditions. Using a proprietary program, thermal analysis results were interrogated to determine microstructural phases and their distribution.

A representative temperature fringe map is shown in Figure 3 for the baseline condition at a quench time of 6 seconds. This part is nearly all martensite after quenching; there are two regions of mixed microstructure in the thickest sections of the part where the cooling rate was the slowest.

A temperature contour map is shown in Figure 4 for the nonuniform cooling conditions at a quench time of 6 seconds. While the bulk of this part is martensite, there are two islands of mixed microstructure that include proeutectoid ferrite, pearlite, bainite and martensite. These islands are located at the two thickest sections of the part, as expected. They are larger than those for uniform cooling, and they are located closer to the bottom of the part where cooling is slowest.

Thermal Stress Model

The thermal stress portion of the finite element code is that portion that calculates the mechanical response of the material to a given temperature history. The large strain, nonlinear code used in this study is capable of modeling nonlinear geometry, nonlinear material properties and nonlinear input and boundary conditions.

The mesh used for this model matches that shown in Figure 1. The material properties applied in this model reflect the phases and mixtures of phases that may be present. For example, to simulate the martensitic transformation of 4140 steel, the coefficient of thermal expansion is modified over the range of temperatures 320° to 300° C (610° to 570° F), and the flow stress is modified to reflect the strength of martensite. Mixed phases are modeled by proportioning the properties of the phases present over a particular range of temperatures.

For the baseline uniform cooling, the radial and axial components of residual stress are shown in Figures 5 and 6. From these it is clear that both high tension and high compression are present. For axial stress, Figure 6 shows that the outer rim has been pulled in tension by the body of the part as it thermally contracts. Consequently, the bottom section of the rim is in compression. The radial stress shows a similar effect, with the bottom section being pulled in radial compression, the top also in compression but at a lower magnitude, and the center section of the rim being in tension. Figure 7 for radial displacement shows that the displacements are not uniform as vertical lines of radial displacement are not present. The locations of greatest radial movement are the left-most positions on these contour lines. This shows that the center of the rim has been displaced more inward than the top or bottom. Thus, the top and bottom are pulled in compression while the center is pulled in radial tension. Figure 8 shows that there is a small amount of plastic deformation, i.e. 0.0025 maximum strain, that accompanies quenching of this material and geometry.

For the case of nonuniform cooling, the stress patterns and magnitudes are somewhat different. Figures 9 and 10 contain radial and axial stress contour maps. From these figures, the rim again is in alternating compression and tension. The cross sections shown in these figures are drawn at 10X distortion of actual shape change for the purpose of viewing real shape distortion. Here, the inner hub has tilted inward toward the centerline at the top of the part, and the bottom of the outer rim has moved inward radially farther than the top. These features are reflected in the contour map of radial displacement shown in Figure 11. The bottom web surface has moved inward farther than the top surface of the web, resulting in general compression of the top web section and tension of the bottom section. For this material, the level of plastic strain is higher, e.g. 0.03 maximum value, with the strain being located at part fillets as shown in Figure 12.

The differences between the case of uniform cooling and nonuniform cooling arise because of differences in the amount and location of the softer mixed microstructure. While present in both forgings after quenching, the regions in the nonuniformly cooled part are larger and located more towards the bottom fillets of the part. This is illustrated by the temperature maps in Figures 3 and 4. The martensitic case is not as uniformly thick in this case as in the case of uniform cooling, allowing a greater degree of bending during quenching. Figures 13 and 14 compare cross sections that have distortions magnified by 30X for comparison. Clearly in Figure 14 for the nonuniform cooling condition, distortion is greater than for uniform cooling.

There is still a great need for R&D programs that include both experimental and analytical tasks. As the needed experimental data become more widely available, analytical models can be validated. In this manner, confidence in the accuracy and capabilities of modeling techniques will allow computer experiments to augment physical experiments to accelerate the knowledge being gained about heat treat processes.

Summary

Thermally induced stresses due to quenching, and associated distortion can be accurately modeled provided the appropriate material property data, and process and boundary conditions are available. In general, the models are extremely sensitive to boundary conditions, which unfortunately include some of the more difficult to measure parameters. As a particular example, a reliable convection/radiation film coefficient as a function of temperature, material, geometry, agitation, and quenchant state (i.e. level of contamination) appears to be absolutely critical to generating an accurate model, and, at the same time, this boundary condition is difficult to obtain. Other key data include mechanical behavior of the various phases present during the quench.

References

1. R.F. Price and A.J. Fletcher, "Determination of Surface Heat-Transfer Coefficients During Quenching of Steel Plates", Metals Technology, pp 203-207, May 1980.

2. R. Brennan, "Evaluating Quenchants in the Heat Treat Facility, Industrial Heating, pp 25-28, January, 1991.

3. R.A. Wallis, et al., "Application of Process Modeling to Heat Treatment of Superalloys", Industrial Heating, pp 30-33, January, 1988.

4. USS, "I-T diagrams", 3rd edition, Pittsburgh, 1963.

5. ASM Handbook V4, Heat Treating, Metals Park, 1991.

6. C. Bates and G. Totten, "Quantifying Quench-Oil Cooling Characteristics", Advanced Materials & Processes, 3/91.

7. B. Buchmayr and J.S. Kirkaldy, "Modeling of the Temperature Field, Transformation Behavior, Hardness and Mechanical Response of Low Alloy Steels during Cooling from the Austenite Region", Journal of Heat Treating 8(2), pp 127-136.

Effect of Process Parameters on Distortion

W.A.J. Moerdijk
Houghton Benelux B.V.
Oosterhout, The Netherlands

Heat Treatment of ferrous and non ferrous metals is nearly always the last process during the production cycle of parts, tools etc. All cost incurred in previous stages of manufacturing, transport and material can turn a part with a positive market value into a scrap weight value, if and when that last step, heat treatment, is not performed according to known parameters. It is therefore rather peculiar that the purchasing department can change from one supplier of material to the other, for the same designated steel, only on an in-house price basis. Quite often material is processed from different sources without performing a few simple tests, except perhaps the Jominy hardenability. This also applies when changing quench oil or quenchant supplier, a quench curve analyses is the minimum a laboratory must do. However it is much better to increase the level of process security by including some simple reproducible tests, like the modified NAVY C and stepped bars. This presentation will compare a limited range of materials and mineral quench oils with polymer quenchants during extended laboratory and practical tests.

The programme, which took 4 years, can be divided into sub frames:

1) A range of commercial available polymers.
2) A range of steels.
3) A range of laboratory tests.
4) Application in sealed quench furnaces at Kleve W. Germany.

Under 1.
Polymers quenchants:

Martensol Inat B.
AQ 251.
Ucon A.
Serviscol 78.

and as comparison water and min. quench oils; Isodur 220: a non additived, medium viscosity cold oil. Durixol 4: an additived medium vis. cold oil. Isomax 166: a high add. low visc. cold oil.

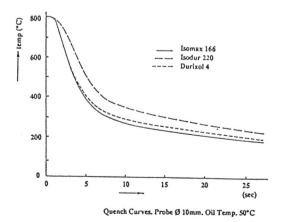

Quench Curves. Probe Ø 10mm. Oil Temp. 50°C

Introduced in a later stage of the programm the following quenchants have been lab tested.
Parquench 60.
Iloquench 500 & 700.
Aquaquench 365.
L.Q.R.

And a higher visc. medium additived oil Isorapid 455. All liquids were checked before and after tests on (where applicable) pH, viscosity, centrifuge, thermal split, bacteria, refractometer and quench curves.

Under 2.
Steel qualities.
Carburizing: C15, 16MnCr5 and 23CrMoB3.
Hardening: CK45, 42CrMo4 and C35BKD.
Tool steels: 90MnCrV8 and 100Cr6.

ST.SOORT	% C	% Si	% Mn	% P	% S	% Cr	% Mo	% V	% B	% Ni	% Cu	% Al
Ck45	0,43	0,26	0,70	0,033	0,037	0,05						0,27
C35BKD	0,37	0,25	0,58	0,005	0,009	0,19	0,02					0,04
42CrMo4	0,38-0,45	0,40	0,80	≤0,035	≤0,035	0,3-1,2	0,15-0,30		0,0034	0,05	0,10	0,04
C15	0,12-0,18	0,10-0,40	0,30-0,60	≤0,035	≤0,035							
16MnCr5	0,13-0,19	0,15-0,40	1,00-1,30	≤0,035	≤0,035	0,80-1,10						
23CrMoB3	0,20-0,25	0,15-0,35	0,70-0,90	≤0,035	≤0,035	0,70-0,90	0,30-0,40		0,003			
90MnCrV8	0,85-0,95	0,15-0,30	1,90-2,10	≤0,030	≤0,030	0,20-0,50		0,05-0,15				

Chemical composition of tool steel quality

This selection was made on the consumption per annum and a direct relation with other steels in the same category.

Under 3.
The following laboratoria results were obtained.

1) On test specimen

a) Hardness
b) Through hardening
c) Crack sensitivity
d) Deformation
e) Corrosion of test panels

In order to obtain meaningful results the following tests were performed.

a) Jominy - hardenability - through hardening.
b) Step bar - hardenability - through hardening - deformation - and crack sensitivity.

c) Modified NAVY C - exact deformation - crack sensitivity.
d) Parts from customers production like fasteners, bolts & nuts, ball bearing races, truck rear axle gears, etc.

Under 4.
Influences of polymers in sealed quench furnaces on; atmosphere control and recovery, corrosion especially in the quench vestibule. Results on production parts normally quenched in mineral oil, and the influence of carbonitriding on the quenchant in relation with the ammonia solvability in water.

Results quenchants & mineral oils

1) Suppliers of polymer quenchants must supply their customers with data and test methods in such a way that it is comparable with information from competitors. During our work we lost time in finding data.

2) The P.A.G. concentrates showed little differences in physical - chemical composition. During our work we discovered 8 brand names in France, 5 of those came from the same manufacturer.

3) Polymer solutions in bath quenching do give a satisfying live cycle, they are operator, environmental and equipement friendly. Whilst in the past bacteria, fungi and yeast had a detrimental effect, the modern products are far less prone to bacteria attacks. An exception must still be made in flame or induction hardening where in general only a section of the metal is heated, the remaining area is flushed with the polymer and washes off the dirt, oils and coolants into the polymer bath, giving rise to contamination.

4) Quench curves supplied by vendors, urgently need an international agreed standard.

The I.F.H.T. quenching committee of Prof. Líscíc does excellent work in this area, and I suggest a working group to be formed between I.F.H.T. - A.S.M. Int. and perhaps MITI.

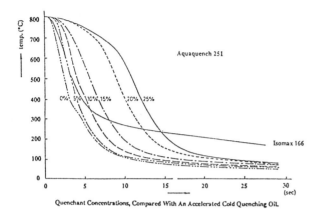

Quenchant Concentrations, Compared With An Accelerated Cold Quenching Oil

5) The majority of polymer solutions quench curves do show a steeper angle from appr. 300°C down to 80°C, compared with quench oils, which does have an effect on martensitic formation and structure, consequently also on distortion, residual stresses and cracking.

6) Despite the small differences in polymer concentrates, dilutions of polymers from a variety of sources do give a wide range in quench time from 800°C to 300°C. Variation measured, run from 1 to 1½ sec at 5 %, to 4 seconds at 25 % concentration. The quench oils Isomax 166 and Durixol 4

take resp. 8 and 9 sec, the Isodur 220 15,5 sec. A practical test showed a very high influence of ammonia on polymer solutions, more about this later.

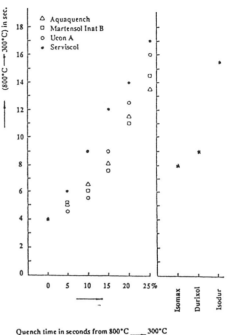

Quench time in seconds from 800°C → 300°C

Selection of test materials

The through hardening effect of Jominy bars during quenching was measured at 5, 20 and 40 mm from face with, water and a P.A.G. in 10, 20 and 30 % with thermocouples at 5 mm from the o.d. We expected the best response from a tool steel so in this case we used 90MnCrV8.

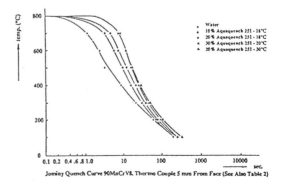

Jominy Quench Curve 90MnCrV8. Thermo Couple 5 mm From Face (See Also Table 2)

Some observations

On the less hardening responsive steel qualities: Ck45, C35BKD, C15

MEDIUM	TEMP.TRAJECT IN °C	IN SECONDS 5 MM FROM FACE	IN SECONDS 20 MM FROM FACE	IN SECONDS 40 MM FROM FACE
	800 - 700	1,3	17,4	43,6
	800 - 600	2,5	36	89
Water	800 - 500	4,5	62	144
temp. 18°C	800 - 400	9,7	100	245
	800 - 300	22,7	174	392
	800 - 200	67,0	333	618
	800 - 100	210	750	1122
	800 - 700	3,6	18,6	43,8
	800 - 600	6,0	37	90
Water +	800 - 500	10,2	66	144
10% AQ251	800 - 400	17,4	107	246
temp. 18°C	800 - 300	33	181	401
	800 - 200	73	360	642
	800 - 100	214	766	1183
	800 - 700	4,4	17,4	46,8
	800 - 600	7,9	42	90
Water +	800 - 500	12,9	67	149
20% AQ251	800 - 400	23,4	108	264
temp. 18°C	800 - 300	42	186	414
	800 - 200	81	375	645
	800 - 100	270	803	1158
	800 - 700	8,4	21,9	43
	800 - 600	12,6	40	87
Water +	800 - 500	17,4	78	147
30% AQ251	800 - 400	27,6	109	264
temp. 20°C	800 - 300	47	185	430
	800 - 200	84	375	636
	800 - 100	258	780	1135
	800 - 700	8,0	25,8	48
	800 - 600	11,8	42	94
Water +	800 - 500	18	72	152
20% AQ251	800 - 400	31	120	258
temp. 30°C	800 - 300	55	191	408
	800 - 200	89	375	633
	800 - 100	327	809	1194

Jominy Quenchtest 90MnCrV8 with thermo couples at 5-20-40mm from face.

and C15 carburized an increase of P.A.G. concentration has a notable effect on the Jominy curve, a higher concentration showed a decrease in hardness.

On the more responding steels like 42CrMo4 and 90MnCrV8 the Jominy curves coincide nearly completely. The carburizing steels 16MnCr5 and 23CrMoB3, though they are in the same group, produced Jominy curves only half parallel. We are convinced that the boron part in 23CrMoB3 is the cause of this difference.

During microscopic observations we noticed different stages of structure within millimeters from the face of the Jominy bar. Conclusion in testing polymers by way of Jominy is not a good method.

A P.A.G. in turbulent spray action is only giving it's cushion effect in the first instant. After the face is cooled down under ± 80°C, the following action is the heat transfer from the bar to the cool face, but the "blanket" of P.A.G. is not occuring anymore due to the too low face temperature. Concentrations of 25 % P.A.G. and higher might give an indication but are certainly not conclusive. Final conclusions about testing polymers employing the Jominy test are:

1) Only at a distance of less than 15 mm from face a clear effect of polymer concentration can be observed.

2) High alloyed steel qualities greater than 15 mm in depth the Jominy test, do not give an effect by changing polymer concentrations.

3) On steel qualities with a Jominy depth of 10 mm and less, the polymer concentration has a clear action. With higher concentration the through hardening reduces.

4) Between 10 and 15 mm Jominy depth the results are inconclusive.

5) The Jominy test does not provide any information about residual stresses or distortion.

6) If and when you want to use the Jominy test for comparison of quenchants employ a high carbon low alloyed steel in order to obtain meaningfull results.

Stepped bar test

Through hardening and residual stresses information can be obtained by using the stepped bar test. It is a cylindrical bar, machined to diameters of 10, 15, 20 and 25 mm with a step length of 50 mm surface roughness 32 Cla. The start diameter of the bar is 30 mm Ø.

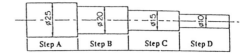

Stepped Bar Test

Since we did expect that with 42CrMo4 and 100Cr6, which are good through hardeners, the diameter of 25 mm Ø would not be enough to obtain a reasonable result, a second larger two stepped was prepared in 50 and 30 mm Ø length of steps 100 mm and 32 Cla. After machining, the bars were stress relieved and measured on total exentricity, with a 0,01 mm clock followed by marking of the bars.

Austinizing at 850°C in a small electric pit furnace under protective atmosphere, time 20 minutes. Total weight between 4 and 5 kg., the test bars were hanged smallest diameter down.

The quench bath volume was 60 ltr. with the polymers at a temperature of 25°C and quench oil at 50°C. Circulation was by means of an impeller with 1000 rpm. For each quenchant, oil and concentration 4 stepped bars have been produced and treated. The given results are the average of four measurements.

Conclusion stepped bar test

CK45 is a typical in between material, partly water partly oil hardening. Quenched in a selected polymer at the right concentration this steel or steels in the same category will provide good hardenability and less distortion compared with water quench.

MEDIUM	%	Ø 25	Ø 20	Ø 15	Ø 10
Water	·	33	35	50	53
Aquaquench 251	5	32	30,5	33,5	38,5
	10	27	27	35	42
	15	24,5	28	32	41,5
Ucon A	5	27,5	33	46	54
	15	29	30	50,5	53
Serviscol	5	29,5	32,5	48,5	55
	15	29	31	37,5	53
Martensol Inat B	5	28	30	44	56
	15	29	29,5	37,5	47,5
Isodur 220	·	22	20,5	21,5	23
Durixol	·	23,5	25,5	27	27
Isomax 166	·	26,5	25,5	28,5	41,5

Average Core Hardness Step.Test Ck45

C35BKD is a good through hardening steel up to 25 mm thickness. The distortion does fluctuate in relation with the selected polymer type and testing is necessary to prevent excessive stresses. This applies especially to complex shapes sharp edges and thickness differences.

MEDIUM	%	Ck45	C35BKD	42CrMo4
Aquaquench 251	1	0,59	0,22	0,11
	2	0,24	0,26	0,17
Ucon A	1	0,61	·	0,15
	2	0,10	·	0,10
Serviscol	1	0,22	0,56	0,40
	2	0,15	0,11	0,27
Martensol Inat B	1	0,16	0,35	0,17
	2	0,16	0,20	0,08
Water	·	1,19	0,60	1,28
Isodur 220	·	0,09	0,14	0,16
Durixol	·	0,12	0,15	0,09
Isomax 166	·	0,05	0,18	0,18

Deformation in m/m of stepped bars after quench

%
Ck45 5 resp. 10
C35BKD 10 resp. 20
42CrMo4 10 resp. 25

42CrMo4 does need slightly higher polymer concentrations. This widely used material (at least in Europe) is a very good hardener but care must be taken to prevent high internal residual stresses.

MEDIUM	CONC.	HARDNESS Ø 30		HARDNESS Ø 50		INCREASE DEFORMATION	
		EDGE	CORE	EDGE	CORE	1	2
Aquaquench 251	10%	51,0	47,0	52,0	38,0	0,29/0,09	0,26/0,06
	20%	51,5	49,0	53,5	38,5	0,06/0,06	0,04/0,06
	30%	51,0	48,5	53,0	40,5	0,04/-0,04	0,02/-0,03
Aquaquench 365	10%	54,5	54,0	54,0	52,5		
	20%	54,5	54,0	54,5	53,5		
Isodur 220		52,0	45,5	52,5	36,0	0,02/0,03	0,01/0,04

Stepped bar test 30-50mm Ø 42CrMo4

100Cr6 the well known bearing material shows excellent through hardening in connection with little distortion. From 10% AQ 365 upwards no problems are expected. One of our observations during the testing procedures was the sometimes critical behaviour of polymer solution and subsequent results. In order to obtain more insight, we made CK45 10 mm diameter, length 100 mm straight bars, starting from 15 mm diameter material. Again for each concentration P.A.G. 4 bars. Twelve bars were turned at 64 Cla and 12 bars centerless grinded at ± 8 Cla.

MEDIUM	CONC.	HARDNESS Ø 30		HARDNESS Ø 50		INCREASE DEFORMATION	
		EDGE	CORE	EDGE	CORE	1	2
Aquaquench 365	10%	62,5	61,8	62,0	60,5	0,07/0,10	0,08/0,11
	20%	62,0	61,5	62,0	59,7	0,36/0,42	0,32/0,38
Isodur 220		61,7	61,6	61,5	59,4	0,28/0,24	0,27/0,24

Stepped bar test 30-50mm Ø 100Cr6

In all bars a thermocouple was introduced, followed by heating, 20 minutes at 850°C in a small horizontal ceramic tube furnace under protective atmosphere, quench bath 40 ltr at 25°C agitated with 1000 Rpm. Results; the grounded bars showed with an increase in polymer concentration a substantial increase in quench drasticity. Quenching in water or oil can benefit from a rough surface finish, polymers do not. As presented on the slide you have observed that the excentricity of the stepped bars small and large diameter fluctuate between water and oil quench.

291

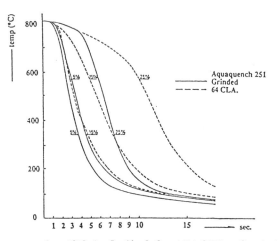

Impact Of Surface Condition On Quench Rate Of Polymer Quenchants

NAVY C test, modified

The NAVY C test is designed for comparison the sensitivity of steels on deformation and cracks during quench in different quench media. In our tests we used a modified type as shown in figure 2.

The advantage of this test is that the deformation can actually be measured at the top slit of the disk. Cracking will occur at the splines, this test did provide us only with quality information. The NAVY C were done on the better through hardening and tool steels 23CrMoB3, 42CrMo4 and 90MnCrV8 in P.A.G.'s AQ 251-365 at several concentrations, quench oils Isorapid 455 and Isodur 220. After machining, with a surface roughness of 32 Cla, the disks were stress relieved, marked, measured an inspected. For the prediction of internal stresses we used the work of Chatterjee-Fischer 1973 and Schröder, Scholtes Macherauch 1984.

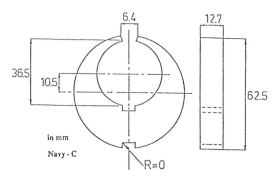

in mm
Navy - C

Modified Navy - C Test Plate

23CrMoB3
Steel quality 23CrMoB3 was tested in carburized and non carburized condition, temp 850°C in protective atmosphere, quench bath 25°C for polymers, 50°C for quench oils and 60 ltr volume. During and after quenching this steel showed no cracks at all, carburized or not. The top slit of the disk narrowed in water, with polymers and oils it widened, appr. with the same distance.

42CrMo4
The disks made of 42CrMo4, quenched 10, 20 and 30% AQ 251, 10 and 20% AQ 365 and the non additived oil Isodur 220 no cracks occured, and the slit widened in all quench media, with the biggest increases in the lower concentrations of quenchants. Higher concentrations and quench oil showed appr. the same results.

MEDIUM	CONC.	HRc			TOP SLIT INCREASE M/M				CRACKS
Aquaquench 251	10%	52,0	53,0	52,0	0,49	0,35	0,27	0,17	NO
	20%	52,0	52,0	52,5	0,36	0,38	0,34		NO
	30%	53,0	52,0	52,5	0,17	0,23	0,17		NO
Aquaquench 365	10%	52,5	53,0	53,0	0,32	0,24	0,30		NO
	20%	52,5	52,0	52,0	0,30	0,24	0,32		NO
Isodur 220	-	51,0	51,5	52,0	0,31	0,38	0,35		NO

Modified Navy C Test on 42CrMo4.
Quenchants compared with non accelerated cold Quench oil

After measuring all disks were cut in order to asses the structure and core hardness, the disks were through hardened and the structure nearly complete (95-100%) martensitic.

90MnCrV8
The treatment of the tool steel 90MnCrV8 has been as follows. Preheat 30 minutes in 400°C, 25 min. austenizing 800°C, quench bath 300 ltr, temp 25°C fast circulation, oil 50°C. Annealing 90 min ± 170°C within 5 min. after quench followed by 120 min. at 200°C. Quenchants used 10, 20 and 30% AQ 251, 15% AQ 365 and the oil Isodur 220. All disks quenched in all concentrations AQ 251 cracked. Though deformation in AQ 251 compared to AQ 365 was nearly the same, zero cracks in AQ 365 and Isodur 220.

Here it was important to check the structure of the metals to understand possible reasons for the cracking. Observations showed a layer structure with coarse carbides. Retained austenite was ± 20%, for this type of material treated at 800°C and annealed at 200°C this is quite normal. Higher austenizing will decrease the layer effect.

Since the cracking is not caused by structural defaults, the obvious reason must be the quench drasticity differences. At martensitic start AQ 251 is still extracting a high amount of energy from the material, whilst AQ 365 does that before the M.S.

The conclusion therefore is that the cracks are caused by the higher temperature gradient during martensitic transformation. In the core, transformation is still in progress, the increase of core volume is larger than the total elasticity of the steel.

The NAVY C disks made of 90MnCrV8 in all quenchants and concentration are through hardened. Quenched in slow oil they don't crack. In AQ 251 all disks cracked up to 30%. In 15% AQ 365 no cracks. The structure of 90MnCrV8 is as expected transformation stresses are probably the cause for cracking in AQ 251.

Tests in practice.

Within the framework of our research, practical application of polymers in sealed quench furnaces was included to study a number of issues, like: The effects of a water containing quench medium on furnace atmosphere and corrosion in the quench vestibule and parts. The testing took place at the R & D facilities of Ipsen at Kleve, Germany, in a TU 1 for 14 days, later followed in a T 7 for one week. The reason for the second trials was that the TU 1 furnace was epuiped with a small impeller mounted in the vertical wall of the quench bath which was not capable in maintaining a good turbulent flow of the quenchant. Reported here are only the tests in the TU 1.

We used as test products standard production parts which usually are treated in S.Q.F. or belt furnaces. A wide range of materials like: 100Cr6, railroad clips springsteel 55Si7, chain fixtures C35, conveyer chain scrapers 30Mn5, fasteners, and carburized rear axle truck gears 20CrMo3 were tested, and after treatment compared with the standard production results. AQ 251 (P.A.G.) in 10 and 25% and Durixol 4 were the selected quenchants. There was a reason for selecting an Ipsen furnace. The inner door is always 2 inches open during processing, in order to obtain the atmosphere also in the quench/cool area, consequently any disturbance of atmosphere, by water vapours etc, can flow in the treating section, including the direct radiation on the surface of the oil or quenchant. The most important factor however was to establish the influence on the atmosphere. Esp. in a carburizing condition water vapour will disturb the carbon potential through the reaction $CO + H_2O = CO_2 + H_2$ so the carbon potential available to the steel is reduced. The atmosphere was controlled through an oxygen probe and is measured by the reaction $2\ CO_2 = 2\ CO + O_2$.

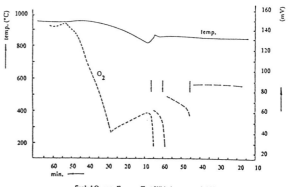

Sealed Quence Furnace Test With Aquaquench 251

One of the processes shown in figure 3, was carburizing. Initial austenizing at 860°C, C potential 0,4%, carburizing at 950°C and C pot. 1,2%. A disturbance would show immediate at such a high C potential. Quenchant a P.A.G.

The quench cycle takes about 15 minutes and the first thing that will happen is the mixing of gasses from atmosphere and quench. Due to the continuous flow of protective gasses to the furnace the atmosphere does recover quickly. A major disruption is the loading/unloading. The decrease in oxygen potential during the first 20 minutes of the carburizing process is the increase in temperature without adjusting the C pot, in order to prevent extreme sootforming until 880°C. Above that temperature and increasing the C pot. to 1,2% the oxygen pot. value is obtained in 20 minutes. This is better than with an oil in the quench. During the second test cycle the construction of the innerdoor threshold was changed in that way that a spring loaded radiation screen protected the surface of the quench medium, this had very good influence on control and results. This last remark does also apply to oil quench systems. With reference to the corrosion aspects of polymer quenchant the following remarks. We did our work in furnaces previous operated on quench oils, the top vestibule was covered with cracked and oxidized oil particules. For this reason we welded a round bar left and right in the top section so we could hang mild steel plates on stainless steel hooks in the upper part.

For the first 10 days, ten Q panels were installed on both sides and at the end of each day these marked and weight plates, one left and right were taken out and secured in silica gel bags for lab testing by means of flushing with N citric acid and weighing. The largest weight loss was 0,2%. Blanks kept aside in corrosion free packaging gave 0,15% weight loss, due to the citric acid treatment.

In conclusion, the tested P.A.G. polymers do contain enough vapour phase corrosion inhibitors to protect the upper parts of the quench chamber. With regard to the hardening results, time does not allow extensive reporting but the exception in the overall good results were the fasteners. Fasteners, nuts & bolts, usually are heat treated in belt or shaker furnaces and not in batch type systems, that was clearly proven in our tests especially the nuts made of C35Pb, C45, C35BKD, 38CrB, KD, 20MnB5 and 20MnB4, provided us with such fluctuating results after quenching in AQ 251 and AQ 365. The results were eratic, though we understand why. Between small parts, like M12 nuts, and with insufficient quenchant turbulence the polymer action is maintained throughout the total quench cycle, the so named "polymer bridges", persistent polymer concentrations, occuring in tight packed batches or small parts covering each other.

No matter the polymer concentration this phenomenon happened, quite in contrary with the Durixol 4 quench oil in which all charges produced satisfactory results. That means that a potential user of polymer quenchants must carefully consider all factors of the heat treatment process, and obtain good information from a well known supplier.

Summary and conclusions

Compared to mineral quench oils, polymers do have a range of disadvantages and advantages like:

-/- Lower unit cost.
-/- Decrease in fire risk.
-/- Increase in processing flexibility.
-/- No or lower part cleaning cost.

-/- Decrease in secondary cost of environment and an increase in working conditions.

-/- Independence of crude oil shortage and lower transport cost.

1) Despite the fact that polymer solutions can replace a major part of oil quenching, currently they can not do it all. Novel developments do show very encouraging advantages and they do need "ad hoc" testing, and good cooperation between supplier and potential user.

2) Development of polymers is continuing by one or two suppliers, this will result in a wider application scope.

3) Criteria of current available polymers in the day to day heat treatment are not entirely clear, standardization is demanded.

4) Polymer solutions do give long life time, they are user friendly.

5) A worldwide standardization of quench curves and physical/chemical information provided by suppliers is necessary.

6) Polymer quench curves below 300°C do sometimes give eratic results compared to quench oil.

7) Laboratory quench curves are not directly representative for the performance in production.

8) Jominy tests with polymer solutions are not representative for their performance with exception of low alloyed high carbon steels.

9) The stepped bar test has itself established as a good and representative test in comparing polymers and quench oils. Deformations and quench stresses are in general higher with current commercial available polymers.

10) The same applies for the NAVY C test modified, which in the area of high alloyed and tool steels will provide more basic information than the stepped bar test.

11) When applying polymers the end results can show a wider band in hardness etc., compared to mineral quench oils. These disadvantages will gradually disappear with the new generation of polymers.

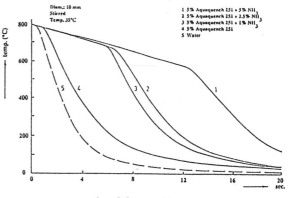

Impact On Quench Rate Of Dissolved NH₃

12) Polymers can be used in all production furnaces without adverse result, if and when parameters are checked before. No vapour phase or fluid corrosion will occur.

13) The application for batch type quenching of small parts with polymer solutions, shall be tested throughout before industrial use.

14) With separated and well stacked batch loads the afore mentioned disadvantage will not occur. However a turbulent quench bath circulation is always demanded.

15) Final conclusion.
In many cases polymer solutions will give an advantage over quench oils. A good, solid

cooperation and understanding
between user and supplier, is
the best down to earth basis for
a successful application. Only a
few suppliers can support you in
all aspects.

Study of Isothermal Heating and Surface Quenching Quality During Laser Heat Treatment

J. Ding, K. Mukherjee, C. Chen, and X. Chen
Michigan State University
East Lansing, Michigan

Abstract

In order to guarantee the original geometrical accuracy and the laser quenching quality of workpieces, the authors established the concept of laser isothermal heating by means of a series of tests and the equation of heat conduction of laser heating condition. Based on the concept of laser isothermal heating, this paper discusses the available temperature range for laser transformation process in a given material and the relationships between dislocation density, residual stress and process conditions of laser heat treatment, and provided an useful method for controlling the temperatures during laser processing (such as laser quenching, cladding and laser-assisted gas jet deposition etc.).

SINCE the laser heat treatment processes have a series of advantages, it is being applied widely to a number of industries. Even though the workpieces treated by laser beam may obtain an excellent geometrical accuracy with lesser distortion than other heat treatment processes, considerable attention has recently been given to the studies of how to control the laser processing quality such as the thermal distortion, residual stress etc. during laser heat treatment. In comparison with the conventional processes, the factor of laser heating temperature is still important to laser heat treatment.

In general, it is believed that the laser heat treatment processes include the laser surface quenching (e.g. laser phase transformation and laser rapid melt-solidification), laser cladding and laser-assisted gas jet deposition etc. The different laser processes have different laser heating conditions and temperatures in accordance with the particular processing quality requirements (i.e. geometrical accuracy, surface hardness, treat depth and strength etc.) and the working condition of the workpieces.

The surface roughness, surface hardness and thermal distortion of the workpieces quenched by laser beam are mainly dependent on the laser heating temperatures i.e. on the laser process parameters (laser power density F, and laser beam irradiating time t). This paper discusses the relationship of the laser process parameters and processing effects during laser energy induced phase transformation process. Based on the equations of heat conduction and related research, the authors established the concept of laser isothermal heating. Under the laser isothermal heating condition of a given material, reproducible surface hardness, surface roughness and residual stress state may be obtained. Consequently, the concept of laser isothermal heating is very useful to the research and production processes of laser surface quenching.

In our experiments, we employed two sets of CO_2 CW lasers, which include a 2.5 KW laser and a 0.6 KW laser. The materials for the laser quenching tests were medium carbon steel, cast iron and automotive exhaust valve steel.

The Laser Isothermal Heating

Based on the equation of heat conduction, previous research papers [1, 2] have introduced the relationships between the laser surface heating temperatures T_{ot}, the laser power density F, the laser beam irradiating time t and the laser hardening depth Z, during laser quenching [2]. The relationships are shown by following equations:

$$T_{ot} = \frac{2F}{K} \times \sqrt{\frac{kt}{\pi}}$$

$$\text{or} \quad F\sqrt{t} = \frac{T_{ot} \times K}{2} \times \sqrt{\frac{\pi}{k}} = M_1 \qquad (1)$$

where K is the thermal conductivity, k is the thermal diffusivity. Equation (1) indicates that for a given material, the surface heating temperature depends on F and t, and when the surface heating temperature is determined, the value of $F\sqrt{t}$ is approximately a constant M_1. It also shown [1, 2]

---

* Visiting scholar from Kunming Institute of Technology, P. R. China that:

$$F \times Z = M_2 \qquad (2)$$

where M_2 is also approximately a constant which is related to the laser surface heating temperature and the thermal property parameters concerned with the material. In other words, when the material is given and the surface heating temperatures are determined in a narrow range, the values of M1 and M2 are determinable.

The Equation (1) for $F \times Z$ = constant may be plotted as in Figure 1. The curve T_m is the upper critical laser heating isotherm. In theory, there are a series of curves which are composed by innumerable points with an identical value of $F \sqrt{t}$ (i.e. an identical laser heating surface temperature). The positions and curvatures of the isotherms curves in the coordinate depend on the values of M_2, i.e. the laser surface heating temperatures and the thermal property parameters concerned with the material. When the laser heating temperature is higher than the melting point of the material, then laser rapid melt-solidification process occurs.

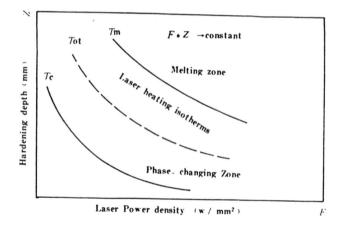

Figure 1. Diagram of F-Z relationship under laser isothermal heating.

The Shoices of Laser Heating Temperature and Corresponding Characteristics

THERMAL DISTORTION: A computer controlled He-Ne laser speckle pattern interferometry was used to measure the thermal distortion of the specimens treated by CO_2 CW laser. The measuring system has the sensitivity of about 0.3 μm to

the distortion and the measurement is non-contacting. The setup is shown in Figure 2.

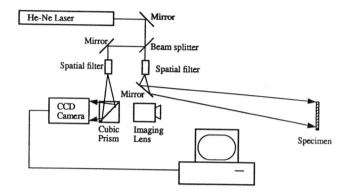

Figure 2. The setup for the measurement of thermal distortion

The size of these specimen was 25x25x1.2 mm. The material used for these specimens is medium carbon steel (AISI 1045). Two sets of the specimens were studied, each of them had four laser quenching strips. One set of specimens was treated so as to produce phase transformation and the other was treated to produce rapid melt-solidification process. The laser heating temperature for the phase transformation was ~ 1200 °C, and it was ~ 1800 °C for the rapid melt-solidification process.

The most of our laser application research work is related to automotive engine parts, so we pay attention to the thermal distortion of specimens treated by CO_2 laser beam. The thermal distortion test was done at a temperature of about 300 °C which is nearly the working temperature of the engine parts. The measured results of the thermal distortion are shown in Figure 3. and Figure 4.

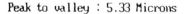

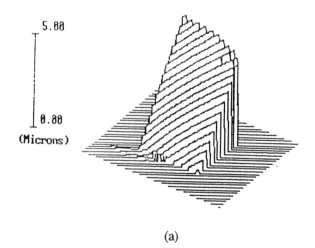

(a)

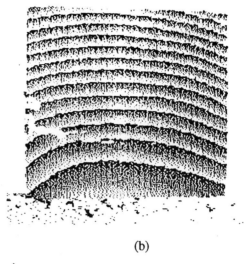

(b)

(μm)

```
1.00
0.00
-1.00
-2.00
-3.00
-4.00
-5.00
-6.00
```

Left   Centre   Right

(c)

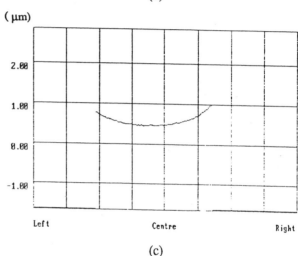

(b)

(μm)

```
2.00
1.00
0.00
-1.00
```

Left   Centre   Right

(c)

Figure 3. The thermal distortion measured by He-Ne laser speckle pattern interferometry for laser phase transformation specimen. (a) 3-D plot of the distortion, (b) the corresponding fringe distribution, (c) a typical cross section displacement.

Figure 4. The thermal distortion measured by He-Ne laser speckle pattern interferometry for laser melt-solidification specimen. (a) 3-D plot of the distortion, (b) the corresponding fringe distribution, (c) a typical cross section displacement.

Peak to valley : 7.39 Microns

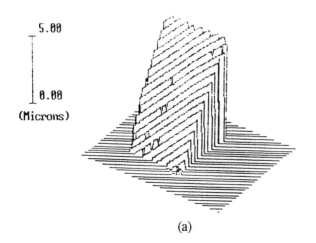

```
5.00
0.00
```
(Microns)

(a)

The distortion value of the peak to valley is 5.33 microns for the specimen with phase transformation and 7.39 microns for the specimen with rapid melt-solidification. The distribution of fringes in both of the cases were approximately uniform. We can see, from the graphical results, the distribution of the fringes and the planeness of the laser phase transformation is better than that of the rapid melt-solidification.

RESIDUAL STRESS: As we know, the residual stress consists of thermal stress and structural stress. In our experiments, when the laser heating temperature is higher than the melting pint of the steel, the values of the residual stress on the laser quenching surfaces were tensile stress, otherwise the values were compressive stress. In some degree, the residual stress is related with laser heating temperatures.

Table 1. shows the residual stress values for the different depths from the surface of a laser hardening layer including melt-solidification zone, phase transformation zone and the substrate.

TABLE 1.

Range	Melt-solidification Zone		Phase transformation zone				Substrate
Position from the primary surface (μm)	0	~200	~500	~800	~1100	~1400	~1700
Value of residual stress (Kgf / mm²)	+16.5	+0.3	-4.8	-12.2	-17.2	-8.4	-6.5
The micro-hardness (Kg / mm²)	1025	1121	1005	750	501	381	325

HARDNESS AND DISLOCATION: The laser heating temperature is important for surface hardness of the workpieces [3], because the carbon diffusivity in austenitizing parent phase influences directly the carbon concentration and dislocation states of the martensites (which are related to the strength and hardness). In Figure 5 (a) and (b), dislocation substructure in the two transmission electron micrographs of laser phase transformation hardening at 1300 °C and 950 °C respectively appears to be different for the two cases. However, further investigation remains to be carried out in order to discover the detail relationship of dislocation density, residual stress and carbon concentration during laser quenching.

(b)

Figure 5. Transmission electron micrographs of the laser phase transformation under different laser heating temperatures: (a) ~ 1300 °C, (b) ~ 950 °C.

Conclusion

In this paper the concept of laser isothermal heating is discussed and it can be expressed in the form of simplified equations or by a diagrammatic sketch. The thermal distortion of the steel specimens treated by laser phase transformation process and laser rapid melt-solidification process was measured by means of the He-Ne laser speckle pattern interferometry. The states of residual stress and dislocation substructure for laser quenching of surface, treated under different laser heating temperatures, were compared. Results indicate that the quality requirements of the workpieces can be met by choosing the suitable laser processing conditions.

References

1. Ding Jianjun et al., Laser Phase-change Heat Treatment. First National Symposium on Laser Heat-treatment. Oct. 1981, China.
2. Ding Jianjun et al., Research on Laser Process Parameters During Phase-change Heat Treatment. Journal of Kunming Institute of Technology, Aug.1991, Vol. 16, pp. 42-48.
3. M.F.Ashby et al., Modelling the Laser Transformation Hardening of Steel. Laser Processing of Materials, edited by K.Mukherjee and J.Mazumder. AIME 1985, pp. 226.

(a)

Proceedings of the First International Conference on Quenching & Control of Distortion, Chicago, Illinois, USA, 22-25 September 1992

An Indirect Method for Quantitative Characterization of the Quenching Power of Quenching Media

T. Réti, I. Czinege, and M. Réger
Bánki Donát Polytechnic
Budapest, Hungary

J. Takács
Recard Company
Györ, Hungary

ABSTRACT

Based on a simplified phenomenological model and computer simulation of transformations taking part during quenching, a computerized technique termed as CTA method (Cooling/Transformation Analysis) has been developed for quantitative characterization of quenching performance of quenching media. Starting with the measured and computer-generated heat flux density given as a function of surface temperature, the developed software is applicable to simulate the cooling/transformation processes occuring in hardenable steel specimens of simple geometry (cylindrical bar, plate). For evaluating and ranking different quenchants, in an indirect manner, the predicted as-quenched characteristics (hardness, martensite content, hardening index) of a preselected reference steel bar with specified size and compositon are used. Experiments based on computer simulation verified that the CTA method enables direct comparison of the results obtained by other similar testing technics (Cooling Curve Analysis, Quench Factor Analysis) devoted to evaluate quenching media.

IT IS KNOWN FROM THE LITERATURE that the testing and characterizing of quenching media used in heat treating practice can be classified into two broad categories [1]:

   * Characterizing in a quantitative manner the heat removal ability of the quenchant,

    * Estimating its hardening power, that is, its ability to develop a specified microstructure and hardness in a given material-section size.

According to experts the most useful methods to characterize the cooling power of a quenchant are the "cooling curve test" and the "quench factor analysis" wherein the change of cooling rates are taken into consideration throughout the complete quenching process.

The examination of quenching performance by cooling curve analysis (CCA method) is a very informative practical procedure. Using a test piece (probe) constructed generally from an austenitic stainless steel, the resulting cooling curves and the corresponding cooling rate functions (which are the first derivative of the temperature-time data with respect to time) sensitively reflects the heat removal characteristics of the quenching fluid. The principal advantage of CCA, that the parameters obtained from a cooling rate curve can be used for ranking and comparing the quenching performance of different quenching oils [2]. One of the limitations in various methods of CCA procedures is that very few metallurgical property correlations have been performed [1].

Another practical concept for evaluating the quenching performance and determining the response of a steel to the imposed quenchant conditions is that of quench factors [3,4]. Quench factor analysis (QFA method) provides a single number, Q_F that interrelates the cooling rate in a part and the transformation kinetics of the alloy. The quench factor reflects the heat extraction characteristics of the quenchant as a function of temperature over the transformation range of the alloy being quenched and permits the direct correlation of the shape of cooling curve with the appropriate property curve for the material of interest [1,4]. The QFA method suffers from the lack of availability of model parameters determining the transformation processes in steel (data of TTT diagrams, thermal properties, material constants).

This paper presents a new-type computerized procedure (called Cooling Transformation Analysis) developed to characterize and compare in a indirect manner the quenching performances of quenchants.

First the concept of the CTA method is outlined, then its possibilities of application are demonstrated by performing computer simulation. Starting with measured and simulated input heat transfer data characterizing the different quenching media, the results obtained by computer simulation of three different quenching performance tests (CCA, QFA and CTA method) are compared and discussed.

MODEL DESCRIPTION

The principle of the CTA (Cooling/ Transformation Analysis) method proposed for evaluating the quenching performance of quenchants is based on the following considerations. Let us assume that the chemical composition, microstructure, geometry of a part to be quenched are specified in advance. In this case the cooling/transformation processes occuring during quenching and the resulting final microstructure are determined only by the surface heat transfer conditions. From this concept the following conclusions can be drawn:

1. For characterizing the quenchant basically the measured heat flux and heat transfer coefficient are the most valuable. They can be determined by direct measurement [5] or in an indirect way, by solving a non-linear transient inverse heat conduction problem, referred to as an ill-posed problem in literature [6].

2. In case the heat flux or heat transfer coefficient characterizing the quenchant is known as input data, and material compositon (hardenable steel or alloy), initial microstructure, and size of the preselected

test specimen (cylindrical bar or plate) are given (that are considered as standardized data), then by proper simulation of the cooling and transformation processes occuring in the part, the as-quenched properties (amount of microconstituents, hardness, and so forth) can be predicted, and in an indirect manner and these quantitative characteristics can be used for evaluating the quenching performance. Briefly, this is the basic concept of the CTA method (see Fig. 1).

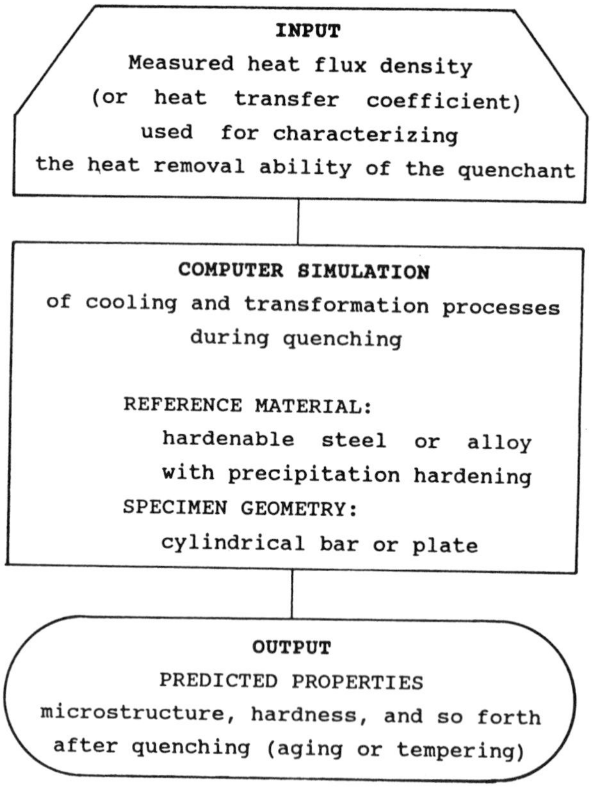

Fig. 1. Basic principle of CTA method

3. The input heat flux (or heat transfer coefficient) function play a key role in the evaluation of quenching performance, since it can be considered as "bridge" between CCA, QFA and CTA methods. In other words, knowing the surface heat flux, there is a practical possibility of comparing directly the results obtained by computer simulation for the three methods mentioned above. As can be stated, the CTA procedure can be regarded as a possible extension and generalization of the QFA method.

For establishing the proper software necessary to apply the CTA method, we used the modified and further developed version of the computerized property-prediction system (PPS) published earlier [7]. The equations forming the base of the PPS can be divided into three main groups as follows:

* The differential equation of heat conduction: By a numerical algorithm (implicite finite difference method), the temperature field in a given part (supposed to be plate or cylindrical bar) is solved.

* The system of kinetic differential equations: for describing the transformation processes occuring in the part,

* Equations devoted to predict the microstructure and properties.

The software developed has been implemented on a personal computer. In what follows, by performing experiments with computer simulation, the applications of the CTA method will be demonstrated.

EXPERIMENTS WITH COMPUTER SIMULATION

The purpose of experiments based on computer simulation was, on the one hand, to verify the correctness of the concept of CTA technique, on the other hand, to demonstrate that the results produced the three testing procedures (CCA, QFA and CTA methods) are mutually compatible. The general methodology of comparative investigations performed by computer simulation is shown in Fig. 2. As a starting point of experiments, 6 different quenchants were selected for analysis. The corresponding heat flux data as a function of the surface temperature are given in Fig 3. Of the six curve the first four (denoted by O6, O7, O8 and W1 respectively) represents measured data taken from literature [1, 8,9], while the last two heat flux functions denoted by S1 and S2 are computer generated (simulated) data. For quantitative characterizing of the heat flux

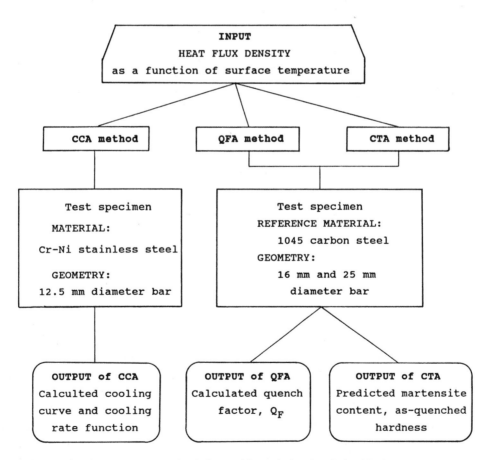

Fig. 2. Methodology of computer simulation used for analysing the relationships between different quenchant testing methods

303

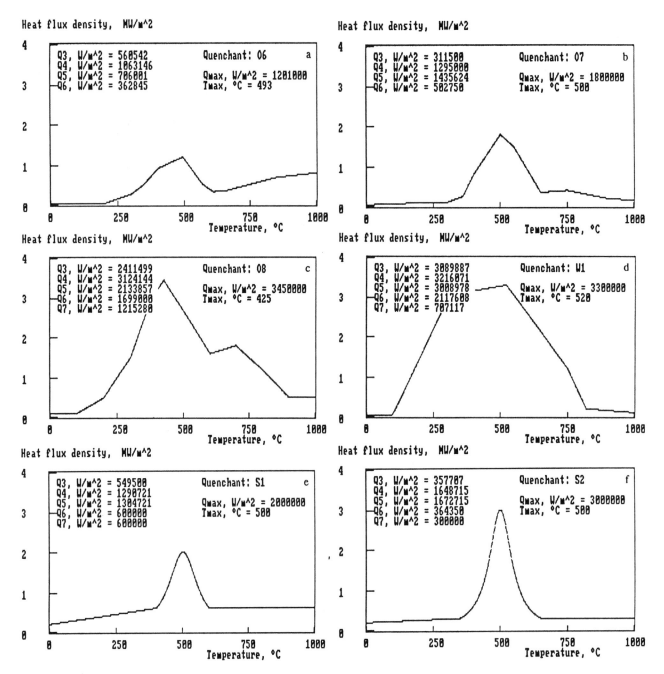

Fig. 3. Heat flux density functions used as input data for experiments by computer simulation

functions, we calculated the quantities Q3, Q4, Q5, Q6 and Q7, which are defined as mean values of heat flux density over the temperature intervals 300-400 °C, 400-500 °C, 500-600 °C, 600-700 °C and 700-850 °C respectively. These data for the six selected quenchants are summarized in Table I.

Simulation of CCA method. Starting with the six preselected heat flux functions, as a first step we have calculated the corresponding cooling curves and cooling rate curves for a

12.5 mm diameter austenitic steel specimen.

As can be seen, the material, size and geometry of the specimen selected was in accordance with the Wolfson laboratory test specification [2]. During investigations the initial specimen temperature T_a was supposed to be 850 °C (1560 °F).

For quantitative description of cooling rate curves relating to the centreline of the probe the numerical characteristics contained in Table II were calculated. T_{cmax} stands for

Table I. Characteristics of heat flux functions for different quenchants

Quenchant		Characteristics of heat flux function, MW/ m^2					
Designation	Type	Q'3	Q4	Q5	Q6	Q7	Q$_{max}$
06	Oil	0.56	1.06	0.71	0.36	0.59	1.20
07	Oil	0.30	1.30	1.44	0.50	0.33	1.80
08	Oil	2.41	3.12	2.13	1.70	1.22	3.45
W1	Water	3.09	3.22	3.01	2.12	0.71	3.30
S1	(Simulation)	0.55	1.29	1.31	0.60	0.60	2.00
S2	(Simulation)	0.36	1.65	1.67	0.36	0.30	3.00

Table II. Numerical characteristics of cooling rate curves

Quenchant	Computed characteristics						
	CR3	CR4	CR5	CR6	CR$_{max}$	T$_{cmax}$	
	K/s	K/s	K/s	K/s	K/s	°C	°F
06	26	52	58	33	64	527	981
07	14	46	77	62	84	590	1094
08	74	109	141	144	154	614	1130
W1	108	144	169	168	176	601	1114
S1	36	47	73	60	82	586	1087
S2	21	42	76	54	86	586	1087

the temperature belonging to maximum cooling rate, while the quantities CR3, CR4, CR5 and CR6 represents the mean values of cooling rate over the temperature ranges 300-400 °C, 400-500 °C, 500-600 °C, 600-700 °C respectively. As an example, four cooling rate curves relating to the quenchants 06, 07, S1 and S2 are illustrated in Fig. 4.

In order to quantify the quenching performance of quenchants on the basis of characteristics points taken from cooling rate curves, the formula derived by the Swedish Institute of Production Engineering Research (IVF) were used [2]:

$$HP = 91.5 + 1.34T_{VP} + 10.88CR5 - 3.85T_{CP}$$

where

HP = the so-called "hardening power" of the quenching oil in relation to unalloyed steels,

T_{VP} = the transition temperature in °C between the vapour phase and the boiling phase,

CR5 = the mean cooling rate in K/s over temperature range 500-600 °C,

T_{CP} = the transition temperature in °C between the boiling phase and convection phase.

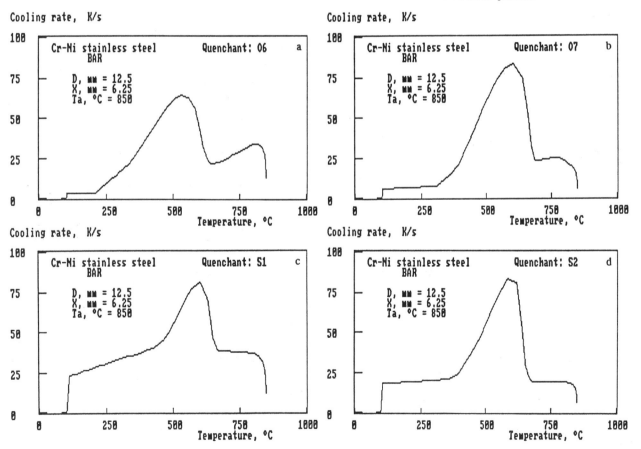

Fig. 4. Calculated cooling rate curves for the centreline of 12.5 mm diameter Cr-Ni austenitic steel bar

305

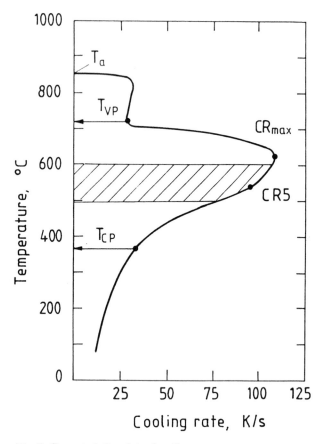

Fig. 5. Characteristic points of cooling rate curve

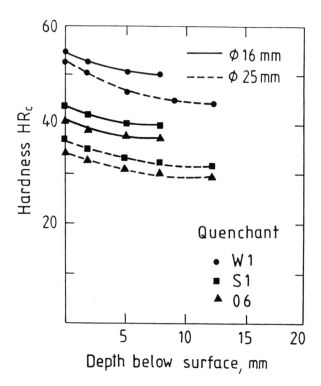

Fig. 6. Some hardness profiles calculated for SAE 1045 after simulated quenching

Table III. Computed characteristic temperatures and hardening powers

| Quenchant | Characteristic temperatures | | | | Hardening power |
| | T_{vp} | | T_{cp} | | HP |
	°C	°F	°C	°F	
06	640	1184	340	644	271
07	695	1283	370	698	436
08	850	1562	275	527	1706
W1	825	1517	100	212	2651
S1	675	1247	435	815	115
S2	675	1247	390	734	321

Fig. 5 shows these three points marked on a cooling rate curve.

Numerical values of the two characteristic temperatures and the calculated hardening power HP for the six quenchants investigated are given in Table III. It should be noted here that the range of HP values for the oils is about 10 - 1000.

Simulation of CTA method. As far as the simulation of CTA method is concerned, cylindrical test specimens of a steel equivalent to SAE 1045 (0.45 % C, 0.65 % Mn, 0.25 % Si), with diameters of 16 and 25 mm (0.64 and 1 in.) were used.

For the six quenchants investigated, the resulting martensite content and the as-quenched hardness were predicted as output data. For the 16 and 25 mm diameter bars the corresponding calculated hardness profiles obtained after simulated quenching with quenchant media 06, W1 and S1 are shown in Fig. 6.

In order to quantify the hardening performance of the quenchants investigated we calculated the so called "hardening index" R_h introduced by Just [10]. This is defined as

$$R_h = 100\ HRc_q\ /\ HRc_{max} \qquad (\%)$$

where HRc_q is the actual value of the as-quenched hardness, and HRc_{max} is equal to the possible maximum steel hardness which is attainable by quenching. This later can be calculated as a function of steel carbon content % C as follows

$$HRc_{max} = 20 + 60\ \sqrt{\%\ C}\ .$$

It has been found that there is a close relationship between the hardening index and the martensite content obtained after quenching [10]. In the case of a fully martensitic microstructure, R_h is equal to 1, while $R_h = 0.75$ corresponds to a martensite content of 50 %. Calculated values of hardening index for the surface and centerline of 16 and 25 mm diameters steel bars are summarized in Table IV.

Simulation of QTA Method. In order to estimate the quench factor on the basis of calculated cooling curves, we used the following formula

$$Q_F = \int_{M_S}^{A_3} \frac{dt}{\tau(T)}$$

where $\tau(T)$ is the critical time for the start of isothermal transformation of austenite at temperature T; A_3 and M_s stand for the composition dependent characteristic transformation temperatures. In practice the start of austenite decomposition (calculated

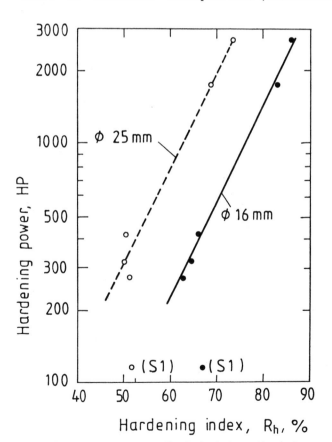

Fig. 7. Comparison of calculated hardening index and hardening power for the centre of 16 and 25 mm diameters steel bars (SAE 1045)

Table IV. Computed hardening indices and quench factors related to 16 and 25 mm diam bars

Quenchant	16 mm diameter bar				25 mm diameter bar			
	Surface		Centerline		Surface		Centerline	
	R_h,%	Q_F	R_h,%	Q_F	R_h,%	Q_F	R_h,%	Q_F
06	66.2	27.8	63.3	29.4	57.7	39.8	51.3	46.7
07	71.1	21.1	65.6	25.2	60.6	32.4	50.4	47.2
08	87.9	6.9	82.8	10.5	85.3	8.9	68.6	21.4
W1	89.7	5.3	85.7	8.3	88.2	6.4	73.9	17.1
S1	71.1	20.5	66.8	23.4	60.4	31.4	52.3	41.2
S2	69.9	23.1	64.2	27.2	58.8	35.3	49.4	49.0

from the stored TTT diagram of the steel) is defined by 1 percent amount of transformed austenite. Computed quench factors related to the surface and the centreline of 16 and 25 mm diameter steel bars are given in Table IV.

DISCUSSION OF RESULTS

The relationship between the computed hardening power, HP and the hardening index, R_h for the centre of quenched steel bars of 16 and 25 mm diameters are illustrated in Fig. 7. It can be seen that, with the exception of quenchant S1, the quenching performance of the tested quenchants are graded identically by both factors. This recognition is supported especially by the computed data regarding the bar of 16 mm diameter. The point related to the quenchant S1 is isolated from the others in the diagram. The reason for this is presumably not the fact that the flux function belonging to S1 is generated by computer, but rather, that the applicability of the formula used for computing HP in case of certain input data is limited. We have to note here that according to our experiences the value of the characteristic temperature T_{CP} can be estimated only with high subjective error, which causes significant scattering when calculating HP. On the other hand, it can be also stated that the same formula gives realistic results for the quenching media of high quenching performance (W1, S2 quenchants).

Analysing the relationships between the hardening index and quench factor a strong agreement are found between the results (see Fig. 8). This is due to the fact, that in the computer prediction of hardening index and

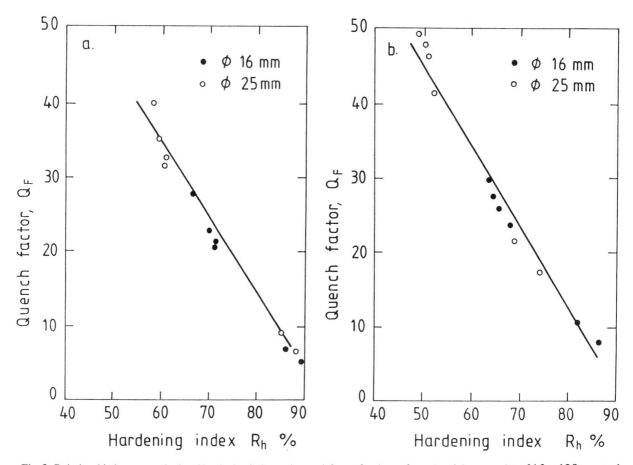

Fig. 8. Relationship between calculated hardening index and quench factor for the surface a.) and the centre b.) of 16 and 25 mm steel bars (SAE 1045)

quench factor the same input data (steel composition, geometry, TTT diagram, etc.) and computational model was used.

CONCLUSION

This paper has been devoted to a new approach of quantifying the quenching performance of quenching media in an indirect way.

By using computer simulation, it was shown how this method (referred as to CTA procedure) could be applied to analyse the relationship between other similar quenching test techniques (CCA and QFA methods). The main conclusion of the experiments based on computer simulation can be summarized as follows:

1. For characterizing and evaluating the quenching and hardening performance of different quenchants and quenching methods, it is advisable to consider the measured heat flux function as the starting point. It would be worth trying to derive a formula similar to equation used for the HP value estimation, which makes it possible to evaluate and grade directly the quenching performance of quenchants on the basis of numerical characteristics related to the measured heat flux or heat transfer coefficient.

2. Starting from the principle of CTA method it would be desirable to elaborate a general and unified testing procedure based on computer simulation, which would be applicable for evaluating every type of quenchant and quenching technique. At present, prediction models based on finite element analysis are available, which form the starting point for establishing a standardized method [11]. For developing a "standardized CTA procedure" the most important model factors (type of computational algorithm, heat transfer parameters, material characteristics, initial microstructure, transformation properties, specimen geometry) should be unambiguously specified.

REFERENCES

1. Bates, C. E., G. E. Totten and R. L. Brennan, ASM Handbook, "Heat Treating", Vol 4, ASM International, 67-120 (1991).

2. Segerberg, S. O., Heat Treating, 20, 30- 33 (1988).

3. Evancho, J, W. and J. T. Staley, Metall. Trans., 5, 43-47 (1974).

4. Bates, C. E., J. Heat Treating, 5, 27-40 (1987).

5. Liscic, B. and T. Filetin, J. Heat Treating, 5, 115-124 (1988)

6. Beck, J. V., B. Blackwell and C. R. St. Clair, "Inverse Heat Conduction: Ill-Posed Problems, Wiley-Interscience, New York, (1985)

7. Réti, T. and M. Gergely, Heat Treatment of Metals, 4, 117-121 (1991).

8. Hildenwall, B.,"Prediction of the Residual Stresses Created during Quenching", Dissertation No. 39, Linköping Studies in Science and Technology, Linköping, Sweden, (1979).

9. Lübben, Th., H., Bomas, H. P.Hougardy, and P. Mayr, Härterei-Technische Mitteilungen, 46, 24-34 (1991).

10. Just, E. and J. Wolff, Zeitschrift für Wirtschaftliche Fertigung, 74, 442-444 (1979).

11. Buchmayr, B. and J. S. Kirkaldy, J. Heat Treating, 8, 127-135 (199)

Proceedings of the First International Conference on Quenching & Control of Distortion, Chicago, Illinois, USA, 22-25 September 1992

The Heat Treating Machine for Advanced and Extreme Vacuum Heat Treating Requirements

P. Heilmann
Leybold Durferrit
Hanau, Germany

W.R. Zenker
Leybold Technologies
Enfield, Connecticut

ABSTRACT

The combination of vacuum, convective heating and high pressure gas quenching at gas pressures of up to 20 bar (290 psi) absolute in a unique vacuum heat treating furnace result in impressive process advantages. This paper reports about the engineering background, equipment concept and process results for vacuum heat treating of tools and dies.

Special results will be reported about large dies, forging materials and distortion control.

Universal and Economical - The Heat Treating Machine for Advanced and Extreme Vacuum Heat Treating Requirements

The heat treatment of tools and parts made of high-alloy steels is being increasingly carried out today in vacuum furnaces with gas, overpressure quenching. The maximum gas quenching pressure is 6 bar abs., and N_2 is usually used as the quenching gas.

This is adequate in order to through-harden tools and parts made of high-speed steel, hot-work steel, and - to a limited dimension - cold-work steel, and to attain a maximum hardness.

Because of too low a quenching speed, the following heat treatments cannot be carried out satisfactorily in these vacuum furnaces:

- tools and parts made of medium and low-alloy steel;
- forging and molding dies with large cross sections and dimensions made of medium alloy steel;
- mass-produced parts with high charging density such as is the case of bulk goods charging;
- parts made of austenitic steels and alloys which, after the solution annealing, require a very rapid quenching;
- parts and tools with dimensions > 100 mm diameter made of cold-work steel

If the gas quenching is not adequate, the quenching must be carried out in the oil bath with all of its disadvantages. These disadvantages are:
- great danger of cracks;
- high warping;
- cleaning of quenching oil
  with alkaline wash water and the complex of problems involving the wasting of this water, or with readily soluble detergents with fluorinated hydrocarbons and the strict regulations imposed on them.

All of this has led to the demand for a universal, economical vacuum furnace which heats up, soaks and quenches more rapidly. For this purpose, a new generation of vacuum furnaces with convective heating and gas high-pressure quenching was developed and built.

The new generation of vacuum furnaces

The experience gained with, and the advantages offered by, the vacuum furnaces with a max. of 6 bars abs. gas quenching were taken over in the design of the new generation vacuum furnaces. These advantages are essentially:
- all-around heating with its homogeneous heating effect;
- cylindrical working space with the possibility of charging the furnace with parts which are even oversize;
- all-around nozzle quenching with its two-dimensional gas flow and homogeneous quenching;
- only few mechanically moving parts in the vacuum furnace which can lead to malfunctions or failures;
- only few lead-ins which can result in leaks in the vacuum furnace.

Drawing 1 shows the construction of the vacuum furnace with the operating conditions for convective heating and gas high-pressure quenching.

A special feature of this new vacuum furnace is the fact that the heating tubes and the quenching nozzle tubes are identical. The tubes are produced from carbon-fiber reinforced carbon (CFC), a new material for furnace construction.

By means of coaxial actuated valve, the gas - in the first place - is blown through the heating tubes onto the charge during the convective heating. A homogeneous and rapid heating in the temperature range of 150 - 800° C or on request to 850° C is thereby possible.

High-pressure gas quenching

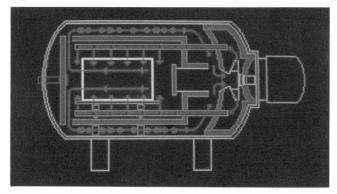

Convective heating

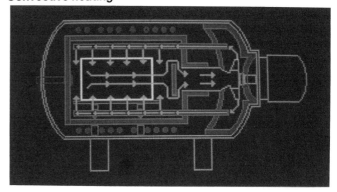

Drawing 1

In the second place, the cooling gas is directed via a ribbed tube heat exchanger during the quenching in order to be blown subsequently by the fan wheel through the nozzle tubes onto the charge.

Figure 1 shows the new vacuum furnace.

Figure 2 shows the internal construction of the heating chamber with the heating/quenching nozzle tubes.

Convective heating

In the temperature range of up to 500 - 600° C, the radiation is not visibly present. For this reason, the heat is transferred under vacuum very poorly or only slowly. Here, the gas helps with its convective heat transfer.

As a result of the accurately directed guidance of the gas in the heating tubes, a rapid and uniform heating is attained.

Experiments have shown that a pressure of about 3 - 4 bar is optimum for the convective heating. At a lower pressure, the rate of heating is too slow. At too high a pressure, the idle running consumption from the furnace rises.

The convection heating offers many advantages such as:
- rapid heating up and soaking in the lower temperature range;
- shorter heat treatment cycles;
- homogeneous heating at high charging density or charging of bulk goods;
- due to homogeneous convection heating, holding and heating stages below the hardening temperature can be eliminated;
- through more homogeneous convection heating, the stresses in the work piece are relieved more effectively, and thereby the warping is reduced;
- utilization of the vacuum furnace plants for the tempering and annealing as well. This, however, is economical only at times when the vacuum plant is not used for other high-grade heat treatment processes, for example in non-manned shifts during the night.

Experiments have shown that the convective heating with high charging density and using dies with large dimensions makes more rapid heating possible.

Graph 1 shows the comparison between convective and radiation heating components of different dimensions.

Comparative Tests
Convection Heating [N$_2$-Gas]
Radiation Heating [Vakuum]

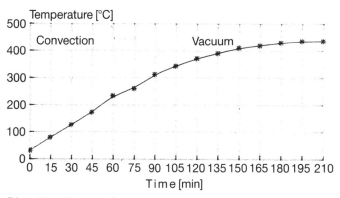

Dim.: 80 × 40 mm −Core−
Weight of Charge 450 kg
Furnace VKSQ 60/60/90

Comparative Tests
Convection Heating [N$_2$-Gas]
Radiation Heating [Vakuum]

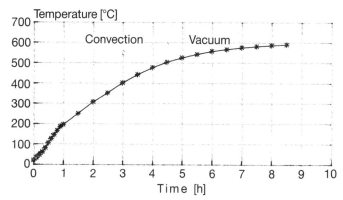

Dim.: 400 × 400 × 600 mm Material X40 CrMoV 5 1
Weight of Charge 600 kg
Furnace VKSQ 60/60/90

Gas high-pressure quenching

In order to increase the rate of quenching, the two influential factors (gas pressure and gas velocity) have been studied. Even with optimum conduct of the gas and maximum fan wheel design, it was no longer possible to increase the gas velocity appreciably. In vacuum furnaces, gas velocities of about 100 m/s are already attained. The gas pressure, however, can be selected as high as desired. For the vacuum furnace, it is only necessary to observe the pressure vessel regulations.

For the gas pressure, the power of the gas recirculation motor is a limitation. As a result of this, gases with other physical characteristics and densities were studied. Here, interesting relationships have been recognized as Table 1 shows.

Physical characteristics of quenching gases

	Chem. symbol	Density at 15 °C 1 bar [kg/m^3]	Density ratio w. resp. to air	Molar mass [kg/K mol]	Gas conditions at 25 °C and 1 bar		
					Specific heat capacity [kJ/kg K]	Thermal conductivity λ W/mk	Dynamic viscosity η Ns/m^2
Argon	Ar	1.6687	1.3797	39.948	0.5204	177×10^{-4}	22.6×10^{-6}
Nitrogen	N$_2$	1.170	0.967	28.0	1.041	259×10^{-4}	17.74×10^{-6}
Helium	He	0.167	0.138	4.0026	5.1931	1500×10^{-4}	19.68×10^{-6}
Hydrogen	H$_2$	0.0841	0.0695	2.0158	14.3	1869×10^{-4}	8.92×10^{-6}

Table 1

For safety considerations, hydrogen can be used only up to 5% by vol. in a gaseous mixture. Argon has physical characteristics which are too poor; helium, on the other hand, has physical characteristics similar to those of H$_2$ and can be used at high pressures because it is not explosive.

If the advantage of He with respect to N$_2$ is compared in Graph 2, advantages are found for the thermal transfer coefficient and - due to its lower density - there is also a lower power requirement for the gas recirculation motor.

Further studies showed that an He gaseous mixture has a better thermal transfer coefficient than a 100% He atmosphere.

In this respect, the following considerations and physical relationships must be observed.

The following physical characteristics of a gas and a gaseous mixture are important for the coefficient of heat transfer:
- dynamic viscosity,
- density,
- specific heat capacity,
- thermal conductivity.

The physical characteristics of a gaseous mixture composed of He/N$_2$ of 0 - 100% by volume were calculated and are represented in Graphs 3 to 6.

Influence of the kind of gas on:
1- Heat Transfer Coefficient
2- Gas blower Rating

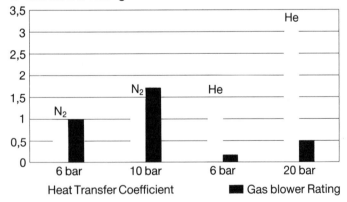

Graph 2

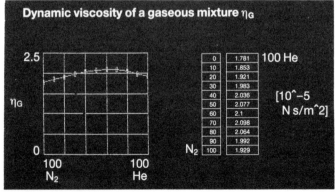

Graph 3

314

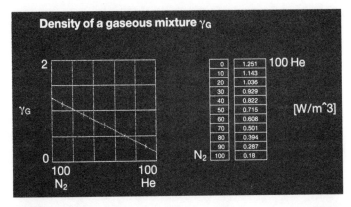

Density of a gaseous mixture γ_G

0	1.251	100 He
10	1.143	
20	1.036	
30	0.929	
40	0.822	[W/m^3]
50	0.715	
60	0.608	
70	0.501	
80	0.394	
90	0.287	
N₂ 100	0.18	

Graph 4

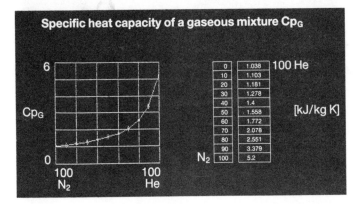

Specific heat capacity of a gaseous mixture Cp_G

0	1.038	100 He
10	1.103	
20	1.181	
30	1.278	
40	1.4	[kJ/kg K]
50	1.558	
60	1.772	
70	2.078	
80	2.551	
90	3.379	
N₂ 100	5.2	

Graph 5

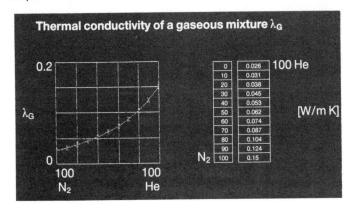

Thermal conductivity of a gaseous mixture λ_G

0	0.026	100 He
10	0.031	
20	0.038	
30	0.045	
40	0.053	[W/m K]
50	0.062	
60	0.074	
70	0.087	
80	0.104	
90	0.124	
N₂ 100	0.15	

Graph 6

The heat transfer of a recirculated stream of gaseous mixture can be determined by means of the Nusselt Relationship (see Formula 1).

$$Nu = C\,Re^{0.8}Pr^{0.43}$$

C = Factor
Re = Reynolds number
Pr = Prandtl number

If the values of the Reynolds and Prandtl numbers are entered into the Nusselt Relationship, the factors of influence for the physical characteristics of the gaseous mixtures in relation to N_2 are found when the formula is solved for the coefficient of heat transfer (Formula 2).

$$\alpha_G = \alpha_{N_2}\left[\frac{\lambda_G}{\lambda_{N_2}}\right]^{0.67}\left[\frac{\gamma_G}{\gamma_{N_2}}\right]^{0.7}\left[\frac{\eta_G}{\eta_{N_2}}\right]^{-0.37}\left[\frac{Cp_G}{Cp_{N_2}}\right]^{0.33}$$

Then, if the physical characteristics of the gaseous mixtures are calculated using this Formula 2, the curves in Graph 7 are found.

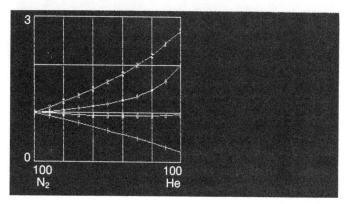

Graph 7

It is clear that the thermal conductivity and the spec. heat capacity have a positive influence on the heat transfer coefficient, while the density has a negative influence on it. The influence of the dynamic viscosity is neutral.

If all values of the curves in graph 7 are multiplied at the same gaseous mixture ratio, an interesting curve is found, as shown in Graph 8.

The optimum gaseous mixture has approx. 70-80% by vol. of He and 20-30% by vol. of N_2. This applies, however, for a gaseous mixture with a temperature of <100° C.

315

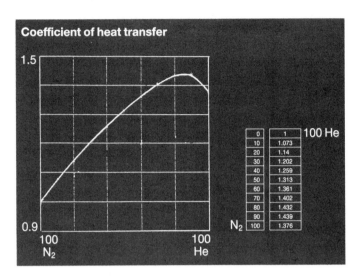

Coefficient of heat transfer

0	1	100 He
10	1.073	
20	1.14	
30	1.202	
40	1.259	
50	1.313	
60	1.361	
70	1.402	
80	1.432	
90	1.439	
100	1.376	N_2

Graph 8

Experiments have shown that an optimum gaseous mixture at higher gas temperatures, such as those which develop in vacuum furnaces during the quenching, is attained at approx. 70-60% by vol. of He and 30-40% by vol. of N_2.

A series of experiments was carried out to determine the quenching intensity.

Graph 9 shows the temperature variation during the quenching process in the core of test samples with various diameters. With a helium gaseous mixture

Gas high Pressure Quenching Vacuum
Furnace VKSQ Quenching 20 bar He/N_2
Core Temperature measurement

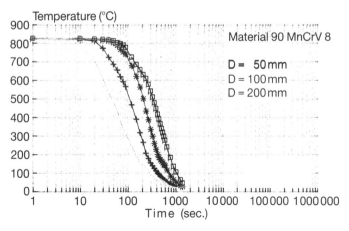

Gas high Pressure Quenching Vacuum
Furnace VKSQ Quenching 20 bar He/N_2
Core Temperature measurement

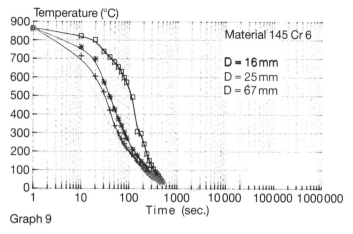

Graph 9

lambda values of about 0.18 to 2 are attained with work pieces with diameters of 12-200 mm.

If the lambda curves for the quenching in oil and in the gaseous mixture N_2/He at 20 bar are compared in Graph 10, it is possible to derive interesting information.

Quench Tests
Material 1.2842 (90 MnCrV8)

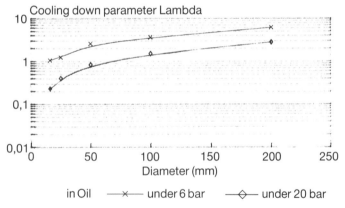

316

Quench Tests
Material 1.2080 (X 210 Cr12)

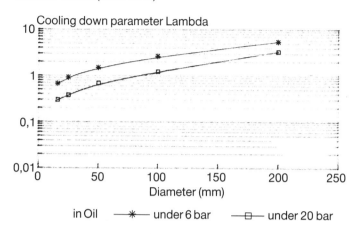

Graph 10

Figure 3

The gas quenching pressure can be programmed to operate at pressures between 1-20 bar. The quenching pressure must be selected depending on the alloy content of the material and the charging density.

Vacuum furnaces can be designed depending on the quenching intensity required for 6, 10 and 20 bar gas quenching pressure.

Figures 3 and 4 show a high-speed steel charge composed of end-milling cutter blanks which were hardened with N_2 at a gas quenching pressure of 10 bar abs. The charge had a gross weight of about 750 kg. Even the blanks in the center had a hardness of 64/65 HRC, and within total load a hardening tolerance of $\pm$ 1 HRC.

Figure 4

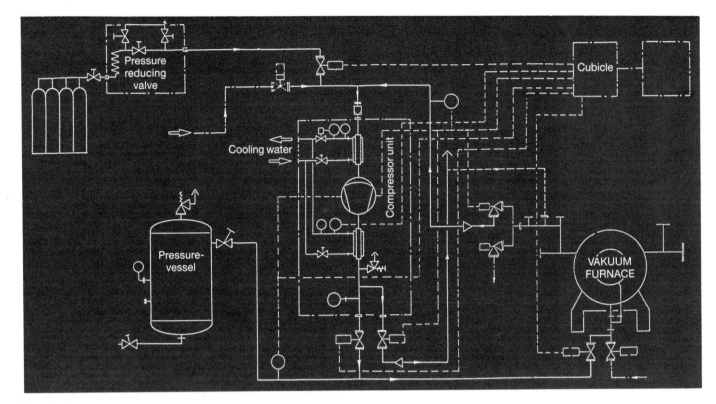

Drawing 2

Helium gaseous mixture recovery system

The price for helium is about 8 to 10 times that for N_2. There are countries in which, for logistic reasons, He is not available or is available only at high expense. In order to make the use of helium gaseous mixtures simpler and more economical, a helium gaseous mixture recovery system was built.

Drawing 2 shows the construction of such a helium gaseous mixture recovery system. The diaphragm compressor can reduce the pressure in the vacuum recipient into the partial vacuum range (0.5 bar). A slight residue (about 50% of the recipient volume) remains as loss. The costs for the loss are low, so the advantage of the gas high-pressure quenching predominates.

If helium should not be available, it is possible to quench at pressures of up to 10 - 12 bar with N_2 or to use an N_2/H_2 mixture.

Literature
Versuchsbericht (Experimental Report):H. Welzig
Gasehandbuch (Handbook of Gases):Messer,Griesheim

Economy

With regard to economy, the following must be said:
- The vacuum furnace of the new generation, in comparison with vacuum furnaces with 6-bar N_2 quenching, is only 10 to 20% more expensive, but still is 30% less expensive than a vacuum furnace with oil quenching.

- Due to the rapid heating up, soaking and quenching with high temperature uniformity, it is possible to heat-treat a broad spectrum of steels and materials with satisfactory results.

- The high gas quenching pressure permits a more densely packed charging, and thus a higher charge weight.

- As a result of the more rapid convective heating, soaking and the quicker quenching at high gas pressures, the heat treatment cycles become shorter.

- By means of the helium gaseous mixture recovery system, the costs for the quenching gas can be considerably reduced.

This versatility of the vacuum furnace of the new generation is the basis for an economical use.

Proceedings of the First International Conference on Quenching & Control of Distortion, Chicago, Illinois, USA, 22-25 September 1992

Martempering and Austempering of Steel and Cast Iron

G. Wahl
Leybold Durferrit GmbH
Hanau-Wolfgang, Germany

ABSTRACT

Martempering and austempering of steel and cast iron give savings in heat treating costs, improved component properties, less distortion and greater economy.
The limits of the process are determined by the composition of the material and the transformation behaviour.

In the austempering of cast iron there are a number of beginnings to be observed, in addition to the application in the transmission field, in mechanical and apparatus engineering and in the automative industry.

1. INTRODUCTION

Essential for a good quality heat treatment is a controlled and very uniform cooling of the whole component. The cooling speed necessary to form martensite or bainite and avoid premature and unacceptable transformations of the structure is determined by the composition of the steel and cast iron used. Oil or martempering salts are usually used for cooling alloyed steels.

Cooling at a temperature below the martensite point Ms causes heavy internal stressing and distortion. Salt baths do not have these disadvantages. When cooling and holding at 180-200°C while martempering a temperature equalization first of all takes place over the whole cross section of the work piece. During the further slow cooling in air to room temperature a low stress martensite transformation of the surface layer takes place.

2. MARTEMPERING IN SALT BATHS

In salt quenching the heat transfer mainly takes place by convection. A boiling period, such as experienced with quenching oils does not occur. Therefore the cooling characteristics of oil are quite different to those of salt martemper baths.

The development of martempering oils also makes it possible to use higher temperatures but, however, there is a visible trend away from oil to salt martemper baths in the treatment of components which tend to distort. A further argument could be that it is easier to control the disposal and treat the effluents.

The cooling intensity of salt martemper baths can be defined by factor λ. As demonstrated in Fig. 1, the cooling effect is influenced by the bath temperature and water content. In order

to use the benefits of marquenching on unalloyed steels also
it is usual practice to improve
the cooling effect by adding
0.5 - 1.0 weight % water. This
measure considerably broadens
the applications for salt martemper baths. Higher water contents do not improve the cooling
effect further and commended for
safety reasons.

Figure 1 shows the influence of
the water content at a bath temperature of 180°C on the cooling
intensity.

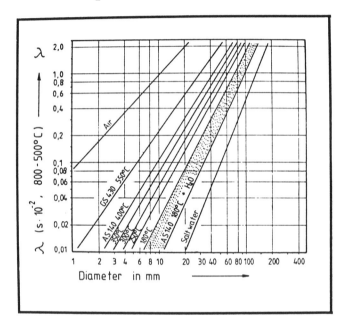

Fig. 1: Cooling parameter for
        AS 140 martempering salt

An automatic probe has been developed for regulating the
cooling intensity of salt martemper baths by adding water.
It replaces the traditional empirical method of adding water.
By continuously monitoring and
controlling the salt martemper
bath, reproducible and defined
cooling conditions to match individual technical needs are
possible within the temperature
range 180-250°C.

The measuring principle is based
on the simulated cooling of a
work piece. Basically the con-

trolling device consists of a
measuring probe, electronic control system and water feeder.

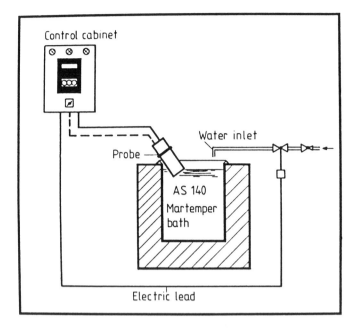

Fig. 2: Salt martemper bath -
         control unit

The sensor is put into the martemper bath and heated up to
200°C above bath temperature by
the built-in heating. Due to the
difference in temperature the
probe gives off a certain amount
of heat to the salt bath, which
is cooler. The amount of heat
given off depends on the cooling
effect of the martemper bath.
Influences due to other salts
being brought into the bath are
also taken into account without
difficulty. The amount of heat
emitted -which is time related -
enables a measured variable to
be derived which is very suitable for regulating the water content and thus the state of an
AS 140 martemper bath.

A measuring cycle runs through
about 20 times/hour and the actual state is shown on the digital display. The desired value
is set and a certain amount of
water added to the martemper
bath via a controlled magnetic

valve when the actual value differs to the set value. This control system has proved to be good under various working conditions.

3. SALT MARTEMPER BATHS COMBINED WITH CONTROLLED ATMOSPHERE PLANTS

In the heat treatment of parts which tend to distort, the benefits of salt martemper baths are gladly used with controlled atmosphere furnaces also. On choosing the appropriate quenching medium, all factors, for example the necessity of employing mandrel hardening, post machining and rejects, omission of meshing times etc. must be taken into account. A practical example of combining controlled atmosphere with salt was carried out with gear components (01).

In order to compare oil hardening with hardening in a salt martemper bath the plant was equipped with two different quenching facilities and operated alternately - oil hardening/salt martempering - throughout the test period. Furnace-related or process-related fluctuations were thereby excluded. Measuring methods used in the comparison are described in Fig. 3.

Two different sized ring gears made from SAE 4027-H were used for the test. Cooling in the oil martemper bath was done at 160°C and in the salt martemper bath at 210°C. The differences in temperature were due to the fact that the martemper oils available for low alloyed steels did not produce constant hardening results at temperatures above 160°C. Fluctuations in core hardness and case depths were clear evidence of this. Furthermore, it was noticed that the martemper oils had a shorter life time which, among other things, resulted in nonuniform hardening results.

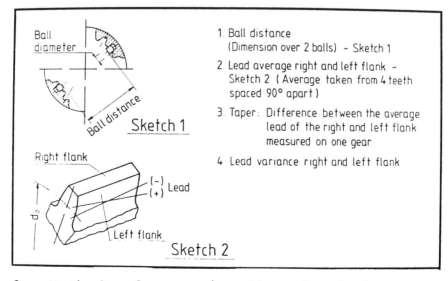

1. Ball distance
   (Dimension over 2 balls) - Sketch 1

2. Lead average right and left flank -
   Sketch 2 (Average taken from 4 teeth spaced 90° apart)

3. Taper: Difference between the average lead of the right and left flank measured on one gear

4. Lead variance right and left flank

Fig. 3: Methods of measuring dimensional changes

A comparison of the changes in dimension which take place during hardening, measured by the distance between two balls mounted in a crown gear, showed distinct advantages in favour of salt quenching. This tendency is also confirmed in Fig. 4 by the tooth lead.

the gas atmosphere after salt martempering.

4. AUSTEMPERING

With this process the desired heat treated state is not obtained by tempering a martensitic

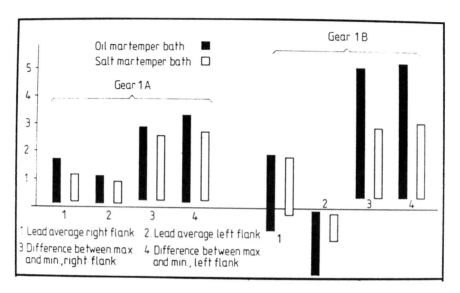

Fig. 4   Changes in dimension during oil martempering and salt quenching

These good results encouraged the user to convert other carbonitriding plants to salt martempering, which too are producing high quality heat treating results. Hardnesses and dimensions are within the specified tolerances. The difference in viscosity between martempering oils and salt martemper baths causes differences in drag out at martempering temperature. Greater losses can be expected with salt martempering baths than with oil.

A real help in minimizing the drag out losses are suitably designed jigs, arrangement of the charges and adequate drip off times over the martemper bath.

Particular attention should also be paid to the cleaning of the jigs. Only careful cleaning will ensure that no trouble occurs in

structure. In actual fact it is a treatment by which, after through heating to hardening temperature, the existing austenite is transformed isothermally into bainite. (02, 03, 04)

This is done by putting the parts, while still at hardening temperature, into a salt bath at 250-450°C. In critical cases the Duplex Quench process (05) can be used to avoid preliminary transformation. First of all the parts are cooled for a short time in a low-temperature salt bath. A second salt bath takes care of the isothermal transformation into bainite.

The transformation of the austenite into bainite does not take place suddenly within this temperature range, as in the formation of martensite. Bainite formation is linked with timerelated precipitations. Therefore work pieces stay isothermic for periods ranging from some minutes to a few hours at the same temperature in the austempering bath. On completion of the transformation, further cooling to room temperature can take place. In practice this is usually done in water to rinse off the salt adhering to the parts at the same time. By using evaporators the salt regained from the water tanks can be returned to the austempering bath.

4.1 Influence of austempering on the component properties

In the broad range of literature available on this subject, austempering is described as having benefits with regard to economy and component properties over normal harding and tempering. Compared with normal hardening and tempering -based on equal strength or hardness -the heat treating time is shorter and toughness values such as elongation, necking, bending behaviour and notch impact strength and fatigue strength are improved.

The toughness properties determined on samples made from SAE 1078 with a strength of approx. 1800 N/mm2 (comparable with approx. 50 HRC) are shown in Fig. 5. In spite of the relatively high hardness, necking and notch impact strength are distinctly above the values of the normal hardened and tempered components. Also very interesting is the maximum bending angle until cracking occurs - it is three times greater after austempering than after normal hardening and tempering.

The dynamic loadability of components is also favourably influenced by austempering. References have been made to increases in the timerelated fatigue strength and fatigue strength. Presumably this is attributable to the low notch sensitivity of the bainite structure.(13)

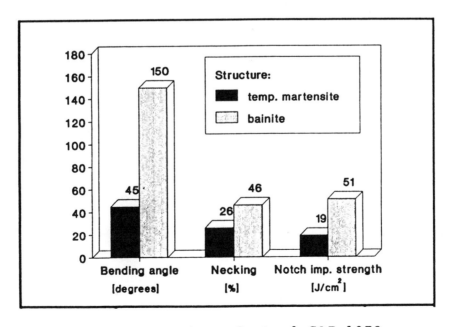

Fig. 5: Properties of steel SAE 1078

4.2 Applications for austempering

T a b l e   1

Work piece	Material	Thickness (mm)	Average HRC
Spring washers, var.sizes	50CrV4	2-5	45-50
Spring rings, var.sizes	Ck60	0.5-2	42-50
Seeger circlips	Ck67	1-3	44-48
Adapter sleeves	C60	1-2	45-50
Notched pins	C90	2-5	55-58
Fasteners	Ck67	0.5-2	46-48
Compensating springs	Ck75	1	46-48
Belt links	67Si7,Ck75	1-1.5	46-50
Chain links	Ck67-75	1-3	44-48
Office machinery parts	Ck55-80	0.5-2	40-50
Type bars	67SiCr5,C55	1	40-50
Concrete nails	41Cr4,50CrV4	5	54-57
Crown nuts	Ck45	6	30-35
Mower plates	Ck67	2-3	52-55
Chain saw parts	Ck67	1-2	50-56
Hinges and fittings	Ck67	1-3	42-48
Hard metal tipped drills	34CrNiMo6,	20	42-48
	50CrV4	15	42-48

This process has proved to be most successful, particularly in the manufacture of small parts in high-volume production. The above Table gives an impression of the range of unalloyed and low alloyed steels treated, with wall thicknesses preferably less than 5 mm but in special cases up to approx. 20 mm thick.

4.3 Austempered nodular cast iron

Austempered ductile iron has become increasingly popular in recent years in the manufacture of components. In the middle of the 1960s investigations were started in the American Automobile Industry to substitute case hardened steels used for transmissions for austempered ductile iron. In a publication on the austempering of ductile iron

John Dodd reports about the use of this material for rear axle gears and pinions in passenger cars instead of case hardened steel parts. There are a number of publications (06,07) on the use of austempered ductile iron for making transmissions. Apart from gear wheels, reference is also made to the use of other components such as bearing shells, crankshafts, wheel hubs and railway axle boxes with the following benefits:

- **Minimizing of mechanical processing by casting the components**

- **less noise due to better damping**

- **savings in costs due to easier heat treatment**

- **good processing properties**

- **savings in weight**

Cast iron is also best austempered within the temperature range of 250 - 450 °C. Depending on their composition, cast iron qualities with wall thicknesses of up to approx. 50 mm are treated.(08, 09, 10)

4.4 Plants for austempering

There are a number of different plant systems - continuous and noncontinuous - available nowadays for carrying out heat treating. This ensures that there are enough possibilities of matching heat treatment to material, component shape and the required throughput. Most common are plants in which austenitizing is done in protective gas and the isothermal transformation into bainite in a salt martemper bath. In many cases salt bath plants alone have also proved themselves.

Salt melts are also being used most successfully for austempering wire coils. For some years now a German steel producer has been operating a salt bath in which 45 t/hour of wire coils each weighing between 400-1200kg can be treated (11).

For treating long slim parts

multi-purpose pit-type furnaces are available in which, after austenitizing in protective gas, the isothermal transformation into bainite can be carried out in salt martemper baths (12,13). Those continuous plants available are shaker hearth furnaces, rotary furnaces and conveyor-belt type furnaces. An effluent free and in many cases also a waste-salt free heat treatment has already been achieved by combining protective gas with marquenching. Evaporator systems for processing the salt-bearing rinsing waters - with the possibility of re-using the regained salt - are available.

Infra red evaporators, which make traditional detoxifying plants superfluous, have proved very successful. The rinse waters enriched with martemper salts are vapourized in the evaporator by infra red rays at temperatures below boiling point within a very short time. This treatment leaves behind solid salts, which can be reused.

The evaporator can be gas-fired or electrically heated. The equipment is compact, easy to handle and requires little space.

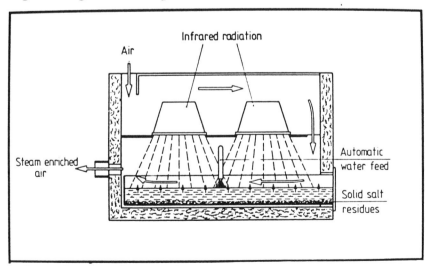

Fig. 6: Infra red evaporator

The evaporator shown in the schematic diagram works fully automatically. An outlet valve facilitates easy removal of the regained salt.

REFERENCES

(01) H. Schlösser   Einsatz von Abschrecksalz AS 140 und Hochtemperaturabschrecköl in einer Getriebehärterei, Durferrit-Hausmitteilungen, Heft 43,(1978), Seite 18-21

(02) K. Falkenmayer   Praktische Anwendung von isothermischen ZTU-Schaubildern, ZwF 75 (1980), H. 9, S. 434-438

(03) J. Motz   Die Kohlenstoffauflösung im Austenit von Gußeisenlegierungen und ihre Bedeutung für die Wärmebehandlung Giesserei, 10/57, H. 18, S.943-953

(04) W. Mannes
      K. Hornung
      H. Rettig   Erprobung von Zahnrädern aus unlegiertem bainitischem Gußeisen mit Kugelgraphit, Konstruieren + Gießen 10 (1985), Nr. 4, S. 19-29

(05) C. Skidmore   Salt Bath Quenching - a Review Heat Treatment of Metals 1986.2, S. 34-38

(06) M. Johnsson   Austenitisch-bainitisches Gußeisen mit Kugelgraphit als Konstruktionswerkstoff im Getriebebau, Antriebstechnik 15 (1976), Nr. 11, Seite 593-600

(07) J.M. Motz   Bainitisch-austenitisches Gußeisen mit Kugelgraphit - hochfest und verschleißbeständig Konstruieren + Gießen 10 (1985), Nr. 2, S. 4-11

(08) J. Dodd   Zwischenstufenvergütung von Gußeisen mit Kugelgraphit Gießerei 65 (1978), S. 73-80

(09) J.W. Boyes
      N. Carter   British Foundry 59 (1966), Nr. 9, S. 379 - 86

(10) W. Scholz
      M. Semchyshen   Mod. Cast. 53 (1968), Nr. 1, Seite 65 - 72

(11) K.-J. Kremer
      K. Neuhaus
      E. Sikora
      M. Wirth   Salzbadbehandlung von Walzdraht Stahl und Eisen 110 (1990) Nr. 6, Seite 51 - 56

(12) K. Heuertz   Mehrzweck-Schachtofen-Automat mit gasdichter beheizter Umsetzvorrichtung, Härterei-Techn.Mitteilungen 42 (1987) 3, S. 169 -173

(13) F.-W. Eysell   Die Zwischenstufenvergütung und ihre betriebliche Anwendung TZ für praktische Metallverarbeitung 66 (1972), Heft 3, Seite 94 - 99

(20) G. Wahl   Development and Application of Salt Baths in The Heat Treatment of Case Hardening Steels Int. ASM-Conference 1989, Colorado USA

Study on UCON® Quenchant A in Application Test

Z. Wen-Shang
Institute of Aeronautical Materials
Beijing, China

ABSTRACT

In this paper the physical properties, cooling characteristics, toxicity and quenching effect of the water-soluble UCON® Quenchant A has been tested to show that this quenching medium possesses excellent performance for heat treatment. The cooling curves of a UCON® Quenchant A aqueous solution are measured by using a model CT-CO$_2$ cooling speed measurement device and the method for determining the correct concentrations of UCON® Quenchant A solution used in industrial quenching bath is provided. By testing the quenching performance of aluminum alloy plates, it has been proven that the distortion of parts is reduced using UCON® Quenchant A, which gains an obvious advantage over other quenching medium This paper proves that UCON® Quenchant A is an ideal quenching medium for aluminum alloys.

INTRODUCTION

A water-soluble organic quenching medium for heat treatment in industrial production has obvious advantages over other conventional quenching media such as water, oil, etc. Typical advantages are: adjustable cooling speed, minimum distortion, prevention of cracking, inflammability, reduction of environmental pollution, energy-savings and cost-effectiveness. Widespread applications of this type of aqueous quenching medium opens up a successful way to solve the distortion and cracking problems encountered during the heat-treatment of aluminum alloy and structural steel products, a common problem in the aerospace industry.

Today, both Union Carbide and E.F. Houghton are the two largest firms in the world producing water-soluble quenching media. Both companies have a large customer base and their products are beginning to be implemented in China. Applications of UCON® Quenchant A, produced by Union Carbide, discussed in this paper are:

A. The physical properties of UCON® Quenchant A

B. The cooling characteristics of UCON® Quenchant A

C. Toxicity testing

D. Concentration determination

E. Effects of quenchant on physical properties of aluminum alloy plates

DISCUSSION

A. Physical Properties of UCON® Quenchant A

Being an organic polymer, poly(alkylene glycol) - (PAG), UCON® Quenchant A is a viscous, colorless liquid with infinite miscibility in water. In practice, according to the requirements of quenching, UCON® Quenchant A is used at solution concentrations of five to fifty percent. The physical properties of UCON® Quenchant A are shown in the Table 1.

UCON® Quenchant A is supplied as an aqueous concentrate containing approximately forty-five percent water. This product has no flash point. In addition, NaNO$_2$ is added as a corrosion inhibitor to protect the quenchant bath from corrosion and the workpieces from rusting.

TABLE 1

Physical Properties of UCON® Quenchant A

	Crude solution	Aqueous solution
Colour	Colourless	Colourless
Specific gravity, d_4^{20}	1.092	increasing with its concentration
Refractive power, h_0^{20}	1.4141	increasing with its concentration
Viscosity $10^6 \ m^2/s$	443	increasing with its concentration
Flash point	none	none
Fire point	none	none
Cloud point, $°C$	none	63
Reverse solubility point, $°C$	none	69
PH value	7	7

B. The Cooling Characteristics of UCON® Quenchant A

The cooling characteristics of a quench medium determines its selection for a particular application. A model JST-830C$_2$ cooling curve apparatus was used to determine quenchant cooling curves at different concentrations. In accordance with standard GB9449, a silver probe is utilized as a sensor. The silver probe is heated to 500°C, which is the temperature of aluminum alloy solid solution heat treatment. Data collection and processing are performed using an IBM computer and the final data and curves are provided by a printer and a plotter.

Figure 1 shows the cooling curves of UCON® Quenchant A at five to fifty percent concentrations. These data show:

1. There are three stages of cooling. The first stage, the A-phase or the vapor-blanket cooling phase, is characterized by slow cooling. The second stage, the B-phase or nucleate boiling phase, is characterized by rapid cooling. The third stage, the C-phase or the convective cooling phase, is the slowest cooling phase. The cooling behavior of each stage will vary with concentration. Generally cooling rates decrease with increasing concentration. The transition for A-phase to B-phase cooling is the "characteristic temperature" which occurs at the

"characteristic time". For ordinary heat treatment, it is desirable to maximize the characteristic temperature, minimize the characteristic time and maximize B-phase cooling rates. These cooling characteristics indicate that the use of UCON® quenchant A should be effective in minimizing distortion and cracking which often accompanies water quenching. The selection of the concentration depends on the material and heat treatment specifications for the alloy of interest.

2. Figure 2 shows that cooling rates decrease with increasing quenchant concentration. These cooling rate decreases are also proportional to increasing solution viscosity. Increasing viscosities exhibit corresponding decreases in heat transfer rates at the cooling metal interface. Therefore, it is possible to control heat transfer rates by varying quenchant concentration. This was true for all three initial solution heat treating temperatures evaluated.

3. Figure 3 shows that both the maximum cooling rates and the temperature where this occurs decreases with increasing concentration. This occurs

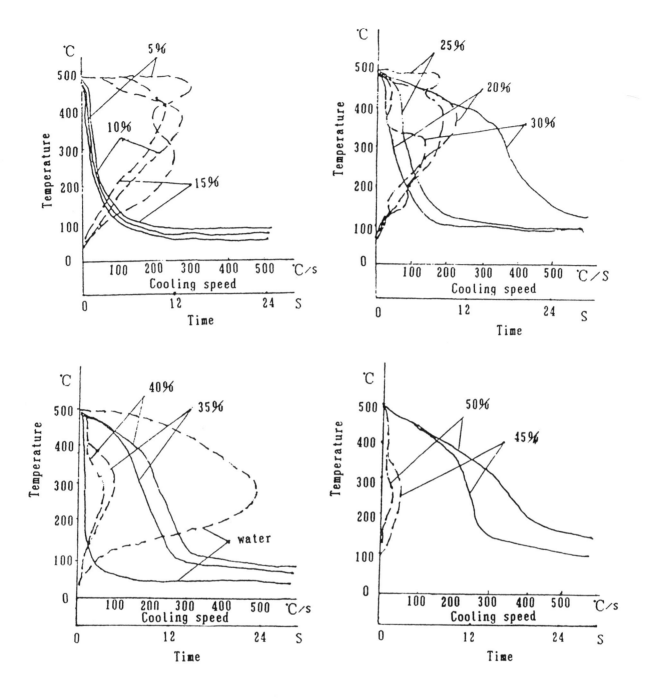

Fig. 1 - Cooling Curves of UCON® Quenchant A With Varying Concentrations

Note: 1) The solid lines are temperature vs. time cooling curves.

2) The dotted lines are cooling speed curves at different temperatures

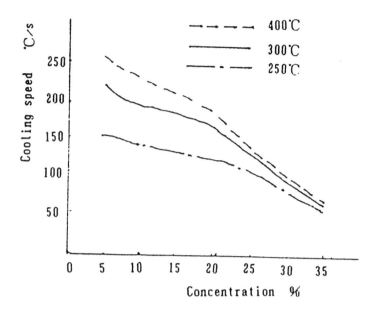

Fig. 2 - UCON® Quenchant A Cooling Speed Changes at 400°C, 300°C and 250°C

because the stability of the vapor blanket surrounding the hot metal decreases with decreasing film thickness and viscosity. Thus the desired transition to nucleate boiling will occur at a higher temperature.

Since, the critical temperature for most aluminum alloys occurs at 420-260°C, it is desirable to maximize cooling rates in this region.

It is desirable to use as high a quenchant concentration as possible while still achieving the critical rate in the 450-250°C transition temperature region in order to minimize cracking and distortion. Figure 3 shows that this corresponds to 15-20% of UCON quenchant A. The maximum cooling rate for water occurs at approximately 280°C. Many aluminum alloys, including aluminum castings, have been quenched in hot water (greater than 60°C) to prevent cracking. However, although cracking is decreased there are large temperature differentials with parts of various

cross-sectional sizes. Therefore, severe cracking and distortion are often unavoidable. UCON® quenchant A has proven to be a viable alternative to hot-water quenching for these types of parts since significant decreases in distortion and cracking are achieved while still meeting the desired physical properties.

C. Test for Toxicity

China Medical University is entrusted with toxicity testing of UCON® Quenchant A. A crude solution enema given to small white mice showed that the half lethal dose (LD50) is greater than 16.2 g/kg. This means that the mice did not die with the dosage tested. According to the International Standard of acute toxicity dosage in industrial products, UCON® Quenchant A may be classified as non-toxic.

D. Determination of Quenchant Concentration

Since some solution is carried away by

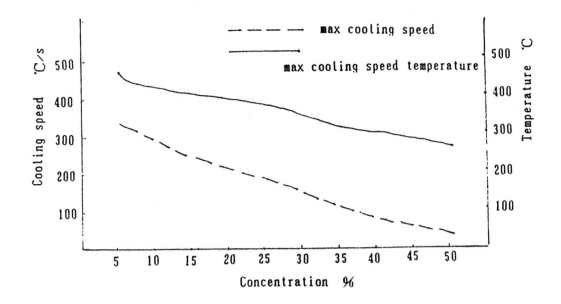

Fig. 3 - UCON® Quenchant A Most Rapid Cooling Speed and the Relationship Between its Corresponding Temperature and Concentration

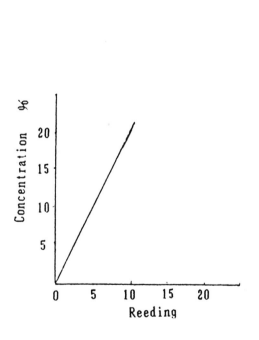

Fig. 4 - Relationship Between Medium Concentration and Reading on Refractometer

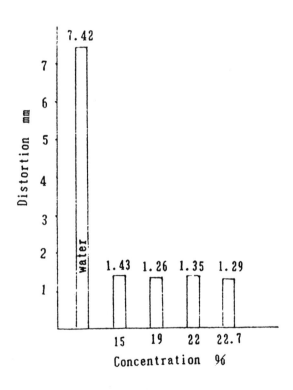

Fig. 5 - Comparison of Quenching Distortion

taking workpieces out of the quenchant bath during the quenching process, UCON® Quenchant A concentration in the bath may vary. If the concentration falls below a fixed range, its cooling characteristics may be effected. Therefore, it is necessary to control the concentration in the heat treatment process.

There are many methods of determining quenchant solution concentration. The hand-held refractometer may be used to determine quenchant concentration. This method is reasonably accurate for uncontaminated quenchant baths.

The relationship between solution concentration and refractive index at a certain temperature is linear. Therefore, a calibration curve such as that shown in figure 4 may be used. (The concentration of UCON® quenchant A can be determined directly from the refractometer scale reading by multiplying the reading by 2.0. As the bath ages, this multiplication factor must be adjusted to account for slight polymer degradation and contamination which may occur.)

E. Effects of Quenchant on Physical Properties of Aluminum Alloy Plates.

1. Distortion
The comparison of the distortion of quenched samples are shown in Figure 5. This data shows that the distortion of water quenched 7075 aluminum alloy plate can be reduced four fold.

2. Strength and Corrosion-Resistance
By using 15,19, 22 and 22.7% UCON® Quenchant A aqueous solution separately, its quenching effect on aluminum alloy plate shows that it has the same effect on their strength and corrosion-resistance as water quenching does.

CONCLUSION

I. UCON® Quenchant A is not smelly or toxic and its aqueous solution is non-flammable and therefore is safe to use in production.

2. UCON® Quenchant A exhibits reverse solubility which facilitates uniform cooling during quenching. The required cooling

characteristics necessary to quench different materials and components are achievable by varying concentration.

3. The quenching test on aluminum alloy plate shows that its required strength and corrosion-resistance can be guaranteed by heat treatment with UCON® Quenchant A, its distortion caused by quenching is minimized, and the man-hours and charge for strengthening after its heat treatment are saved. UCON® Quenchant A is an ideal quenching medium for aluminum alloys.

Proceedings of the First International Conference on Quenching & Control of Distortion, Chicago, Illinois, USA, 22-25 September 1992

Residual Stresses in Laser Heat Treatment of Plane Surfaces

J. Grum and P. Zerovnik
Faculty of Mechanical Engineering
Murnikova, Ljubljana, Slovenia

ABSTRACT

The research work was focused into the study of residual stresses after quenching plane surfaces at different kinematic and optical conditions. Varying the kinematic conditions, we achieved that the laser beam conditioned different cooling conditions, resulting in the occurrence of different hardness profiles along the depth of the trace and different geometric relationships between the individual traces. Thus the laser beam was guided to follow a zig-zag path, and a circular or cornered spiralling line. Besides these variations in guiding the laser beam, we observed the effects of changes in the focal distance of two adjacent traces.

In this way we obtained: partial overlapping of the traces, contact, and different distances between the traces. It was found out that the size and shape of the quenching trace and the distances between the traces exert an influence on the size and direction of residual stresses. Alternatingly distributed quenched and unquenched bands frequently create preferential conditions in sliding of parts mostly as a result of their geometric features.

INTRODUCTION

Lasers represent one of the most important inventions of the twentieth century. With their development it was possible to get a highly intensive, monochromatic, coherent, highly polarized light waves. The first laser was created in 1960 in Californian laboratories with the aid of a resonator from an artiticial ruby crystal. Dated in this period is also the first industrial application of laser which was used to make holes into diamond materials extremely difficult to machine. First applications of laser for metal machining were not particularly successful mostly due to low capabilities and instability of laser sources in different machining conditions. These first applications, no matter how successful they were, have however led to a development of a whole number of new laser source types. Only some of them have met the severe requirements and conditions present in metal cutting. As the most successful among them, CO_2-laser should be mentioned. The high intensity CO_2-lasers have proved to be extremely successful in various industrial applications so from the point of view of technology as well as economy. A great number of successful applications of this technique have stimulated the development of research activities which since 1970 have constantly been increasing.

Some of the most notable advantages of laser sources are:

- high adaptability to small-series of workpieces of simple and complex shape and various size.

- by adapting the laser source, very different materials can be machined or very different machining operations performed.

For this purpose, besides CO_2-lasers, Nd-YAG and Excimer-lasers with a relatively low power and a wave length between 0.2 to 1.06 μm have been successfully used. A characteristic of these sources is that, besides a considerably lower wave length, they have a smaller focal spot diameter and much higher absorptivity than CO_2-lasers.

This paper reports on the investigations in surface heat treatment which were carried out on the laser machining system, LPM 600 with a capacity of 600 W, made by a Slovene laser-manufacturer ISKRA - Centre for Electro-optics. Important parameters of heat treatment with a laser beam are: laser power, laser trace width, and workpiece speed. **Figure 1** is a photo of the ISKRA laser machining system with a built-in continuous difusion-cooled CO_2-laser, directing the light via a metallic mirror and a focusing lens onto the workpiece surface. A very important component of this system is a numerically controlled positioning table with a built-in BOSCH-ALPHA3 computer unit. This computer unit is linked through an interface to the laser control system, which enables laser ignition, opening and closing of the nozzle and opening and closing of the process gas etc. **Figure 2** presents a schematic of the ISKRA -CO_2 laser source with all the important specifications.

A combination of PL 600 with the CNC-controlled positioning table KS-2 for precise cutting and welding.

Positioning table:
— X × Y travel   360 × 510 mm
— working area   420 × 700 mm
— positioning accuracy   ±0.01 mm
— minimum increment   0.001 mm
— maximum speed   6 m/min
— floor dimensions   1900 × 1550 mm
— CNC control   Bosch Alpha 3
                        with linear and
                        circular interpolation
— power requirements   5 kVA, 3 × 380 V, 50 Hz

Figure 1: Laser machining system, ISKRA - Centre for Electro-optics, Slovenia

In heat treatment using laser light interaction, it is necessary to achieve the desired heat input which is normally determined by the hardened layer depth. Cooling and quenching of the overheated surface layer is in most cases achieved by self-cooling since after heating stops, heat abstracted into the workpiece material is so intensive that the critical cooling speed is achieved and thus also the wanted hardened structure. **Figure 3** illustrates the dependence of power density on laser light interaction time on the workpiece surface in order to carry out various metalworking processes. Diagonally,

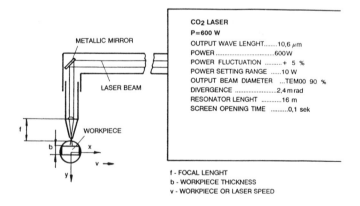

Figure 2: A schematic of LEM 600 ISKRA CEO CO_2 laser source with specification

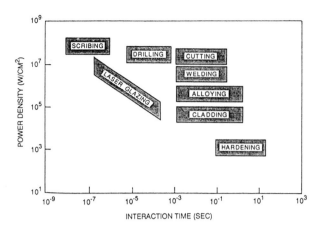

Figure 3: Power density and interaction time in laser metalworking processes /13/

there are two processes, i.e. scribing and hardening, for which quite the opposite relationship between power density and interaction time has to be ensured. In scribing material vaporization at a depth of a few microns has to be achieved, and with it the alphanumeric recording of prescribed quality and character resolution. On the other hand, for hardening a considerably lower power density per workpiece surface unit is required, but the interaction times are the longest among all the mentioned metalworking processes.

SURFACE HEAT TREATMENT

Heat treatment, or in other words the conditions in heating, are described by the equation of heat conductance in a 3-D body /1/. The solution of this equation has enabled graphical presentations of the temperature cycles on the surface and in the workpiece interior with respect to the heating time with a given source. In this equation, the power density can be changed, which enables the determination of the temperature cycle on the surface, and thus solid state hardening of the workpiece material, or hardening by melting the material slightly, and then rapidly cooling it down.

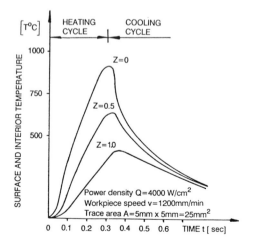

Figure 4a: Temperature cycle on the workpiece surface and in its interior versus interaction time /1/

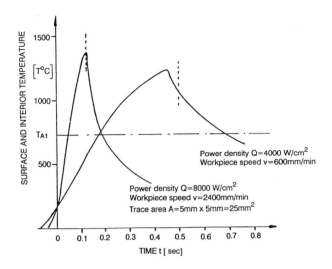

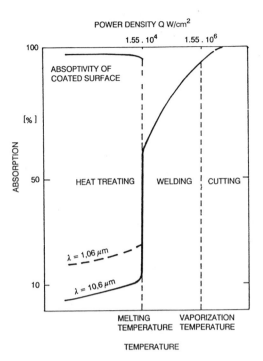

Figure 4b: The effected of laser light interaction time on the temperature cycle during heating and cooling at various power densities and workpiece speeds /1/

Figure 4a shows the temperature cycle on the surface and in the workpiece interior during heating up and cooling down using the above laser source, and **Figure 4b** the temperature cycle on the surface at various power densities and required heating up times. In this way it is possible to regulate the thickness of the hardened surface layer. Thus surface hardening is to a great extent dependent on power density, which allows the conclusion that the most favourable region for surface hardening is that with laser power P_2 or P_3, enabling a considerable variation in workpiece speed and corresponding laser trace area, **Figure 4c.** A similar description of the conditions in hardening is given in reference /2/ where the emphasis is put on the experimental work and graphical presentation enabling an easier selection of the heat treatment conditions. In references /3/ and /4/, heat treatment/hardening of different kinds of gear drive housing made of cast iron or malleable cast iron is reported. Another interesting application of laser for heat treatment is reported in /5/ where laser was used for the heat treatment of the engine cylinder inner wall and piston rings for which various laser light density profiles were applied /5/. Another interesting study /6/ is a comparison between various heat treatment procedures such as flame hardening, induc-

Figure 5: The effect of temperature on laser light absorptivity /9/

tion hardening with laser hardening of a nodular cast. The results have confirmed that laser-hardening offers many advantages thanks to very low losses in wear testing. A theoretical and partly experimental study of induction hardening and laser hardening /7/ was carried out for a simple cylindrical steel specimen with the objective to determine the temperature cycles and residual stresses. A comparison of highly energy intensive electron beam heat treatment and laser-beam treatment on a number of specific machine parts is reported in /8/, assessing both methods and stating the main advantages offerred by these two laser metalworking processes.

EXPERIMENTAL ARRANGEMENT

Selection of Absorbers

With the interaction of the laser light and its movement across the surface, very rapid heating up of metal workpieces can be achieved, and subsequent to that also very rapid cooling down or quenching. The cooling speed, which in conventional hardening defines quenching has to ensure martensitic phase transformation. In laser hardening the martensitic transformation is achieved by self-cooling, which means that after the laser light interaction the heat has to be very quickly abstracted into the workpiece interior. While it is quite easy to ensure the martensitic transformation by self-cooling, it is much more difficult to deal with the conditions in heating up. The amount of the disposable energy of the interacting laser beam is strongly dependent on the absorptivity of the metal. The absorptivity of the laser light with a wave length of $10.6\,\mu$m ranges in the order of magnitude from 2 to 5 % whereas the remainder of the energy is reflected and represents the energy loss. By heating metal materials upto the melting point, a much higher absorptivity is achieved with

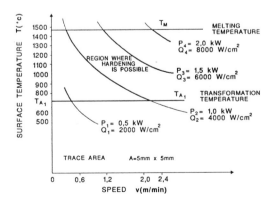

Figure 4c: The effects of power density and workpiece speed on hardening /1/

335

an increase of upto 55% whereas at vaporization temperature the absorptivity is increased even upto 90% with respect to the power density of the interacting laser light.

Figure 5 illustrates the relationship between laser light absorptivity on the metal material surface and temperture or power density /9/. It is found that, from the point of view of absorptivity, laser beam cutting does not pose any problems, as the metal takes the liquid or evaporated state, and the absorptivity of the created plasma can be considerably increased. Therefore it is necessary to heat up the surface, which is to be hardened, onto a certain temperature at which the absorptivity is considerably higher and enables rapid heating up onto the hardening temperature or the temperature which is for safety reasons lower than the solidus line. This was successfully used in heat treatment of camshafts as reported in /10/.

Another possibility of surface preparation is to apply an absorber. Additional coating consisting of an absorber layer has to have the following properties:

- high ability to absorb laser light

- fast and easy application onto the metal surfaces

- simple preparation and low cost

- uniform spreading on the surface with a film thickness ranging between 20 to 40 μm

- high heat conductivity ensuring heat transfer onto the base material

- high adhesion onto metal surfaces regardless of the deposition method, e.g. submerging, spraying, painting

- easy removal of the remainders from the surface.

Absorbers suitable for this kind of application are various metal oxide powders, carbide graphite, zink phosphate, manganese phosphate and also black paint.

Since heat treatment/hardening goes on at considerably lower temperatures than the hardening temperature (Th/Tm = 0,5...0,6) is, the desired efficiency of the laser beam interaction is not achieved. Therefore we chose three different types of surfaces prepared in different ways, i.e.:

- surface machined by grinding

- surface treated with absorber A

- surface treated with absorber B

Absorber A named MIOX.PVK MEDIUM, Slovene manufacturer, is a product made on polymer basis with additions. It was applied onto the surface by submerging. The thickness of the absorber A layer was not more than 20 μm which was achieved by hardening at room temperature.

Absorber B named MELIT EMAIL, also manufactured in Slovenia, was deposited by submerging. Hardening was carried out at a temperature between 120 -150°C for 30 to 40 min. MELIT EMAIL is made on the basis of alkide and alumine resins which have

to be air-dried before hardening for a time of 10 -15 min for the solvents to evaporate and the paint deposit to set. The instructions given by the manufacturer have to be strictly observed, and special attention should be paid to the cleanliness of the surface, paint deposition technique and drying method.

SELECTION OF HEAT TREATMENT CONDITIONS

Experiments were carried out using the above mentioned laser system in the following heat treatment conditions:

- laser power P = 350 W

- workpiece speed with respect to laser beam position "v" v = 1000 mm/min using absorber A or B and v = 400 mm/min without absorber

- focal distance of the focusing lens f = 2.5 " or 63.75 mm

- focus distance from the workpiece surface w = 11 mm

- workpiece surface area 20 x 10 mm = 200 mm^2

- laser beam travel: along a square spiralling line starting in the middle following

- a zig-zag line starting on the edge of the longer workpiece side line.

Both ways of laser beam travel and workpiece dimensions are shown in **Figure 6**. After hardening two adjacent traces overlap by a third of their length, which affect the hardening trace profiles and surface properties.

WORKPIECE MATERIALS

To study the successfulness of laser beam heat treatment, the following workpiece materials were chosen: - Č 1531 JUS or 1042 AISI structural carbon steel for heat treatment and surface harden-

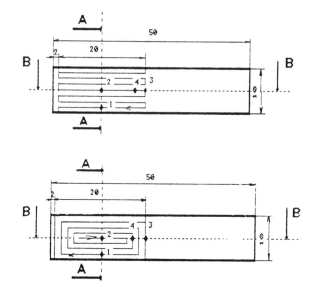

Figure 6: Workpiece geometry and different ways laser beam travel

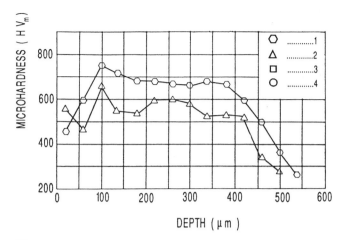

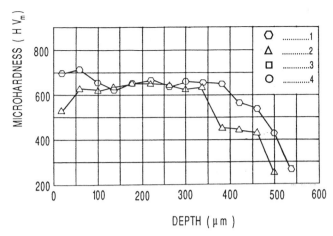

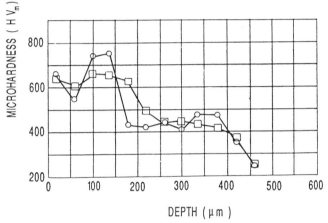

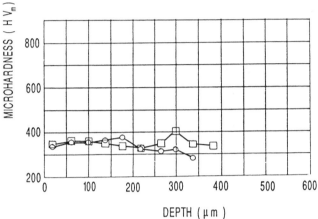

Figure 7: Microhardness of Č 1531 steel after zig-zag hardening, using absorber A as a function of depth - in transverse direction Fig. 7a and in longitudinal direction Fig. 7b

Figure 8: Microhardness of Č 1531 steel after zig-zag hardening, using absorber B - in transverse direction Fig. 8a and in longitudinal direction Fig. 8b

ing - Č 4732 JUS or 4142H AISI chromium-molybdenum structural steel - Č 6445 JUS tungsten-chromium steel with a vanadium addition, suitable for hot or cold work tools.

The chemical composition of the above steels is:

Č 1531   0.46 % C - 0.65 % Mn

Č 4732   0.42 % C - 0.65 % Mn - 1.05%Cr - 0.22%Mo

Č 6445   0.80 % C - 0.40% Mo - 1.o% Cr - 2.0% W - 0.4% Si

The dimensions of the workpieces were adapted to the chosen heat treatment procedure and the requirements of subsequent residual stress measurements. They had a length of 50 mm, a width of 10 mm and a thickness of 4 mm.

EXPERIMENTAL RESULTS

Microhardness Analysis

After heat treatment the workpieces were cut in the transverse and longitudinal direction. In Fig. 6 the transverse cross-section is

denoted by A-A and the longitudinal cross section by B-B. On these sections microhardness was measured, and micro- and macro analysis were made. Based on microhardness measurements, the influences that might have an effect on the results of surface heat treatment by a laser beam were observed and divided into the following groups:

-   influences of absorber A or B

-   influences of alloying elements or types of steel on the size of the hardening trace, and hardness

-   influences of heating-up and cooling-down speeds on heat treatment as a result of different laser beam travelling line across the workpiece surface

-   effects of mass on heat treatment

Transversely, on A-A cross-section, microhardness was measured on two positions, denoted by point 1 and 2 in Figure 6. Along the length of the workpiece, on cross-section B-B microhardness measurements were made at two positions denoted in the figure by numbers 3 and 4. Microhardness was measured on the location of the largest trace, while 10 to 15 measurements were made at different depth layers.

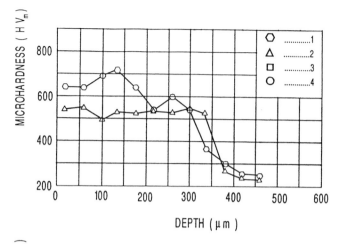

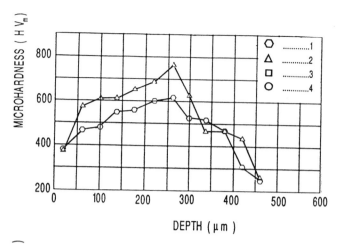

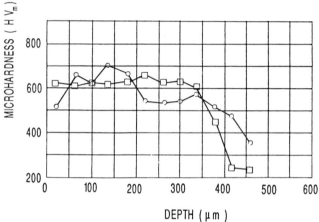

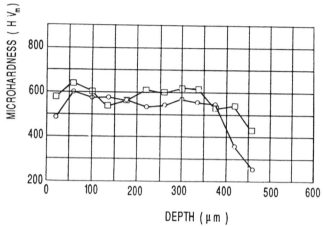

Figure 9: Microhardness of Č 1531 steel after square spiral hardening, using absorber A - in transverse direction Fig. 9a and in longitudinal direction Fig. 9b

Figure 10: Microhardness of Č 1531 steel after square spiral hardening, using absorber B, - in transverse direction Fig. 10a, in longitudinal direction Fig. 10b

Figure 7 shows a microhardness profile of Č 1531 subsequent to laser beam hardening in transverse and longitudinal cross section, from 1 to 4 using absorber A, and Fig. 8 using absorber B. In both cases a zig-zag laser beam travel was chosen. The variation of hardness in depth (Fig.9, absorber A) is in agreement with the speed of the heat abstraction at points 1 and 2 whereas at points 3 and 4 it is a bit surprising since it has two transition areas. The first transition area lies at a depth of 150 to 200 μm and the second at a depth of 400 to 500 μm. Both transition areas are a result of temperature phase transformations

$$1. \ T \geq T_{A3} \ \text{and}$$

$$2. \ T_{A1} \leq T < T_{A3},$$

which speak for an amount of martensitic transformation and thus confirm the difference in microhardness. The variation of microhardness as a function of depth **Figure 8**, absorber B, confirms that the selected absorber is less successful, since the microhardness profile is straight as far as down to 400 μm across the longitudinal cross section in points 3 and 4. This means that the surface did not reach the temperature of the austenitic phase, or that due to the heating method the stress relieving of the surface was considerable.

For reasons of comparison let us inspect the hardening of the same steel where the laser beam travels along a square spiral, using absorber A, **Figure 9**, and absorber B, **Figure 10**. In Figure 9 microhardness is very uniform upto the transformation region at a depth of 300 μm. The highest microhardness is achieved on the surface at point 1, then at point 3 and 4. The differences in microhardness at point 3 and 4 are minimal, higher deviations lying at point 2 in the middle of the square spiral. At point 2 the microhardness reaches a value of between 500 and 530 HVm, which is due to the slowest heat abstraction rate and effects of slight stress relieving. In **Figure 11** and **12**, differences in microhardnesses at measuring position 1 are illustrated for all three kinds of steel, using absorbers A or B after hardening with a laser beam travelling along a square spiralling line. **Figure 13** and **14**, in addition, show the differences in microhardnesses using abosorbers A or B at measuring position 1, with laser beam travelling along a zig-zag line. On the basis of the microhardness data from figs.11 and 14, one can note the following:

- differences in surface hardness

- effects of surface stress relieving

- different microhardness gradients with respect to depth

- differences in the hardened layer thickness.

338

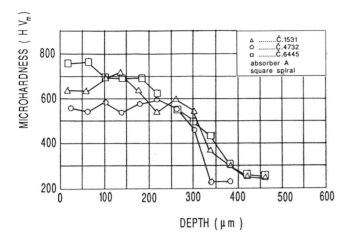

Figure 11: Comparison of microhardnesses at measuring position 1 for all the investigated steels, - square spiral hardening, absorber A

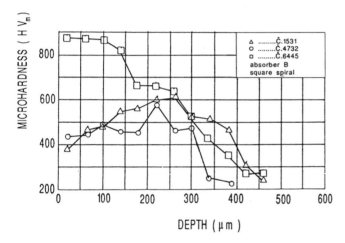

Figure 12: Comparison of microhardnesses of all the investigated steels, measuring position 1 - zig-zag hardening, absorber B

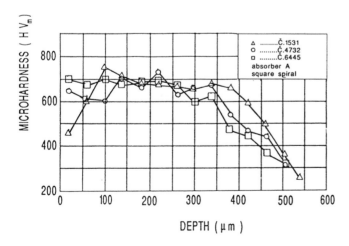

Figure 13: Comparison of microhardnesses of all the investigated steels, measuring position 1 - zig-zag hardening, absorber A

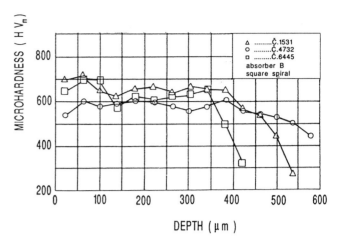

Figure 14: Comparison of microhardnesses of all investigated steels, measuring position 1 - zig-zag hardening, absorber B

For these reasons we prepared the instructions for an appropriate selection of heat treatment regime which would suit the kind of workpiece material and size and shape of the laser trace.

RESIDUAL STRESS ANALYSIS

Yang in /11/ report on the simulation of thermal conditions in the workpiece material in laser heating and cooling, and calcution of residual stresses by the 2D-finite-element method. Grevey et al in /12/ determined the magnitude of residual stresses on the basis of Orlisch diagrams and assessment of thermal conditions.

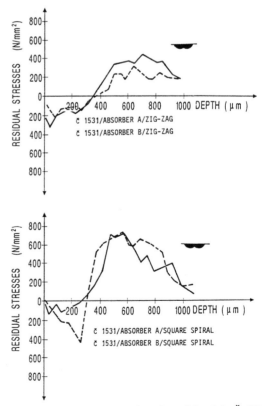

Figure 15: Residual stresses as a function of depth in Č 1531 steel

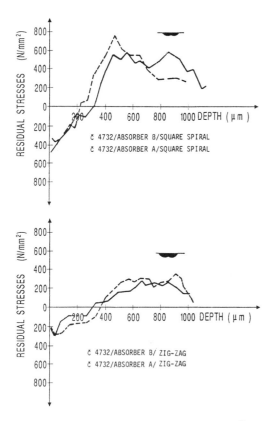

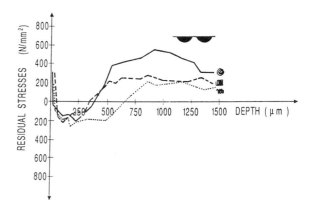

Figure 18: Residual stresses in Č 6445 steel for inter-spaced hardening traces, concentric circle, square spiral and zig-zag hardening methods

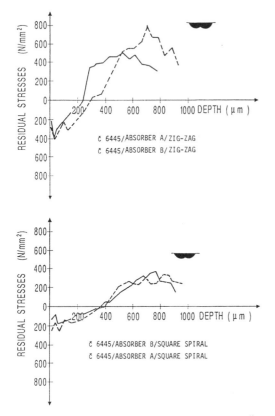

Figure 16: Residual stresses as a function of depth in Č 4732 steel

This study based the identification of residual stresses on the relaxation method measuring workpiece deformation. The relaxation

Figure 17: Residual stresses as a function of depth in Č 6445

was induced by electro-chemical removal of the stressed surface layer. The removal causes a breakdown in the existing equilibrium state. The restoration of the equilibrium is in this case accompanied by workpiece strains. The strains were measured by means of resistance strain gauges and calculated into stresses using a mathematical model with the corresponding software. **Figure 15** shows a comparison of residual stress size as a function of hardening trace depth for Č 1531 using absorbers A and B for both ways of laser beam travel. The results have shown that in zig-zag laser beam travelling, a hardened and stress relieved structure is obtained because of an overlapping trace. Such a post-heat treatment state results in low residual stresses in the surface layer and in tensile residual stresses in the deeper layers down to 400 μm. Of great significance is also the stress gradient since at slight changes of compressive into tensile stresses, the dynamic external loads on the workpiece are lessened. The exactly opposite behaviour of residual stresses is found when the laser beam travels along a square spiral. Here, especially in the case of absorber B, compressive stresses are highest, with an extreme increase in stresses at the compressive -tensile transition, and a considerably high tensile stress in deeper layers at approx. 600 μm. **Figures 16** and **17** show residual stress profiles for Č 4732 and Č 6445 as a function of different ways of laser beam travel and different abosorbers. From the results it can be noted that, using an adequate procedure, it is possible to achieve low residual stresses and slight transitions between the compressive surface zone and the tensile lower layer zone. **Figure 18** illustrates the variation of residual stresses with an intermittent laser trace. The residual stresses in the compressive and tensile zone are low, a special feature being a thin surface layer with tensile stresses. Since after laser heat treatment grinding should be applied, in each of the discussed cases compressive stresses in the surface are a usual result.

CONCLUSIONS

The contribution presents the results of surface heat treatment with a laser beam, studying the effects of the changes in: surface preparation with different kinds of absorbers, kinematic conditions of laser beam travel across the workpiece surface, and different types of traces (overlapping/intermittent). The experiments were made for three very typical steels. On the basis of metallographic analysis, microhardness measurements in various depth layers, and residual stress measurements, instructions for the selection of conditions used in laser heat treatment of the discussed steeels were prepared.

REFERENCES

1 - Kawashumi, H., Source Book on Applications of the Laser in Metalworking, American Society for Metals, Metals Park, Ohio, U.S.A., 185-193, 1981.

2 - Courtney, C and W. M. Steen, Source Book on Applications of the Laser in Metalworking, American Society for Metals, Metals Park, Ohio, U.S.A., 195-208, 1981.

3 - Wienman, J. A., Source Book on Applications of the Laser in Metalworking, American Society for Metals, Metals Park, Ohio, U.S.A., 209-217, 1981.

4 - Yessik, M. and R. P. Scherer, Source Book on Applications of the Laser in Metalworking, American Society for Metals, Metals Park, Ohio, U.S.A., 219-226, 1981.

5 - Seaman, F. D. and D. S. Gnanamuthu, Source Book on Applications of the Laser in Metalworking, American Society for Metals, Metals Park, Ohio, U.S.A., 179-184, 1981.

6 - Fukuda, T., M. Kikuchi, A. Yamanishi and S. Kiguchi, Proc. of the Third Int. Congress on Heat Treatment of Materials, The Metals Society, London, England, 2.34-2.39, 1983.

7 - Melander, M., Y. Shanchang Y. and T. Ericsson, Proc. of the Third Int. Congress on Heat Treatment of Materials, The Metals Society, London, England, 2.75-2.85, 1983.

8 - Sayegh, G., Proc. of the Third Int. Congress on Heat Treatment of Materials, The Metals Society, London, England, 2.63-2.74, 1983.

9 - Engel, S. L., Source Book on Applications of the Laser in Metalworking, American Society for Metals, Metals Park, Ohio, U.S.A., 149-171, 1981.

10 - Mordike, S., Dr. Puel and H. Szengel, Zbornik radova međunarodnog savetovanja, Nove tehnologije toplinske obrade metala, Zagreb, Croatia, 1-12, 1990.

11 - Yang, Y. S. and S. J. Na, Surface and Coating Technology 38, 311-324, 1989.

12 - Grevey, D., L. Maiffredy and A. B. Vannes, Journal of Mechanical Working Technology 16, 65-78, 1988.

13. Sandven, O. A., Laser Surface Transformation Hardening, American Welding Society and Manufacturing Productivity Centre, New Orleans, Louisiana, U.S.A., Proc. from the 1988 Conf. on Laser Surface Modification, 1983., p.p. 8-42.

DATE DUE

2/15	9	2		
3/2				
2/5				